Settling Velocity, cm/sec

Temperature	20°C	24°C	20°C	24°C	20°C	24°C
Shape Factor	0.5	0.5	0.7	0.7	0.9	0.9
$-1.00\ \phi$	18.2	18.4	20.8	20.8	25.0	25.0
-0.75	16.6	16.8	19.0	19.0	22.6	22.6
-0.50	15.0	15.2	17.2	17.2	20.0	20.2
-0.25	13.4	13.6	15.4	15.6	17.6	17.8
0	11.8	12.0	13.6	13.8	15.2	15.6
0.25	10.4	10.6	11.8	12.0	13.0	13.4
0.50	9.0	9.2	10.2	10.4	11.0	11.2
0.75	7.7	7.9	8.7	8.9	9.2	9.5
1.00	6.4	6.6	7.3	7.5	7.6	7.9
1.25	5.3	5.5	6.0	6.2	6.3	6.5
1.50	4.4	4.6	4.9	5.1	5.1	5.3
1.75	3.6	3.7	3.9	4.1	4.2	4.4
2.00	2.9	3.0	3.2	3.4	3.3	3.5
2.25	2.3	2.4	2.5	2.7	2.6	2.8
2.50	1.8	1.9	2.0	2.1	2.0	2.2
2.75	1.4	1.5	1.5	1.6	1.6	1.7
3.00	1.1	1.2	1.1	1.2	1.1	1.2

Sources

H. G. Page, 1955, Phi-millimeter conversion table, *J. Sed. Pet.*, 25, 285–292.

American Society for Testing Materials, 1968, Wire-cloth sieves for testing purposes (E 11-61), *in* 1968 Book of A.S.T.M. Standards, Pt. 10, 535–540.

W. S. Tyler Co., 1963, Testing sieves and their uses, Handbook 53, W. S. Tyler Co., Cleveland, Ohio, 48 pp.

J. M. Zeigler and Barbara Gill, 1959, Tables and graphs for the settling velocity of quartz in water, above the range of Stokes' Law, Woods Hole Oceanographic Institution Ref. No. 59-36, Woods Hole, Massachusetts, 67 pp.

Procedures in Sedimentary Petrology

PROCEDURES IN SEDIMENTARY PETROLOGY

Edited by

ROBERT E. CARVER

University of Georgia
Athens, Georgia

WILEY-INTERSCIENCE

A Division of John Wiley & Sons, Inc.
New York • London • Sydney • Toronto

PREFACE

Procedures in Sedimentary Petrology is intended to be a guide to currently standard, or accepted, methods of measurement and analysis in sedimentary petrology. Contributors were asked to avoid extensive discussion of historical development and modification of procedure in order to concentrate on concise description of the most widely accepted method, or methods.

My original chapter outline included only subjects that are considered to be exclusive to sedimentary petrology and to be well tested, or standardized. On consultation with the contributors, the outline was modified, and expanded to cover a few subjects that overlap other well-established fields of geology or are relatively untested. The result is a manual of procedure, including both technique and analysis, which should be of value to all sedimentary petrologists from graduate students, for whom the book is primarily intended, to senior investigators.

With two exceptions, every chapter was reviewed by at least two contributors and myself. Dr. James D. Howard examined two chapters and is the only reviewer who is not also a contributor. Normally, chapters were reviewed by authors of the most closely related chapters, so that reviews were authoritative and critical in the best sense of the word.

I suggested, in the original notes on chapter content, a few topics of primary importance for each chapter. Beyond these suggestions, chapter content was left entirely to the authors. A reasonable overlap between chapters was encouraged in order to avoid gaps in the treatment and to present more than one view of critical areas between techniques, or between technique and analysis. I take full responsibility for omission or inclusion of major procedures, but the contributors are fully responsible for the content of their chapters.

Considerable effort was devoted to standardizing the form of references that appear at the end of each chapter and in the General References. A format considered to be suitable to the subject, in which the most frequently cited sources are most abbreviated, was adopted. A list of abbreviations follows the contents. As a general rule, journal volume numbers appear in boldface type, page numbers in normal type, without further designation.

Final drafts of manuscripts were completed in 1969. Literature published (as opposed to dated) in 1968 and early 1969 was, in most cases, available to the contributors prior to submission of the final drafts.

Athens, Georgia   Robert E. Carver
June 1970

CONTENTS

CONTENTS

ABBREVIATIONS

Abh.	=	Abhandlungen
Abs.	=	Abstract
Abt.	=	Abteilung
Acad.	=	Academy
Akad.	=	Akademie
Am.	=	American
Ann.	=	Annual, annals, annales
A.P.I.	=	American Petroleum Institute
Assoc.	=	Association
A.S.T.M.	=	American Society for Testing Materials
Bull.	=	Bulletin
Circ.	=	Circular
Chem.	=	Chemistry
Co.	=	Company
Comm.	=	Committee
Congr.	=	Congress
Dept.	=	Department
Devel.	=	Development
Econ.	=	Economic
ed.	=	Editor, edition
eds.	=	Editors
Eng.	=	Engineers
Fig.	=	Figure
Figs.	=	Figures
Geol.	=	Geological, geologique, geologisch
Ges.	=	Gesellschaft
Inf.	=	Information
Inst.	=	Institute

Inter.	=	International
Inv.	=	Investigations
J.	=	Journal
Jahrb.	=	Jahrbuch
Lab.	=	Laboratory
Mag.	=	Magazine
Min.	=	Mineralogy, Minéralogie
Misc.	=	Miscellaneous
Mitt.	=	Mitteilungen
ms.	=	Manuscript
Nat.	=	Natural
No.	=	Number
Paleon.	=	Paleontology
Pet.	=	Petrology
Petrol.	=	Petroleum
Pl.	=	Plate
P.P.	=	Professional Paper
Proc.	=	Proceedings
Prof.	=	Professional
Publ.	=	Publication
Quart.	=	Quarterly
Rept.	=	Report
Ref.	=	Reference
Res.	=	Research
Rev.	=	Review
Sci.	=	Science
Sed.	=	Sedimentology
S.E.P.M.	=	Society of Economic Paleontologists and Mineralogists
Ser.	=	Series
Soc.	=	Society, Société
Surv.	=	Survey
Symp.	=	Symposium
Tech.	=	Technical
Trans.	=	Transactions

u.	=	und
Univ.	=	University
U.S.	=	United States
Verh.	=	Verhandlungen
Wiss.	=	Wissenschaft, wissenschaftlich
Zeits.	=	Zeitschrift

Procedures in
Sedimentary Petrology

SECTION
I

ANALYSIS
OF SEDIMENTARY
STRUCTURES

CHAPTER 1

MEASUREMENT OF SEDIMENTARY STRUCTURE ORIENTATION

JOHN H. HOYT

University of Georgia Marine Institute, Sapelo Island, Georgia

A variety of types of sedimentary structures can be measured to give information about the origin, deposition, and history of sediments. This chapter is concerned chiefly with megascopic features that are measured and described in the field and that formed contemporaneously with deposition.

STRATIFICATION

An important and universal characteristic of sediments and sedimentary rocks is layering or *stratification* which occurs in several sizes, orientations, and relationships. Accurate knowledge and description of stratification are vital in structural, environmental, and paleogeographic studies.

STRATIFICATION TERMS

Campbell (1967) has discussed the following terminology and utilizes natural subdivisions of rock units which differ principally in areal extent and interval of time for formation. A layer of rock or sediment that can be distinguished from layers above and below is a *stratum*. It is separated from adjacent layers by *stratal surfaces*. The smallest megascopic stratum is a *lamina*. Comformable laminae in a distinctive structure are grouped to form a *laminaset*. The stratum that reveals the principal rock layering is a *bed*, and a number of superposed, similar beds make a *bedset*. Depositional surfaces that reveal the principal layering are *bedding surfaces*.

McKee and Weir (1953) proposed another system of stratification terminology that has been in use for several years. In this system a *lamina* is a stratum less than 1 cm thick, and a *bed* is greater than 1 cm. Strata of similar characteristics are grouped in a *set*, and similar sets in a *coset*.

Layers oblique to bedding surfaces are referred to as *cross-strata*. A *formation* is a stratum or group of strata that form a unit for geologic mapping. *Blanket* deposits are widespread sheets of sediment with a width-to-thickness ratio greater than 1000:1 (Krynine, 1948). *Tabular* units have width-to-thickness ratios between 1000:1 and 50:1, and *prisms* are between 50:1 and 5:1. Krynine has also proposed that sedimentary bodies with a volume of more than 500 cu miles be termed large; between 500 and 1 cu miles, medium; and those less than 1 cu mile, small.

DIP AND STRIKE

One common field observation is the inclination of stratification. The basic measurements are *dip*, which is the maximum angle of inclination measured from the horizontal, and *strike*, which is the direction of a horizontal line in the plane of the inclined unit, measured perpendicular to the dip.

The most common method of measuring dip, in the field, is by a Brunton compass containing a clinometer for this purpose. If measurements are made at a distance from the outcrop, it is important to sight in the plane of stratification (along the strike), as determined by observations in a third direction, to get a *true dip* rather than an *apparent dip*. True dip may be determined from two apparent dips graphically as shown in Fig. 1 or by using stereographic diagrams as described in Chapter 2, page 29. It is necessary to record the apparent dip angle and direction of dip to make these computations.

The compass may also be placed directly on a stratification surface to measure the dip and strike. A flat piece of wood or clip board is sometimes useful in measurements on a stratification surface. Care must be taken that the surface does not contain irregularities or foreign objects that prevent the board from lying directly along the bedding. A compass is used in slightly different directions to observe the dip several times to get a maximum reading. Alternately, the strike may be determined first by centering the bull's-eye level and marking or noting the direction on the strata. The dip is then measured perpendicular to the strike. Various specialized pieces of apparatus can be used to give a suitable surface for measurement (Pryor, 1958; Armstrong, 1967; Parizek, 1967).

Dip and strike can be measured with plane table and alidade. This measurement is particularly useful, and sometimes necessary, with low-angle dips or when the dip must be determined more accurately

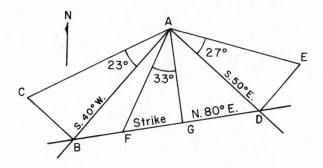

Fig. 1 Diagram showing method of converting two apparent dips to true dip. Apparent dips: $23°$ in direction $220°$ (S. $40°$ E.); $27°$ in direction $130°$ (S. $50°$ E). Solution: angle $BAC = 23°$ (rotated about line AB); angle $DAE = 27°$ (rotated about line AD); $BC \perp AB$; $DE \perp AD$; $BC = DE$; BD = strike = $N.$ $80°$ E; $AG \perp BD$; $DG = BC = DF$; angle FAG = true dip = $33°$ S.

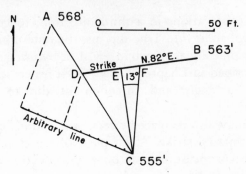

Fig. 2   Diagram showing method of determining strike and dip with plane table and alidade. Elevation of three points A, B, and C are determined (or relative elevations). Line AC connecting high and low points is divided into 13 divisions (568' - 555' = 13'). Position of 563' is noted on line AC. Line BD is strike = N. 82° E. EC is direction of dip. DF = 8' (rotated about line EC) is plotted from scale. ECF = dip = 13° S.

than with a compass. The plane table and alidade are used to determine the locations and elevations of three points on the stratum to be measured. Dip and strike are then found graphically (Fig. 2). Maps contoured on a stratification surface are also useful for determining dip and strike if the surfaces are sufficiently large. Strike at a particular location is parallel to the adjacent contours. Dip may be obtained graphically (Fig. 3).

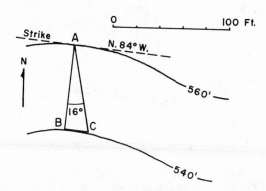

Fig. 3   Method of determining dip and strike from a structure contour map. Find dip and strike at A. Line AB is perpendicular to contours; strike = N. 84° W. BC = 20' (rotated about line BC). BAC = dip = 16° S.

MEASUREMENT OF STRATIFICATION ORIENTATION

Usually, it is not sufficient to determine only the dip of cross-strata; the structural dip of the unit or formation containing the strata must also be known. In addition, some estimate of the initial dip or depositional dip of the formation is necessary. Unfortunately, it is seldom possible to determine initial dip accurately, and the best that can be done is to assume that conformable, continuous beds are within a few degrees of horizontal when deposited. Although this is true of some shelf sediments and perhaps flood-plain deposits, it is not always so for continental slope, nearshore, dune, and continental deposits. The dip of the formation must be corrected first for initial dip (if it can be determined). The corrected structural dip is used to adjust the inclination of the strata to find the relation to horizontal at the time of deposition (Fig. 4). The necessary computation can also be made on a stereographic net.

In setting up procedures for measuring cross-strata it is advisable to consider how many measurements will be taken from the same set. If the inclination within a set is uniform, one measurement per set is sufficient. If, however, the stratification planes are curved or trough-shaped, several measurements are necessary to determine the orientation and shape of the set so that a reading can be taken down the plunge of the trough, for the plunge is usually the most useful measurement in orientation studies.

Special problems are encountered when determining regional formation dip in areas of rapid facies change. If dip is determined from widely spaced measurements, lithologic change may place the formation contact above or below the lateral equivalent position that would represent an initial deposition surface. Facies changes are particularly misleading in subsurface studies where well control may be inadequate to determine precise facies relationships. It is then necessary to use a lithologic change or "marker" above or below the formation. If this is done, particular attention should be paid to unconformities.

Dip can also be determined by mechanical devices, such as well-logging equipment in bore holes and from oriented cores. In addition to the corrections for structural and initial dip, the inclination of the bore hole should be considered.

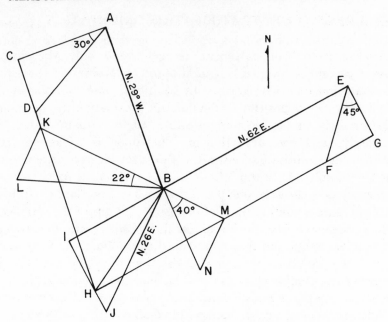

Fig. 4 Graphic method of removing secondary tilt. Stratification striking N. 29°
W. and dipping 30° W. are within a formation striking N. 62° E. and
dipping 45° S. Assume zero initial formational dip. Find original dip and
strike of stratification. AB = strike of stratification. Angle CAD = dip of
stratification = 30° W. BE = strike of formation. Angle FEG = dip of
formation - 45° S. $CD = FG$. BH = horizontal projection of intersection
of stratification and formation planes. Rotate plane $BEGH$ about line
BE. IJ - EF. BJ = strike of stratification prior to tilting N. 26° E.
$KL = CD$. KB is original dip direction of stratification. Angle KBL = dip
of stratification in direction of original dip after secondary tilt.
$MN = CD$. Angle MBN = dip of formation in direction of original dip of
stratification. Original dip of stratification = MBN plus KBL =62°. (After
McLaughlin, 1948)

The measurement of stratification inclination in unconsolidated
sediment presents some special problems. Natural exposures are not
readily available; excavating or trenching is generally necessary. Best
results are obtained if the sediment is wet but not saturated. In dunes
the area may be dampened by flooding, whereas in fluvial and
marginal marine localities the zone of saturation may be lowered by
pumping. Stratification should be examined on two perpendicular
faces to determine inclination. The use of peels (Chapter 10) and

x-ray photography (Chapter 11) is helpful in sediment lacking obvious stratification. Stratification can be determined from cores, but coring must be done carefully to avoid disruption of strata. Best results are obtained by pushing or driving thin-walled tubes into the sediment, then freezing, splitting, and impregnating the sample core. Coring in subaqueous environments has proved particularly bothersome. A box corer, developed by Reineck (1958), shows promise for near-surface sediment, and vibrocorers may eventually provide information to a depth of several feet.

In consolidated beds that have abundant small-scale current-ripple cross-bedding it is difficult to measure dip directions accurately in the field. These beds are commonly very fine grained and have weathered surfaces that obscure bedding. Accordingly, hand samples on which the strike and dip of the sample are marked can be brought into the laboratory, sawed to provide fresh intersecting surfaces, and reoriented to their original attitude. Cross-strata outlined in ink can then be measured.

DESCRIPTION OF CROSS-STRATIFICATION

Two systems are currently used for classifying and describing cross-stratification. One proposed by McKee and Weir (1953) is simple and easy to remember, but it is less complete than a classification by Allen (1963) that attempts to include all major factors that may be encountered.

McKee and Weir recognize three main types of stratification (Fig. 5).

1. *Simple cross-stratification* consists of sets whose lower bounding surfaces are nonerosional, but are represented by nondeposition or a change in character.

2. *Planar cross-stratification* consists of sets whose lower bounding surface is a plane surface of erosion.

3. *Trough cross-stratification* consists of sets whose lower bounding surface is a curved surface of erosion.

The main types of stratification are classified further by considering the shape of each set of cross-strata: *lenticular* — set bounded by converging surfaces, at least one of which is curved; *wedge* — set

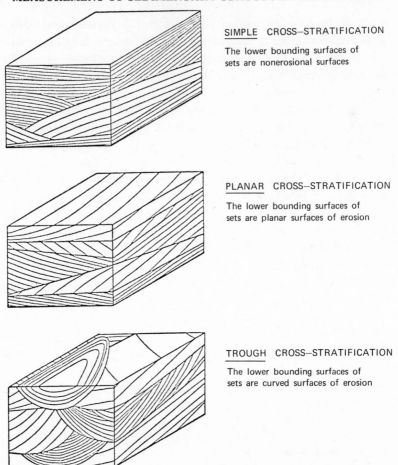

SIMPLE CROSS—STRATIFICATION

The lower bounding surfaces of sets are nonerosional surfaces

PLANAR CROSS—STRATIFICATION

The lower bounding surfaces of sets are planar surfaces of erosion

TROUGH CROSS—STRATIFICATION

The lower bounding surfaces of sets are curved surfaces of erosion

Fig. 5   Classification of cross-stratification. (From McKee and Weir, 1953)

bounded by planar, essentially parallel surfaces. Lenticular and wedge-shaped cross-strata are further described on the attitude of the axis of the set as being *plunging* or *nonplunging*, and as being *symmetric* or *asymmetric*, depending on whether or not the cross-strata on opposite sides of the axial plane correspond in size and shape. Additional differentiation is made on the curvature of individual cross-strata: *convex*, if they arch upward; *concave*, if they arch downward, and *straight*, if they are not arched. The cross-stratification is *high angle* if it has an inclination of 20° or more, and *low angle* if less than 20°. The magnitude of the cross-stratification is described as *small scale* if less than 12 in. (30 cm) in length, *medium scale* if between 1 and 20 ft (30 cm to 6 m) in length, and *large scale* if more than 20 ft (6 m) in length.

Allen (1963) attempts to cover all possible combinations of naturally occurring cross-stratification and to include them in a descriptive classification based on six parameters: grouping, magnitude, environment, lower bounding surface, angular relation, and lithology (Fig. 6). These criteria are used to distinguish 15 types of cross-stratification which, for the sake of brevity, are given designations of the Greek alphabet: alpha-cross-stratification, beta-cross-stratification, and so on (Figs. 7 and 8). To avoid confusion the letter delta is not used.

Alpha-cross-stratification consists of solitary sets that are typically large in scale. The surface beneath each set is nonerosional, planar, and the sets are discordant with respect to lower bounding surface. The sets are lithologically homogeneous; the cross-strata are straight or concave-upward and, in vertical sections, are parallel to maximum dip.

Beta-cross-stratification is similar to alpha-cross-stratification except that the lower bounding surface is essentially a planar erosional surface.

Gamma-cross-stratification is similar to beta-cross-stratification, but is distinguished by having a irregular, erosional lower bounding surface.

Epsilon-cross-stratification is similar to alpha-cross-stratification, but the cross-strata are lithologically heterogeneous.

Zeta-cross-stratification is represented by solitary sets that are large in scale and bounded underneath by an essentially cylindrical erosional surface with a horizontal axis. The cross-strata are concordant with the lower bounding surface and are lithologically homogeneous.

Eta-cross-stratification is similar to zeta-cross-stratification except the lower boundary is a scoop-shaped, erosional surface and the cross-strata are discordant to the lower boundary surface and are lithologically heterogeneous.

Theta-cross-stratification is similar to zeta-cross-stratification, but is situated under a trough-shaped erosional surface that plunges at both ends. The cross-strata are discordant to the lower boundary and are lithologically homogeneous.

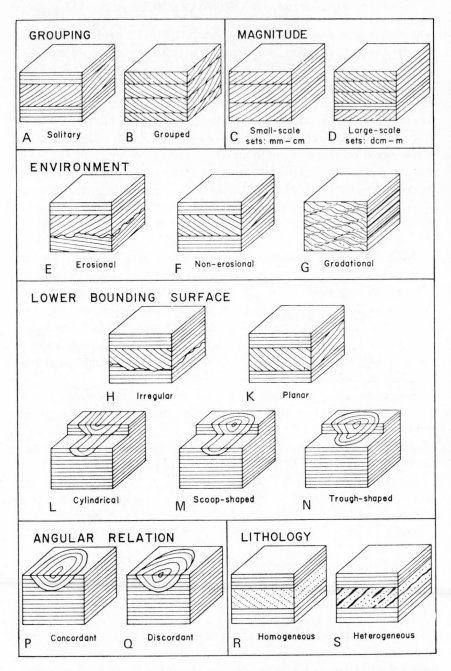

Fig. 6 Descriptive terms applicable to cross-stratified units. (From Allen, 1963)

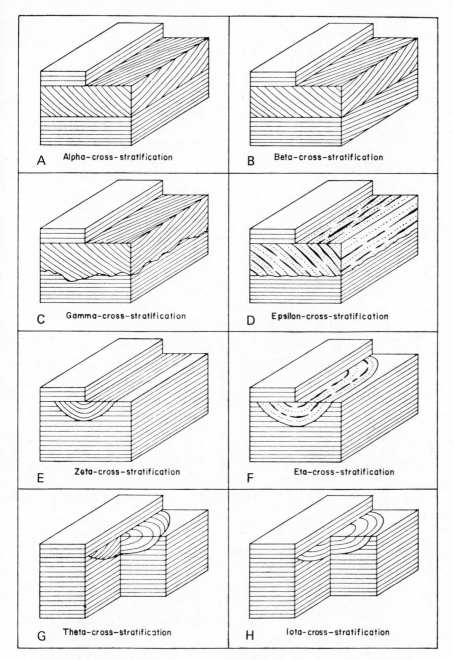

Fig. 7 Generalized block diagrams illustrating types of cross-stratification. (From Allen, 1963)

13

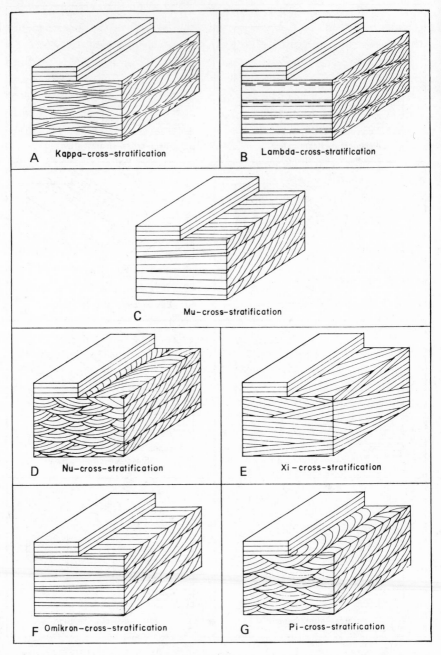

Fig. 8 Generalized block diagrams illustrating types of cross-stratification. (From Allen, 1963)

14

Iota-cross-stratification is similar to theta-cross-stratification, but the cross-strata are concordant with the lower bounding surface.

Kappa-cross-stratification consists of cosets of grouped sets that are individually small in scale. The cosets are bounded by imaginary, irregular gradational surfaces and defined by changes in attitude of the cross-strata. The cross-strata are continuous, but are mostly discordant across boundary surfaces. The cross strata are lithologically homogeneous.

Lambda-cross-stratification is similar to kappa-cross-stratification, but the sets are divided by gradational and planar imaginary surfaces that are defined by pronounced changes in attitude of cross-strata and are discordant to imaginary surfaces. The cross-strata are lithologically homogeneous.

Mu-cross-stratification consists of grouped sets that are individually small in scale. The lower surface of each set is a planar erosional surface. The cross-strata of a set are discordant with lower boundary and are lithologically homogeneous.

Nu-cross-stratification is similar to mu-cross-stratification except that each set is bounded below by a scoop-shaped, plunging erosional surface.

Xi-cross-stratification is represented by grouped sets that are individually large in scale. Each set is underlain by a planar nonerosional surface. The cross-strata are discordant to lower surface and are lithologically homogeneous.

Omicron-cross-stratification is similar to mu-cross-stratification, but has large-scale sets.

Pi-cross-stratification is similar to nu-cross-stratification, except the sets are individually large in scale.

This classification is probably too complicated for field use without notes. It does, however, emphasize essential relations that should be recorded to permit classification in the laboratory and should lead to increased knowledge and a better understanding of stratification.

RIPPLE MARKS

Ripple marks are undulatory bed forms produced in unconsolidated sediments by the action of wind and water currents and by waves. Ripple marks, which occur in a variety of sizes and shapes, are useful in paleocurrent, paleogeographic, and environmental studies. The two most common shapes, symmetrical and asymmetrical, have served as a basis for classification for several years. Recently, however, Tanner (1967) has outlined numerous parameters that may be useful in distinguishing a greater variety of ripple marks. These include: ripple index (RI, wavelength divided by height); ripple symmetry index (RSI, horizontal distance from highest point on crest along the gentler slope to the deepest point in the trough divided by the horizontal distance between crest and trough taken along the steeper slope); continuity index (CI, crest length divided by mean spacing between crests); straightness index (SI, distance parallel to crest, along which curvature can be seen, divided by departure of crest from a straight line); bifurcation index (BI, distance between two bifurcations, along a single crest, divided by mean spacing between crests); parallelism index No. 1 (PI_1, distance parallel to crest, along which curvature can be seen, divided by the term mean spacing between crests, multiplied by maximum crest spacing divided by minimum spacing); and parallelism index No. 2 (PI_2, the difference between maximum and minimum crest spacing divided by mean spacing between crests).

Tanner reports the following results in distinguishing environments:

CI   Values above 15:   almost invariably wind or wave types
Values above 10:   swash, wind, or wave formed
Values below 10:   about 85% current formed
Values below 6:   about 95% current formed
Values below 4:   almost invariably current formed

SI   Values above 100:   almost invariably wind or deep-water waves
Values above 15:   almost invariably wind or waves
Values between 9 and 15:   swash, wind or waves
Values below 6:   almost invariably wave or current formed
Values below 4:   almost invariably current formed

BI   Values above 10:   about 80% wave formed
Values above 6:   about 75% wave formed
Values less than 1.0:   typically wind formed

PI_1 Values above 7:   98% wave formed
Values between 2 and 7:   98% wind, wave or swash formed

Values below 1:   98% current formed

PI$_2$ Values above 0.4:   90% current or swash formed

Values below 0.2:   90% wind or waves

The size classification of ripple marks is arbitrary. Forms with amplitudes to 1 ft are called ripple marks, from 1 to 3 ft, megaripples and greater than 3 ft, sand waves. Analysis of ripple marks should include direction of crest trend and orientation of asymmetry as well as dimensional measurements.

Ripple marks with distinctive shapes and features have the following characteristic names: *backwash* ripple marks—broad, rounded crests with shallow troughs formed parallel to the shoreline by backwash on the beach, wavelength commonly about 18 in., amplitude 3/4 to 1/2 in.; *flat-topped* ripple marks—shallow water ripple marks planed off during falling water level, variable size and symmetry; *linguoid* ripple marks—lobate ripple marks with tongue-like projections formed by current action; *rhomboid* ripple marks—diamond-shaped ripple marks formed on the beach by swash and backwash wavelength commonly 2 to 6 in., amplitude less than 1/8 in.

IMBRICATION

Pebbles and cobbles of unequal dimensions deposited by currents usually have a preferred orientation that produces a shingling or imbrication. In streams and channels with unidirectional flow, pebbles are commonly aligned with the long dimension parallel to the current and dipping upstream. Gravel on beaches is less well oriented, but the long dimension tends to dip toward the ocean.

Sand grains are also deposited by currents in preferred orientations with the long dimension dipping upcurrent. On the beach, however, the imbrication, in some instances, is toward the shore (Curray, 1956), and backwash appears to control deposition. With cobbles and pebbles, wave swash orients the particles because the permeability of the beach reduces markedly the surface backwash.

Orientation of disk-shaped pebbles is described in terms of the short axis, wheras the orientation of triaxial forms is based primarily on the long axis. Dip and direction of the defining axis, as well as size and other parameters, such as composition, should be recorded. As with stratification, structural attitude of the containing strata should be measured as well as initial dip, which may be significant in some alluvial deposits. Sand grain orientation can be measured in thin sections of carefully collected samples of known orientation (Chapter 12). Unconsolidated or slightly consolidated sediments should be impregnated with plastic to facilitate sectioning (Chapter 9).

LINEAR STRUCTURES

The orientation of linear structures in flat-lying strata is measured by simply aligning a compass parallel with the structure and recording the direction.

Several techniques are used to measure linear structures in folded strata, but all employ some way for correcting the tilt of strata. One of the simplest is to measure the strike of the bed; then measure the rake (acute angle between strike and lineation in the plane of bedding) of the lineation. The orientation is found by adding (or subtracting) the rake to the strike.

Another method is to (1) scratch a pencil line on the bedding plane parallel to the strike, (2) align the long edge of a notebook along the strike line, (3) clamp a pencil to the notebook (using an elastic band) so that it is parallel to the lineation, and (4) rotate the notebook to horizontal and measure the orientation of the pencil with a compass. If the beds are overturned, the notebook must be rotated through an angle greater than 90°. Using a clear sheet of plexiglass about 4 x 8 x 1/8 in. in place of the notebook aids in aligning the pencil parallel with the structure. Specialized devices for speeding up data collecting by this field-tilt-correction technique are described by Cloos (1933) and ten Haaf (1959).

Both techniques described compensate for the rotation of bedding around the line of strike only. If the directional structure occurs on the limb of a plunging fold, the bed has also undergone rotation around a vertical axis. If the plunge of the fold is known, a correction for the orientation can be made using the tables of Norman (1960) or Ramsey (1961). Corrections may be ignored for simple folds that plunge less than 25°.

DISCUSSION

Symmetrical structures, such as groove molds, parting lineation, and oriented casts, define a line of orientation only and not a direction or sense. For example, for a structure oriented east-west, it is immaterial whether the orientation is recorded as east ($90°$) or west ($270°$). In contrast, flute molds, cross bedding, and current ripple marks define a sense of motion along a line. By convention, the direction that the current moved *toward* is recorded. Direction determination is facilitated by using an azimuth compass (marked $1°$ to $360°$), although a quadrant compass is suitable if the compass is read carefully. For most work a compass need be read only to the nearest multiple of $5°$.

Generally, only one directional reading should be taken from each sedimentation unit at an outcrop. This enables the data to be interpreted from the viewpoint of frequency of events (currents). An exception is made when the orientation of different structures within the same stratigraphic unit is compared. A careful estimation of the preferred orientation is adequate in a sedimentation unit in which individual elements (such as groove molds) show considerable scatter, but have a distinct, preferred orientation. If, however, two modes are visible, each should be recorded; if no mode is apparent, it is necessary to record the orientation of many single elements, compute a mean and then test the data for significance.

REFERENCES

Allen, J. R. L., 1963, The classification of cross-stratified units with notes on their origin, *Sedimentology,* 2, 93–114.

Armstrong, G. C., 1967, An instrument for measuring planes and vectors in space, *J. Sed. Pet.,* 37, 1241–1243.

Campbell, C. V., 1967, Lamina, laminaset, bed and bedset, *Sedimentology,* 8, 7–26.

Cloos, H., 1933, Primare Richtungen in sedimenten der rheinischen Geosynkline, *Geol. Rundschau,* 29, 357–367.

Curray, J. R., 1956, Dimensional grain orientation studies of Recent coastal sands, *Bull. Am. Assoc. Petrol. Geologists,* **40,** 2440–2456.

Haaf, E. ten, 1959, Graded beds in the northern Apennines, Ph.D. thesis, Groningen, Holland, 102 pp.

Krynine, P. D., 1948, The megascopic study and field classification of sedimentary rocks, *J. Geol.,* **56,** 130–165.

McKee, E. D., and G. W. Weir, 1953, Terminology for stratification and cross-stratification, *Bull. Geol. Soc. Am.,* **64,** 381–390.

McLaughlin, K. P., 1948 Secondary tilt: a review and a new solution, *J. Geol.,* **56,** 72–74.

Norman, T. N., 1960, Azimuths of primary linear structures in folded strata, *Geol. Mag.,* **97,** 338–343.

Parizek, R. R., 1967, Strike-dip indicator, *J. Sed. Pet.,* **37,** 1249–1250.

Pryor, W. A., 1958, Dip direction indicator, *J. Sed. Pet.,* **28,** 230.

Ramsay, J. G., 1961, The effects of flooding upon the orientation of sedimentation structures, *J. Geol.,* **69,** 84–100.

Reineck, H. E., 1958, Kastengreifer und Lotrohre "Schnepfe" Gerate zur Entnahme ungestorter, orientierter Meeresgundproben, *Senckenbergiana Lethaea,* **39,** 42–48, 54–56.

Tanner, W. F., 1967, Ripple mark indices and their uses, *Sedimentology,* **9,** 89–104.

CHAPTER 2

MATHEMATICAL TREATMENT OF ORIENTATION DATA

LEE R. HIGH, JR.

Oberlin College, Oberlin, Ohio

M. DANE PICARD

University of Utah, Salt Lake City, Utah

Sedimentary structures are used extensively in a variety of geologic problems. Because they are commonly primary and are formed at the site and time of deposition, sedimentary structures are among the most useful criteria for interpreting depositional environments. Furthermore, because many sedimentary structures have directional symmetries, considerable study has been devoted to determining the relationship between structure orientation and either the current system of the depositing medium or the geography of the basin. The environmental significance of sedimentary structures is beyond the scope of this chapter; the reader is referred to recent studies for further information (Klein, 1962; Lane, 1963; McKee, 1957 and 1964; Ore, 1964; Pettijohn, et al., 1965; Picard, 1957 and 1967; van Straaten, 1954 and 1959; and Visher, 1965). In this chapter are discussed some of the common methods for analyzing orientation data from sedimentary structures. Although we emphasize sedimentary structures, the statistical methods described here can be applied to other types of directional data, such as oriented fossil debris and sedimentary grains. The treatment is not exhaustive and only standard methods are described. For special techniques and applications, the reader is referred to appropriate papers.

DATA EVALUATION AND RETENTION

Data Collection

The collection of orientation data from sedimentary structures is described in the preceding chapter. Structures yielding orientation data are generally of two classes, linear and planar. In addition, each class yields single-ended data and double-ended data. Double-ended data describe an orientation, but not the sense. For example, linear features, such as parting lineation, aligned bryozoa, or parallel grain axes, can have an orientation of 150° and 330°. Without further information, the actual direction of transport is indeterminate. Single-ended data, such as asymmetric ripple marks (linear), cross-stratification (planar), or imbricate grain fabrics, indicate a preferred direction. The techniques and illustrative examples described here utilize only single-ended data. Most of the methods can be modified, however, for double-ended data by multiplying all angles by two and substituting 2θ for θ in all equations.

In collecting orientation data, each structure is recorded separately. If data from different structures are grouped indiscriminately, serious mistakes will occur if the structures do not record the same direction events. Since the relative orientations of sedimentary structures are not consistent in all settings, the orientation relationships must be determined. For example, ripple marks and cross-stratification can yield similar directions in one geographic setting but occur at right angles to one another in another setting. Equivalence (or non-equivalence) of different structures therefore must be demonstrated. Once equivalence is established, data can be grouped for analysis. In most rock units, only two or three structures are abundant enough to yield reliable paleocurrent directions. Minor structures are omitted and the equivalence of only major structures is demonstrated.

Tests for Equivalence

Several techniques can be used to demonstrate the equivalence of data derived from different structures. Each structure represents a population whose members are directions, usually grouped by compass intervals (i.e., 0° to 29°, 30° to 59°, and so on). The data consist of the number of orientation measurements (or percentage of measurements) within each interval. In a first approximation, the data are inspected visually and separate structures are said to have "similar" or "dissimilar" distributions of orientation measurements. For the data in Table 1 cross-stratification and ripple marks appear to be similarly oriented.

Several statistical tests are suitable for determining equivalence by testing if two or more samples are drawn from a single population.

TABLE 1   Paleocurrent directions in Crow Mountain Formation, Wyoming (after Tohill and Picard, 1966, p. 2563)

	Section Numbers					
	12		13		16	
			Structure			
Compass Intervals	Cross-strat.	Ripple Marks	Cross-strat.	Ripple Marks	Cross-strat.	Ripple Marks
0–29	0	0	0	1	1	0
30–59	5	4	3	0	2	2
60–89	4	2	5	0	3	3
90–119	2	0	0	0	1	4
120–149	2	0	1	0	0	2
150–179	1	0	1	0	0	1
180–209	0	0	1	0	3	1
210–239	0	1	5	2	0	4
240–269	0	2	1	3	1	1
270–299	1	1	1	0	0	0
300–329	0	0	1	2	0	0
330–359	0	0	0	0	1	1
Total	15	10	19	8	12	19
Arithmetic Mean	97°	135°	157°	232°	135°	151°
Standard Deviation	±65°	±105°	±91°	±95°	±101°	±82°

The most frequently used are Student's t test and the χ^2 test, which are discussed in standard statistics texts (Dixon and Massey, 1951, pp. 97-109, 184-191; Miller and Kahn, 1962, pp. 52-133; Krumbein and Graybill, 1965, pp. 132-134; and Griffiths, 1967, pp. 321-364). A short and useful description of these tests is also included in Folk (1968, pp. 57-61). However, the usefulness of these two tests is limited by several factors. The t test assumes normal distribution which is not always justified for orientation data. Furthermore, unimodality is required for the t test. However, in most paleocurrent studies, so few observations are made at each locality that modality is not obvious and cannot be assumed (Table 1). Rather, modality must itself be determined statistically. Furthermore, the t test can be used to compare only two populations at a time. Thus the t test can rarely be used for paleocurrent data. With the χ^2 test the expected number in each interval must exceed five for the test to be valid. Thus a minimum of 60 observations are required if 30° intervals are used. Few paleocurrent studies have data this extensive. For the data in Table 1 these limitations do not permit the use of either the t test or the χ^2 test. These objections are overcome, however, by using an additional statistic, the Kolmogorov-Smirnov test, to test orientation populations for equivalence.

The Kolmogorov-Smirnov test is a graphic means of measuring the spread between two or more frequency curves and estimating the probability that the two curves are samples of an homogeneous population. For a given sample size, as the spread decreases, the probability that the samples are from a single population increases. The following example illustrates the use of this statistic.

Table 1 presents data from the Crow Mountain Formation (Triassic) of central Wyoming (taken from Tohill and Picard, 1966, p. 2563). This sandstone and siltstone unit contains abundant cross-stratification and asymmetric ripple marks. Cumulative frequency plots of the orientation data are shown in Fig. 1. The curves for ripple marks and cross-stratification are similar. In the Kolmogorov-Smirnov test, the null hypothesis is that the separate frequency curves describe samples drawn from different populations. To test the probability of the null hypothesis, the maximum spread of the two cumulative frequency curves is measured ("d_n" in Fig. 1). The minimum expected spread, assuming each curve is a different population (null hypothesis), is calculated from the formula:[1]

$$d_{.05} = 136\sqrt{1/n_1 + 1/n_2}$$

where n_1 and n_2 are the number of measurements for each curve, and the subscript indicates the probability of error. Thus, if the maximum observed spread (d_n) is less than the expected value $(d_{.05})$, the null hypothesis is rejected at the .05 significance level, and it is concluded that the samples are drawn from a single population. As shown in Fig. 1, at each of three Crow Mountain sections, ripple marks and cross-stratification can be considered to comprise a single population. Thus orientations of cross-strata and ripple marks can be grouped for further analysis.

Stratigraphic Variation

In addition to recording separately the orientations of different structures, the stratigraphic position of orientations should be recorded. Current patterns may have changed during deposition, especially if the sequence contains unconformities or marked facies changes. If the current pattern did change, directions obtained from different positions within a rock unit will represent different populations. If data are not recorded stratigraphically, these changes may not be recognized and the value of paleocurrent determinations is questionable. To test for the existence of stratigraphic variation in orientation, rock units are divided into subequal members, either natural or arbitrary, and orientation directions for each unit are evaluated for equivalence with the other units. In the remainder of this chapter, we assume that all data at a locality are homogeneous.

CORRECTION OF DATA FOR TECTONIC TILT

When strata are flat-lying, structure orientations measured in the field are the same as when formed; they can therefore be analyzed

---

[1]The value of $d_{.05}$ can be read from tables and graphs prepared by Dixon and Massey (1951, pp. 256–258) and Miller and Kahn (1962, pp. 464–470). These sources should be consulted for more detailed descriptions of the Kolmogorov-Smirnov test. The actual probability for any observed d_n can be calculated from tables in Smirnov (1948, pp. 279–281).

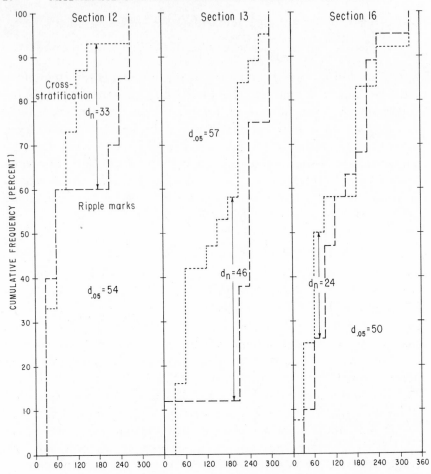

Fig. 1 Cumulative frequency curves for ripple mark and cross-stratification directions at three Crow Mountain localities, central Wyoming. See text for discussion of results. Data are from Tohill and Picard (1966, p. 2563). Ripple marks measured perpendicular to crests in direction of foreset laminae. Cross-stratification measured in direction of maximum foreset inclination.

without further treatment. However, if the beds have been tilted subsequent to formation of the structures, the orientations may have changed. Thus in areas of local or regional tilting, the data may require correction before they can be analyzed. The methods used generally rely on rotation of the beds about the strike through the angle of dip. Necessity and methods differ, however, for linear and planar structures; each is discussed separately.

Linear Structures

Orientations of linear structures are less disturbed by tectonic tilting than are orientations of planar structures. For dips less than 25°, the maximum possible error between the measured and original azimuths is 3° (Ramsay, 1961, p. 86). Thus, based on the errors of field measurement of structure orientation and the deviation from straightness of linear structures, tilt corrections can be ignored at localities where the structural dip is less than 25°. However, where the dip is more than 25°, possible errors become large and data require correction.

Tilt correction is achieved by rotating the bed, keeping the angle between the structure orientation and the strike of the bed (axis of rotation) constant. When the bed has been rotated to the horizontal, the original orientation is restored. Descriptive geometry or stereographic projections can be used for bed rotation. These techniques are described in standard texts (Donn and Shimer, 1958; Higgs and Tunell, 1959) as the simple rotation of tilted beds. The data can be corrected, however, in the field at the time of collection. This procedure, which is usually more convenient than making later corrections, is the general practice.

To correct linear orientation data in the field, it is necessary to remember that the angle, in the plane of bedding, between the strike of the bed (axis of tilt) and the structure orientation (in tectonic nomenclature this angle is the "rake") is independent of the amount of folding or dip. The attitude of the bed does not affect this angle. This can be demonstrated by marking lineations on an index card. If the card is rotated about an axis, the angle between the axis and the lineation remains constant. The angle is a function of the orientation of the axis of rotation (strike) and not the amount of rotation (dip). This angle can be determined in the field by marking the strike of the beds on the bedding surface and, with a protractor or Brunton clinometer, measuring the angle between the strike and the structure orientation. The original orientation of the structure is then determined by adding the rake angle to the strike of the beds if the angle opens clockwise relative to the strike line, and subtracting the rake angle if the angle opens counterclockwise. The result is the orientation of the structure before tilting.

To summarize, linear orientations require no correction if the structural dip is less than 25°. In areas of greater dip, the original orientation is determined by adding or subtracting the rake angle to the strike of the beds.

Planar Structures

The orientation of planar structures is more sensitive to tilting than that of linear structures and the data should always be corrected for dip. Cross-stratification is the most abundant type of planar structure from which orientations are determined. The axis of trough or festoon cross-stratification is linear and treated as other linear structures. If cross-stratification orientations are determined by measuring the direction of maximum foreset inclination, other techniques are necessary. The correction is based on the "two tilt problem" described in standard texts (Donn and Shimer, 1958; Higgs and Tunell, 1959). Although the problem can be solved trigonometrically, stereographic projections are easier. Fig. 2 illustrates the steps involved in the stereographic correction.

Special Cases

In the foregoing discussion it was assumed that structural tilt was the product of flexural folding about a single horizontal axis. In many areas these assumptions are justified. Exceptions are found, however, and Ramsay (1961) has developed methods for correcting orientation data for plunging folds and shear folds.

ANALYSIS OF DATA

Many techniques are used to analyze paleocurrent data. The numerous methods partly reflect general dissatisfaction with available techniques, since widely used methods have failed to resolve individual cases. To a large degree the mathematics now in use were developed in the 1950's during a period of considerable interest in circular or periodic statistics and their application to geology. Analytical techniques described here are based mainly on theory

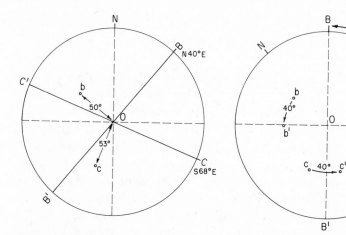

Step 1: <u>Given</u>:

Bedding: Strike N40°E (BOB'), Dip 50°SE
Cross-stratification: Strike S68°E (COC'),
    Dip 53°NE
<u>Plot</u>:
Pole of bedding (b)
Pole of cross-stratification (c)

Step 2a: Rotate bedding to horizontal

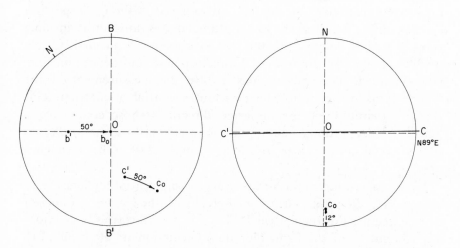

Step 2b: Rotate bedding to horizontal
         (cont.)

Step 3: Read original orientation of
        cross-stratification.
        Strike: N89°E (COC'), Dip 12°N
        Current direction: N1°W

Fig. 2   Stereographic correction for tectonic tilt.

developed by these workers (see Jones, 1968, for a discussion of this problem and for references) and the interested reader is referred to these papers.

In general, the analysis of structure orientations involves two problems. The first problem is to establish the preferred orientation (or orientations) at each locality. Some measure of the orientation and the confidence in that measure must be determined. The second problem is to synthesize the calculated direction (or directions) at each locality into a regional pattern. From this pattern are drawn conclusions about the paleogeography and environment. It should be apparent from the foregoing that complications arising from possible or real polymodality must be considered. This final aspect, unimodal versus polymodal distributions, is commonly ignored, thereby reducing the validity of particular studies.

MEASURES OF CENTRAL TENDENCY

The primary purpose of paleocurrent studies is to determine and to interpret preferred orientations. To this end, methods that generate preferred directions from the data are required. Although inspection can determine the direction, a procedure that is satisfactory in some general studies where only the overall direction is needed, formal methods that yield the preferred direction and confidence limits are recommended. These techniques involve the measurement of the tendency of the data to cluster about a central orientation. This central orientation is the preferred direction whose confidence is determined by testing a null hypothesis based on the assumption that the observed cluster arises by random sampling of an homogeneous population.

A convenient and rapid means of measuring central tendency is by determining the modal class, the interval with the greatest number of measurements. For the data in Table 1, the modal classes for sections 12, 13, and 16 of the Crow Mountain Formation are 30 to 59°, 210 to 239°, and 60 to 89°, respectively. Midpoints of these intervals are preferred paleocurrent directions.

A common method of measuring central tendency is to calculate the mean of the data. Arithmetic means, which were calculated in Table 1, are not generally suited to circular data. If observations span the origin, the calculated mean can be opposed to the actual mean. As an extreme example, the mean (average) of 45° and 315° is 180°, rather than the actual value of 360°. If the data are properly transformed, by adding a constant to all orientations, this problem can be overcome and arithmetic means calculated. However, this procedure greatly increases the risk of computational error. In addition, the selection of the proper transform value is difficult when the data are scattered over the compass. If an inappropriate transform value is used, the arithmetic mean is unpredictably biased.

Because orientation data are circular, rather than linear, techniques of vector analysis are generally used. Single-ended orientation measurements are vectors that indicate both direction and magnitude. For each observation the magnitude is unity, whereas repeated directions or measurements within an interval have correspondingly higher magnitudes. All measurements together form a population that generally has one or more modes. Each mode is described by two features, orientation and dispersion. Since the mean or resultant vector describes both modal orientation and dispersion, it is an adequate description of the entire population. The orientation of the resultant vector is a measure of the modal orientation, and the magnitude of the resultant is a measure of the dispersion of data within the mode. Thus the mathematical determination of central tendency within a population requires three steps: calculation of the resultant vector; calculation of the magnitude of the resultant vector; and testing of the calculated vector against random distributions.

Graphic Techniques

Orientation vectors can be analyzed graphically for the vector resultant. For ungrouped data each observation is plotted as a unit vector. For grouped data (Table 1) the length of the vector is proportional to the number of observations within each interval. Fig. 3 shows the vector resultant for section 16 of the Crow Mountain (Table 1). The consistency ratio is calculated by dividing the length of the resultant by the sum of the individual vectors (Reiche, 1938, p. 913) and is inversely related to the standard deviation (Reiche, 1938, Table 1, p. 915). Thus a consistency ratio of 1 represents perfectly preferred orientation, and a value of 0 represents random orientations.

Section 16

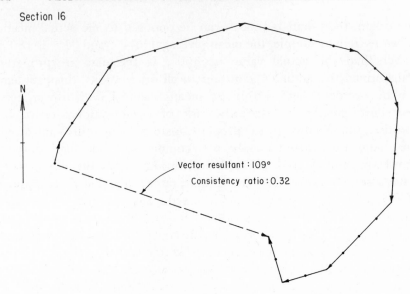

Fig. 3 Graphic determination of vector resultant. The consistency ratio is calculated by dividing the length of the vector resultant by the sum of the lengths of the component vectors.

Calculation of Vector Resultant

The vector resultant can also be calculated from the formula

$$\tan \theta = \frac{\Sigma \, n_o \sin \theta_i}{\Sigma \, n_o \cos \theta_i},$$

where θ = azimuth of vector resultant,

n_o = number of observations per interval (grouped data),

θ_i = midpoint of ith interval (grouped) or direction (ungrouped) in degrees.

The magnitude r of the vector resultant is given by

$$r = \sqrt{(\Sigma \, n_o \sin \theta_i)^2 + (\Sigma \, n_o \cos \theta_i)^2} \; .$$

A modification of the χ^2 test known as the Tukey χ^2 test (Harrison, 1957; Rusnak, 1957; Middleton, 1965, 1967) is widely used. The orientation of the vector resultant is calculated from the formula

$$\theta = 0° + \alpha \quad \text{if} \quad C \geqslant 0$$

or

$$= 180° + \alpha \quad \text{if} \quad C < 0$$

where

$$\tan \alpha = S/C,$$

$$C = \frac{\Sigma N \cos \theta_i}{(\Sigma \cos^2 \theta_i)^{1/2}},$$

$$S = \frac{\Sigma N \sin \theta_i}{(\Sigma \sin^2 \theta_i)^{1/2}},$$

$$N = \frac{n_o - n_e}{\sqrt{n_e}}$$

and where θ = azimuth of vector resultant,

θ_i = midpoint of ith interval in degrees,

n_o = number of observations per interval,

n_e = expected number of observations per interval (that is, average).

The resultant vector, θ, is an estimate of the central tendency whose significance is tested by the formula

$$\chi^2 = C^2 + S^2 \quad \text{for 2 degrees of freedom.}$$

A table showing general significance of various values of χ^2 is given in Dixon and Massey (1951, p. 308) and is reproduced partially here (Table 2). If the calculated value of χ^2 exceeds the value in the table, the preferred orientation, θ, is significant at that level. The table gives therefore the probability that a minimum value of χ^2 calculated from the observations is not the product of random orientations. If the calculation of χ^2 yields an unacceptably low general significance value, the null-hypothesis (random distribution) is accepted. Intermediate χ^2 values may reflect polymodal distributions. Middleton

(1965, p. 548; 1967) has noted that this use of the χ^2 test avoids the usual restriction that each class contain at least five measurements. The classes in the Tukey χ^2 test are the dimensions C and S, not the compass intervals. Thus the Tukey χ^2 test is useful for localities where ten or more directions are measured.

TABLE 2   Chi-square probability for two degrees of freedom

			Probability		
	90	95	97.5	99	99.5
χ^2	4.61	5.99	7.38	9.21	10.60

Table 3 is a work sheet, modified after Harrison (1957, p. 103), in which the data from section 16 of the Crow Mountain (Table 1) are analyzed by the Tukey procedure. The preferred direction of combined ripple marks and cross-stratification at this locality is 111°; the general significance of this direction is 0.975. There is a probability therefore of less than 0.025 that the observed concentration of measurements about 111° was produced from an isotropic population. It is appropriate to mention here that the Tukey χ^2 test for grouped data and the Rayleigh test for ungrouped data are essentially equivalent (Durand and Greenwood, 1958, pp. 230-231).

The methods just described for measuring central tendency should be applied only to unimodal distributions. If the distribution is polymodal, single estimates of mean direction yield intermediate directions that do not reflect the actual distribution. Because many paleocurrent patterns are unimodal, this restriction is not serious. Furthermore, each of the above methods is applicable to the individual modes of a polymodal distribution, yielding modal vector resultants and modal vector means. The Kolmogorov-Smirnov test can then be used to demonstrate the independence of the several modes. For extensive studies, however, the calculations are lengthy and time consuming. Additionally, it is difficult to define separate modes in many polymodal distributions. Individual modes can overlap and, if both are non-normal, they cannot be differentiated. Additional techniques are needed, therefore, for polymodal distributions.

TABLE 3   Work sheet for Tukey χ^2 test (after Harrison, 1957, p.103)
Data from Crow Mountain Formation (Tohill and Picard, 1966, p. 2563)

Intervals	θ_i	n_o	$n_o - n_e$	N	$\cos \theta_i$	$N \cos \theta_i$	$\sin \theta_i$	$N \sin \theta_i$
0−29	15	1	−1.6	−1.2	.97	−1.2	.26	−0.3
30−59	45	4	1.4	1.1	.71	1.1	.71	0.8
60−89	75	6	3.4	2.6	.26	0.7	.97	2.5
90−119	105	5	2.4	1.8	−.26	−0.5	.97	1.4
120−149	135	2	−0.6	−0.5	−.71	0.4	.71	−0.4
150−179	165	1	−1.6	−1.2	−.97	1.2	.26	−0.3
180−209	195	4	1.4	1.1	−.97	−1.1	−.26	−0.3
210−239	225	4	1.4	1.1	−.71	−0.8	−.71	−0.8
240−269	255	2	−0.6	−0.5	−.26	0.1	−.97	0.5
270−299	285	0	−2.6	−2.0	.26	−0.5	−.97	1.9
300−329	315	0	−2.6	−2.0	.71	−1.4	−.71	1.4
330−359	345	2	−0.6	−0.5	.97	−0.5	−.26	0.1
Total		31				−2.5		6.5

$n_e = 2.6$
$C = -1.0$
$S = 2.6$
$\tan \alpha = -2.6$
$\alpha = -69°$
$\theta = 111°$
$\chi^2 = 7.8$

For 30° intervals

$$(\Sigma \sin^2 \theta_i)^{1/2} = (\Sigma \cos^2 \theta_i)^{1/2} = 2.46$$

Polymodal Distributions

Polymodal distributions can range from simple, bimodal opposed patterns to complex, nonsymmetric patterns. Bimodal opposed patterns can be analyzed by a modification of the Tukey χ^2 test proposed by Middleton (1965, 1967) or by a graphic method devised by Selley (1967). Complex polymodal distributions are conveniently analyzed by a method developed by Tanner (1955).

To analyze polymodal distributions, the compass is divided into intervals and the number of observations within each interval recorded. We recommend 30° or 45° intervals. Larger intervals are insensitive to trends and smaller intervals are not justified in view of field measurement errors. Table 4 lists measurements taken from three localities in the Red Peak Formation of Wyoming (Picard and High, 1968). The standard deviations are calculated from the formula

$$s = \sqrt{\frac{\Sigma(n_o\theta_i)^2 - \dfrac{(\Sigma n_o\theta_i)^2}{\Sigma n_o}}{\Sigma n_o - 1}},$$

where n_o is the number of observations per interval, and θ_i is the midpoint of the ith interval.

TABLE 4   Paleocurrent orientations, Red Peak Formation, Wyoming
(after Picard and High, 1968)

Compass Intervals	Stratigraphic Section		
	Red Grade	Horse Creek	Rawlins
0–29	22*	7*	9*
30–59	20*	6	10*
60–89	6	1	4
90–119	1	2	5
120–149	1	2	3
150–179	1	0	6
180–209	4	5	5
210–239	3	8*	5
240–269	7	8*	5
270–299	12	4	0
300–329	24*	6	5
330–359	17	3	6
Total	118	52	63
Average (n_e)	9.8	4.3	5.2
Standard Deviation (s)	8.8	2.7	2.1

*Significant concentrations $(n_o \geqslant n_e + s)$.

Those intervals containing an observed number of measurements (n_o) within one standard deviation of the expected or average number (n_e) are considered indistinguishable from random distribution. Those intervals that are more than one standard deviation above or below the expected number represent significant concentrations or voids in the distribution. For one standard deviation the general significance is 0.67. Intervals that exceed the expected value by more than two standard deviations are significant at the 0.95 level. For this test a high level of significance should not be demanded. When three or more broad modes are distributed over eight or twelve intervals, overlapping and interference are to be expected. Consequently, the 0.67 level of significance is considered acceptable and indicative of a preferred direction.

REGIONAL ANALYSIS

Once preferred directions at localities are determined, the data are analyzed for regional paleocurrent patterns. Aside from subjective evaluation (visual inspection of data), the most frequently employed technique uses moving averages. The procedure is illustrated in Fig. 4. An arbitrary grid is superimposed on the map showing preferred directions at each locality (Fig. 4a). All directions within mutually adjacent blocks are averaged, and the result plotted at the common corner (Fig. 4b). The net effect is to smooth local irregularities and to distribute evenly the current symbols across the map without regard to outcrop control.

Alternatively, trend surface analysis can be used to determine the regional paleocurrent pattern. Agterberg, Hills, and Trettin (1967) recently described use of this procedure in the analysis of paleocurrent data.

For markedly polymodal distributions, it may not be possible to analyze quantitatively regional paleocurrent patterns. When it is difficult to differentiate regional systems among several modes, subjective evaluation is necessary. For example, neither the simple bimodal pattern in the Sirte Basin (Selley, 1967) nor the complex quadramodal pattern of the Red Peak Formation (Picard and High, 1968) can be reduced by moving averages or trend surface analysis. Rather, inspection determines regional paleocurrent patterns. Consequently, it is difficult to recognize variations in the pattern across the map area.

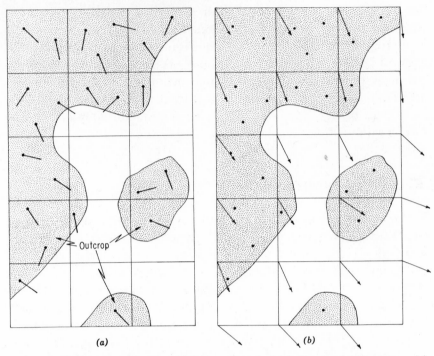

(a)   (b)

Fig. 4 Determination of regional paleocurrent trends by moving averages. (*a*) Preferred direction is plotted at each locality. (*b*) Mean or average directions for adjacent squares are plotted at common corner.

PRESENTATION OF DATA

Presentations of paleocurrent data must satisfy at least two requirements, the actual observations should be given and the presentation should be graphic to enable visual comprehension of major trends. Within the framework of these restrictions several methods are possible.

A tabular listing of observations (Tables 1, 3, and 4) satisfies the first requirement. All data are presented and available to other workers. These tables should be included. If the paleocurrent distributions are unimodal and vector means or resultants are calculated, these can be represented on a map by an arrow. At each locality the direction of the arrow indicates the direction of preferred orientation, and the length is proportional to the consistency ratio or to some other measure of vector strength. If standard deviations are calculated, these can be shown as arcs through which the arrow passes.

Alternatively, data can be shown directly on rose diagrams (Fig. 5a). Its major advantage is that the total distribution is given graphically. Thus, for distributions that are not normal and unimodal, rose diagrams are preferable to arrows. Although rose diagrams present actual observations that are grouped into intervals, tabular listing of the data is still necessary. Rose diagrams tend to be difficult to read for data and are commonly inadequate for presenting observations.

For complex polymodal distributions, rose diagrams are ambiguous. Overlapping modes tend to be obscured and difficult to recognize. For these distributions, compass diagrams (Fig. 5b) are recommended. They present the data and indicate major modes and nodes in an easily interpreted, visual manner. For most polymodal distributions, compass diagrams are satisfactory for presenting graphically the data and the initial level of interpretation.

MODELS AND INTERPRETATION

The final stage in an analysis of paleocurrents is evaluation of the regional pattern. At present the interpretation of pattern is largely subjective. Klein (1967, Table 1, p. 373) and Selley (1968) have summarized patterns associated with various environments with respect to modality and relations to slope and source directions. Their disagreements emphasize the need for additional information from modern environments if paleocurrent patterns are to be related firmly to specific environments. For example, based on a few Recent studies, it is tentatively concluded that shallow marine currents are quadramodal and the onshore and offshore directions are more abundant than longshore directions. This pattern can be confirmed only by additional studies. If complex polymodal distributions are to be interpreted with any degree of confidence, additional information on modern and ancient environments is required.

An interesting example of the possible interpretive use of quantitative paleocurrent data was suggested by Hamblin (1958). Hamblin suggested that the standard deviation in fluvial sedimentary rocks is a measure of the degree of stream meandering (1958, pp. 51-57).

Regional increases in the standard deviation of cross-stratification should correspond therefore to decreases in the stream gradient as base level is approached. Hamblin confirmed this relationship by examining modern streams. Thus the degree of paleocurrent dispersal in fluvial units is a measure of distance from the stream mouth. The discovery of similar relationships for other paleocurrent patterns would add to the interpretive value of the patterns.

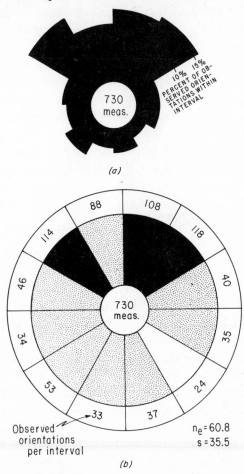

Fig. 5   Presentation of paleocurrent data. Data are from Red Peak Formation (Picard and High, 1968). (a) Rose diagram. (b) Compass diagram. Stippled segments are within one standard deviation of the mean number of occurrences. Solid segments are more than one standard deviation above, and blank segments below, the mean.

SUMMARY OF PALEOCURRENT TECHNIQUES

Fig. 6 presents a review of the suggested techniques for analyzing paleocurrent data. Orientations of sedimentary structures are corrected first for tectonic tilt. The Kolmogorov-Smirnov statistic tests the corrected orientations for equivalence and stratigraphic homogeneity. The tests need not be performed for each locality. Rather, if localities with the best control show no significant differences among data from different sedimentary structures or diverse stratigraphic units, the tests are considered successful. The data at each locality should be listed in a table and inspected for modality. Vector resultants and magnitudes should be calculated for all data and the vectors tested by the Tukey χ^2 (or Rayleigh) test. If the test is successful, the distribution is unimodal and the calculated preferred directions are plotted as an arrow. Regional patterns can then be determined by moving averages or trend surface analysis. If the test indicates that the vector is not significant, the distribution is either random or polymodal. Polymodal distributions are plotted on compass diagrams and preferred directions determined by Tanner's modal analysis. Regional trends of polymodal distributions are generally determined subjectively. Final interpretation of paleocurrent data is made by analogy to Recent environments and current systems.

ACKNOWLEDGMENTS

We thank C. D. McCormick, G. V. Middleton, S. S. Streeter, and W. R. Vroman for critical review and suggestions for improvement of this chapter. V. R. Picard typed the manuscript; the illustrations were drafted by K. J. Braun. Partial financial support from the National Science Foundation (Picard, grant GA-1002) and the Petroleum Research Fund (Picard and High, grant 3217-A2) for other projects was helpful. This chapter is a joint effort, shared equally by us.

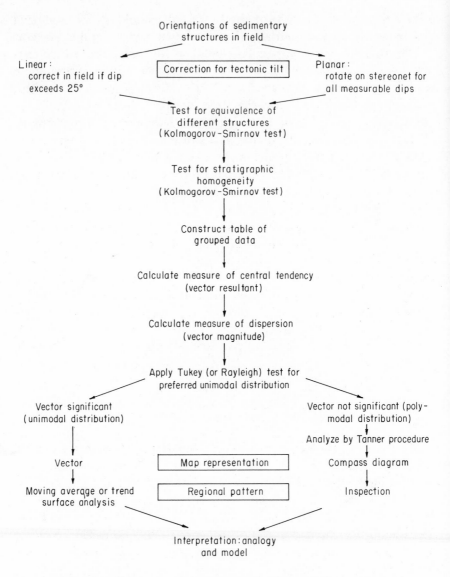

Fig. 6   Summary of techniques for analyzing paleocurrent data.

REFERENCES

Agterberg, F. P., L. V. Hills, and H. P. Trettin, 1967, Paleocurrent trend analysis of a delta in the Bjorne Formation (Lower Triassic) of northwestern Melville Island, Artic Archipelago, *J. Sed. Pet.,* **37**, 852–862.

Dixon, W. J., and F. J. Massey, Jr., 1951, Introduction to statistical analysis, McGraw-Hill Book Co., 370 pp.

Donn, W. L., and J. A. Shimer, 1958, Graphic methods in structural geology, Appleton-Century-Crofts, 180 pp.

Durand, D., and J. A. Greenwood, 1958, Modifications of the Rayleigh test for uniformity in analysis of two-dimensional orientation data, *J. Geol.,* **66**, 229–238.

Folk, R. L., 1968, Petrology of sedimentary rocks, 2nd ed., Hemphill's, 170 pp.

Griffiths, J. C., 1967, Scientific method in analysis of sediments, McGraw-Hill Book Co., 508 pp.

Hamblin, W. K., 1958, Cambrian sandstones of northern Michigan, Michigan Dept. Conservation, Geol. Surv. Division Publication 51, 149 pp.

Harrison, P. W., 1957, New techniques for three-dimensional fabric analysis of till and englacial debris containing particles from 3 to 40 mm in size, *J. Geol.,* **65**, 98–105.

Higgs, D. V., and G. Tunnell, 1959, Angular relations of lines and planes, W. C. Brown, Dubuque, 43 pp.

Jones, T. A., 1968, Statistical analysis of orientation data, *J. Sed. Pet.,* **38**, 61–67.

Klein, G. deV., 1962, Sedimentary structures in the Keuper Marl (Upper Triassic),*Geol. Mag.,* **99**, 137–144.

_____, 1967, Paleocurrent analysis in relation to modern marine sediment dispersal patterns, *Bull. Am. Assoc. Petrol. Geologists,* **51**, 366–382.

Krumbein, W. C., and F. A. Graybill, 1965, An introduction to statistical models in geology, McGraw-Hill Book Co., 475 pp.

Lane, D. W., 1963, Sedimentary environments in Cretaceous Dakota Sandstone in northwestern Colorado, *Bull. Am. Assoc. Petrol. Geologists,* **47**, 229–256.

McKee, E. D., 1957, Primary structures in some Recent sediments, *Bull. Am. Assoc. Petrol. Geologists,* 41, 1704–1747.

———, 1964, Inorganic sedimentary structures, pp. 275-295, in John Imbrie, and Norman Newell, eds., Approaches to paleoecology, John Wiley and Sons, 432 pp.

Middleton, G. V., 1965, The Tukey Chi-Square test, *J. Geol.,* 73, 547–549.

———, 1967, The Tukey Chi-Square test: a correction, *J. Geol.,* 75, 640.

Miller, R. L., and J. S. Kahn, 1962, Statistical analysis in the geological sciences, John Wiley and Sons, 483 pp.

Ore, H. T., 1964, Some criteria for recognition of braided stream deposits, Wyoming University *Contributions to Geol.,* 3, 1–14.

Pettijohn, F. J., P. E. Potter, and R. Siever, 1965, Geology of sand and sandstone, Indiana University Printing Plant, Bloomington, Indiana, 208 pp.

Picard, M. D., 1957, Criteria used for distinguishing lacustrine and fluvial sediments in Tertiary beds of Uinta Basin, Utah, *J. Sed. Pet.,* 27, 373–377.

———, 1967, Stratigraphy and depositional environments of the Red Peak Member of the Chugwater Formation (Triassic), west-central Wyoming, Wyoming University *Contributions to Geol.,* 6, 39–67.

———, and L. R. High, Jr., 1968, Shallow marine currents on the Early (?) Triassic Wyoming shelf, *J. Sed. Petrol.,* 38, 411–423.

Ramsay, J. G., 1961, The effects of folding upon the orientaton of sedimentation structures, *J. Geol.,* 69, 84–100.

Reiche, P., 1938, An analysis of cross-lamination of the Coconino sandstone, *J. Geol.,* 46, 905–932.

Rusnak, G. A., 1957, A fabric and petrologic study of the Pleasant-view sandstone, *J. Sed. Pet.,* 27, 41–55.

Selley, R. C., 1967, Paleocurrents and sediment transport in near-shore sediments of the Sirte Basin, Libya, *J. Geol.,* 75, 99–110.

———, 1968, A classification of paleocurrent models, *J. Geol.,* 76, 99–110.

Smirnov, N., 1948, Table for estimating the goodness of fit of empirical distributions, *Annals Math. Statistics,* 19, 279–281.

Straaten, L. M. J. U. van, 1954, Composition and structure of Recent marine sediments in the Netherlands, *Leidse Geologische Mededelingen,* **19,** 1—110.

———, 1959, Minor structures of some Recent littoral and neritic sediments, *Geol. Mijnb.,* **21,** 197—216.

Tanner, W. F., 1955, Paleogeographic reconstructions from cross-bedding studies, *Bull. Am. Assoc. Petrol. Geologists,* **39,** 2471—2483.

Tohill, Bruce, and M. D. Picard, 1966, Stratigraphy and petrology of Crow Mountain Sandstone Member (Triassic), Chugwater Formation, northwestern Wyoming, *Bull. Am. Assoc. Petrol. Geologists,* **50,** 2547—2565.

Visher, G. S., 1965, Use of vertical profile in environmental reconstruction, *Bull. Am. Assoc. Petrol. Geologists,* **49,** 41—61.

SECTION
II

SIZE ANALYSIS

CHAPTER 3

SIEVE ANALYSIS

ROY L. INGRAM

University of North Carolina, Chapel Hill, North Carolina

The distribution of sizes of sedimentary particles with intermediate diameters in the range of 1/16 to 16 mm (sand and fine gravel) is most commonly determined by sieving. In the United States, the United States Standard sieves or the Tyler Standard sieves (Table 1) are used by most workers.

TABLE 1 Sieve openings

Wentworth Scale, mm	Phi Scale	$\sqrt[4]{2}$ Scale, mm	U. S. Standard [a]				Tyler[b] Mesh
			Opening, mm	Mesh	Permissable Variation		
					Average ± %	Maxi. + %	
16	−4.00	16.000	16.0		3	6	
	−3.75	13.454	13.5		3	6	
	−3.50	11.314	11.2		3	6	
	−3.25	9.514	9.51		3	6	
8	−3.00	8.000	8.00		3	6	2½
	−2.75	6.727	6.73		3	6	3
	−2.50	5.657	5.66	3½	3	10	3½
	−2.25	4.757	4.76	4	3	10	4

49

TABLE 1   Sieve openings (Continued)

Wentworth Scale, mm	Phi Scale	$\sqrt[4]{2}$ Scale, mm	U. S. Standard [a]				Tyler [b] Mesh
			Opening, mm	Mesh	Average ± %	Maxi. + %	
					Permissable Variation		
4	−2.00	4.000	4.00	5	3	10	5
	−1.75	3.364	3.36	6	3	10	6
	−1.50	2.828	2.83	7	3	10	7
	−1.25	2.378	2.38	8	3	10	8
2	−1.00	2.000	2.00	10	3	10	9
	−0.75	1.682	1.68	12	3	10	10
	−0.50	1.414	1.41	14	3	10	12
	−0.25	1.189	1.19	16	3	10	14
1	0.00	1.000	1.00	18	5	15	16
	0.25	0.841	0.841	20	5	15	20
	0.50	0.707	0.707	25	5	15	24
	0.75	0.595	0.595	30	5	15	28
1/2	1.00	0.500	0.500	35	5	15	32
	1.25	0.420	0.420	40	5	25	35
	1.50	0.354	0.354	45	5	25	42
	1.75	0.297	0.297	50	5	25	48
1/4	2.00	0.250	0.250	60	5	25	60
	2.25	0.210	0.210	70	5	25	65
	2.50	0.177	0.177	80	6	40	80
	2.75	0.149	0.149	100	6	40	100
1/8	3.00	0.125	0.125	120	6	40	115
	3.25	0.105	0.105	140	6	40	150
	3.50	0.088	0.088	170	6	40	170
	3.75	0.074	0.074	200	7	60	200
1/16	4.00	0.062	0.063	230	7	60	250
	4.25	0.053	0.053	270	7	60	270
	4.50	0.044	0.044	325	7	60	325
	4.75	0.037	0.037	400	7	60	400
1/32	5.00	0.031					

[a] A.S.T.M., 1966, pp. 447–448.
[b] W. S. Tyler Co., 1967, p. 10.

Before a sediment is sieved, the individual particles must be separated from one another. Clay, cementing agents, and soluble salts must be removed. Sediments containing cementing agents that cannot be removed without altering the individual particles cannot be analyzed by sieving (e.g., most silica-cemented sandstone or most carbonate-cemented clastic limestones).

For sieve analysis, most sediments can be placed in one of three categories: (A) sediment contains clay and cementing agents, (B) sediment contains clay, but does not need to be freed of cementing agents or pigments, and (C) sediment contains no clay, cementing agents, or soluble salts. All the steps in the procedure below should be done for Type A sediments. Only those steps marked with a □ need be done for Type B sediments. Only those steps marked with a △ need be done for Type C sediments.

As the purposes and requirements of sieve analysis and the nature of sediments are variable, however, all of the steps should be considered and decisions made on each sample or suite of samples as to which steps will be performed or eliminated.

PRELIMINARY TREATMENT

□ △ 1. Dry sediment in air or in a 40°C oven.

At higher temperatures any clay present may be baked into a bricklike substance, thus making dispersal of the clay difficult. Temperature-sensitive minerals such as halloysite will also be altered at higher temperatures.

□ △ 2. Break all clumps. Mash with fingers; use wooden or rubber pestle in a mortar; or use a wooden rolling pin. Use enough force to separate the grains, but avoid breaking individual grains.

□ △ 3. Mix sediment thoroughly and split to get the desired weight of sample. The exact size of sample that should be used depends on several factors: the size and sorting of the sediment, the shape and roundness of the grains, the number of screens that will be used, the shaking time. Consequently exact weights cannot be specified. As a preliminary guide the following approximate weights are suggested: fine gravel—500 gm; coarse sand—200 gm; medium sand—100 gm; fine sand—25 to 50 gm. About 15 gm (range of 5 to 25 gm) of material finer than 1/16 mm is needed if a pipette or hydrometer analysis is to be made. Another split may be necessary for the analysis of this fine material. Any of several alternate procedures may be used.

Coning and Quartering

Pour the sample on a flat surface so as to form a cone (Fig. 1). Use a straight edge and separate the cone into four quarters. Push aside two of the alternate quarters. Mix the remaining two quarters and form another cone. Continue the quartering process until the desired size sample is obtained. Do not attempt to get a sample of an exact size. For example, if a 100 gm sample is wanted, a sample within the range of 75 to 125 gm will usually be acceptable.

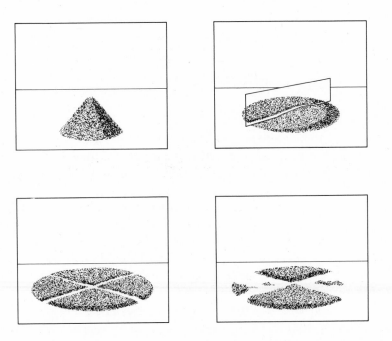

Fig. 1 Splitting sample by coning and quartering. (After W. S. Tyler Co., 1967, p. 13)

Mechanical Sample Splitters

A variety of types of mechanical sample splitters are available; the most common type is shown in Fig. 2. Pouring the sample through the splitter will divide the sample in half. Replace one of the pans containing one-half of the sample with a clean pan, and pour this half through the splitter again. The clean pan will then receive one-fourth of the original sample. Repeat as needed to get the desired weight of sample.

Overloaded sieves do not separate efficiently. If the weight on any screen exceeds the values shown in Table 2, a smaller size split of the sample should be obtained and re-sieved.

□ △ 4. Weigh the split sample to the nearest 0.01 gm. Record on form for recording sieve analysis, line 6 (Fig. 3).

Some workers use a small (about 10 gm) split of the sample and determine the moisture content of the sample by heating in a 110°C

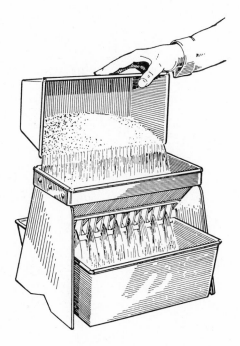

Fig. 2 Sample splitter (From W. S. Tyler Co., 1967, p. 13)

oven for 2 hours and cooling in a desiccator. The weight of the sample to be sieved is then corrected for the moisture content. This supposedly more accurate method, however, may well result in a less precise analysis because of the rapid absorption of moisture by the sample between the desiccator and the balance (Folk, 1968, p. 38).

TABLE 2   Maximum sieve loads on 8-in. diameter sieves.
Based on the conservative interpretation, extrapolation,
and combination of the works of Shergold (1946), Whitby (1958),
and McManus (1965).

Sieve Size		Adjacent Sieves Differ By		
Phi	mm	1/4 Phi	1/2 Phi	1 Phi
−2.00	4	40 g	80 g	160 g
−1.75		36		
−1.50		34	68	
−1.25		31		
−1.00	2	28	56	110
−0.75		26		
−0.50		24	48	
−0.25		22		
0.00	1	20	40	80
0.25		18		
0.50		17	34	
0.75		16		
1.00	1/2	14	28	60
1.25		13		
1.50		12	24	
1.75		11		
2.00	1/4	10	20	40
2.25		9		
2.50		8	17	
2.75		8		
3.00	1/8	7	15	
3.25		6		
3.50		6	12	
3.75		5		
4.00	1/16	5	10	20
4.25		5		
4.50		4	8	
4.75		4		
5.00	1/32	3	7	15

When using the nonmoisture correction technique, all materials should remain exposed to the air until equilibrium is reached with the moisture in the air.

REMOVAL OF SUBSTANCES THAT INTERFERE WITH DISPERSAL

The decision as to whether or not to remove carbonates, organic matter, iron oxides, or soluble salts depends on the purpose of the analysis and the composition of the sample.

SIZE ANALYSIS BY SIEVING

1. Sample No. _____ 2. Analyst_____ 3. Date_____ 4. Summary of Preliminary Treatment, Dispersal, etc. _____

5. Dispersant added_____; Concentration_____; Vol._____
6. Weight of untreated sample _____
7. Weight of sample after removal of carbonates, organic matter, iron oxides_____

| 8. Sieve Opening | | | Grade Size | Wt. Retained | Weight % | Cumulative % |
Phi	mm	Mesh				

Fig. 3   Form for recording sieve analysis.

Removal of Carbonates

This procedure is not recommended if mineralogical studies are to be made on the sample.

5. Place the sample in a 250- to 600-ml beaker. Add 25 ml distilled or deionized water and stir.

6. Add 10% HCl slowly until effervescence stops.

If carbonate material is abundant, the addition of 10% HCl will eventually result in a very large volume of liquid. When the beaker is nearly full, concentrated acid may be added very slowly or the excess liquid may be decanted or siphoned off.

7. Heat to 80 to 90°C. Add HCl until effervescence stops. A more exact procedure is to add HCl until a pH of 3.5 to 4 is reached and maintained. The pH can be checked by using: (a) a pH meter, (b) a pH indicator solution on a spot test plate (e.g., brom phenol blue indicator solution), (c) pH paper. Methyl orange indicator paper which changes from yellow in a neutral solution to orange at pH 3.1 to 4.4 and red below pH 3.1 is recommended.

8. If much carbonate material is present, the dissolved calcium ions will interfere with the dispersal of the sample, will hinder the removal of organic matter with the H_2O_2 treatment, and will precipitate as calcium oxalate in the iron removal treatment. Wash sample with very dilute HCl (about 0.1%). Repeat washing two or three times. The liquid can be tested for calcium by making a small amount of the liquid in a test tube alkaline to litmus paper with ammonium oxalate. A white precipitate of calcium oxalate will form if much calcium is present.

Washing can be done in several ways: (a) transfer sediment to one or more centrifuge tubes using the wash solution in a wash bottle and a rubber policeman. Stir thoroughly. A rubber stopper smaller than the inside of the centrifuge tube attached to a glass stirring rod makes a good stirrer (Fig. 4). Centrifuge and decant liquid above the sediment cake. (b) If the material is essentially all sand or coarse silt, let the sediment settle in the beaker and decant or siphon off the liquid. (c) Place porcelain filter candle in beaker and remove liquid with suction or vacuum pump (Fig. 5). The sediment that cakes on the filter is removed by applying back pressure.

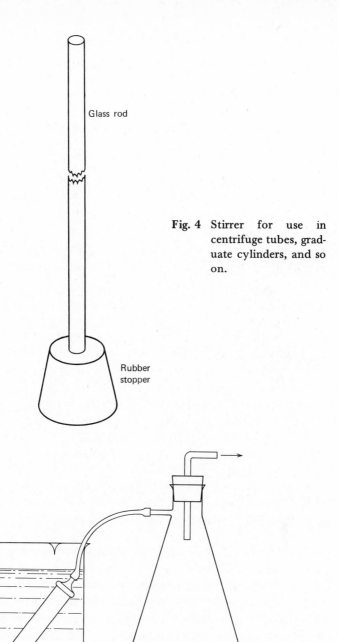

Glass rod

Rubber
stopper

Fig. 4 Stirrer for use in centrifuge tubes, graduate cylinders, and so on.

Fig. 5 Filtrations using porcelain filter. (From Krumbein and Pettijohn, 1938, p. 67)

Removal of Organic Matter (Jackson, Whitting, and Pennington, 1949, pp. 77–81)

This procedure will seldom remove all the organic matter, but is still very helpful in dispersing the sediment. This procedure may be stopped after any step, when most of the organic matter has been removed.

□ 9. If little organic matter is present, place the sample in a 400-ml beaker and add 100 ml of 6% H_2O_2 slowly and with constant stirring. Cover and heat to 40°C for 1 hour. Bring to a brief boil at the end of the heating period to remove excess H_2O_2.

10. If much organic material is present, do the following:

(a) Remove excess clear liquid by decantation after gravity settling or centrifuging.

(b) Add 30% H_2O_2 *very* slowly while stirring until frothing stops. Do not let sample froth over. Avoid contact of skin with 30% H_2O_2, for this reagent will cause burns.

(c) Heat to 40°C on a hot plate for 10 minutes. It may be necessary to remove sample from heat and to cool with a jet of cold water to prevent frothing over. Use a larger beaker if samples consistently froth over.

(d) Evaporate to a thin paste but not to dryness. Add 10 to 30 ml 30% H_2O_2, cover with watch glass, and digest at 40 to 60°C for 1 to 12 hours. Repeat until organic matter is removed.

(e) Bring to a brief boil to remove excess H_2O_2.

Removal of Iron Oxides (Leith, 1950, pp. 174–176)

11. Place sample in a 400-ml beaker and add water to make a volume of about 300 ml.

12. Place aluminum (a cylinder of sheet aluminum is preferable, but any form of recoverable aluminum will do) in beaker.

13. Add 15 gm oxalic acid (powder or concentrated solution containing 15 gm) and boil gently for 10 to 20 minutes. Add more oxalic acid as needed to remove all the iron.

Removal of Soluble Salts

14. Remove excess liquid by decanting after centrifuging or gravity settling, or by filtering. If the liquid is turbid with suspended

clay, digest by placing tube or beaker containing the sediment in a boiling water bath to cause flocculation. For suspensions that resist flocculation, add a very small amount of NaCl.

15. Wash (see step 8) two to five times. Stop if clay starts to disperse. Recent marine sediments should be washed until chloride-free. A drop of 4% silver nitrate in the filtrate will form a white precipitate of silver chloride if chlorine is present.

If a large number of marine samples are to be processed, removal of the salts by dialysis is recommended (Müller, 1967, pp. 33–34) so that many samples may be processed at once. Place each sample in a dialyzer bag and place in a large container filled with water. The water in the container should be changed frequently. Fastest results are obtained if water flows continuously through the container. The process may take several days for marine clays.

Drying and Weighing

16. Dry sample in air or in a 40°C oven. If no clay is present, the drying may be done in a 100°C oven. The drying process will be speeded up if the sample is spread out in a thin layer on a large watch glass, aluminum plate, and so on.

17. Let sample come to equilibrium with the moisture in the room air (about 1 hour).

18. Weigh sample to nearest 0.01 gm and record on line 7 of Fig. 3. This is the weight that is used in calculating percentages after sieving.

DISPERSAL

If the sample contains no clay and little silt, the dispersal procedure may be eliminated. For accurate work, however, the sample should be dispersed, for sand grains often have an almost invisible coating of clay particles. The dispersal procedure should result in the replacement of all exchangeable cations held by the clay with sodium ions and in the removal of other ions that hinder dispersal.

☐ 19. Place sample in a 400-ml beaker and add 200 ml of distilled or deionized water.

☐ 20. Add a volume of 10% Calgon equal to wc, where w is the weight of the sample and c is the estimated percent of clay. In other words, add 1 ml of 10% Calgon for each gram of estimated clay. If in doubt, overestimate the percent clay. Record on line 5 in Fig. 3. Until a person has some experience in estimating the percentage of clay, it is perhaps advisable to follow the recommendations of ASTM Specification D422-63 (1963, p. 208) and add a constant 50 ml of 10% Calgon to each sample.

The amount of and the kind of dispersing agent that will give the best dispersal depend on the amount of clay, the type of clay mineral, and the kinds of adsorbed ions. Usually this information will not be known. The work of Rolfe, Miller, and McQueen (1960) has shown that some clays need an amount of dispersing agent equal to 10% by weight of the amount of clay. For best dispersal the optimum amount of dispersing agent to add must be determined by experimentation for either too little or too much will decrease the amount of dispersal. Fortunately, the dispersal curve for Calgon has a rather broad, flat top, so. that not using the optimum amount of dispersing agent will not introduce a very large error.

Dispersing agents commonly used are sodium hexametaphosphate (Calgon), sodium tripolyphosphate, tetrasodium phosphate, sodium carbonate, sodium hydroxide, sodium oxalate, sodium silicate, and ammonia. Ten percent solutions of any of these may be used, but Calgon is usually considered the best overall dispersing agent. Calgon has the added advantage of complexing calcium ions that may be in solution rather than forming insoluble calcium precipitates, which the carbonate and oxalate dispersing reagents do.

☐ 21. Let soak overnight. Then pour into the dispersion cup of a mechanical analysis stirrer (Fig. 6) and mix for 1 to 5 minutes. A faster but less efficient alternate is to boil the suspension gently for 15 minutes and then to mix in the stirrer for 5 minutes (sand) to 30 minutes (clay).

For samples that resist dispersal, centrifuge, decant clear liquid, and repeat steps 20 and 21. The decanted liquid will often contain the substances that interfered with dispersal so that the clay will disperse merely by shaking it in distilled or deionized water.

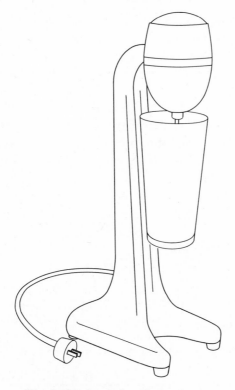

Fig. 6   Mechanical analysis stirrer.

SEPARATION OF FRACTION TO BE SIEVED

The fraction to be sieved may be separated by wet sieving (steps 22–24) or by decantation (steps 25–28).

Separation by Wet Sieving

☐ 22. Pour the dispersed sediment onto a wet 1/16-mm screen set over a large funnel. Make certain that all the sediment has been washed from the dispersion cup (use a wash bottle). If the silt and clay are to be analyzed, collect the sediment that passes through the screen in a 1000-ml cylinder.

☐ 23. Wash the residue on the screen with distilled or demineralized water from a wash bottle. Continue washing until nearly all fines are washed through screen. Do not exceed 1000 ml if the fine fraction is to be analyzed.

☐ 24. Dry sand on sieve in a 110°C oven, over a hot plate or under an infrared drying lamp. The sieve may be damaged if heated higher than 150°C.

Often the 1/16 mm screens need to be made available for other uses as soon as possible. An alternate drying procedure is to use a wash bottle and transfer the sand on the sieve into a large funnel lined with rapid-filtering filter paper. The sand is then dried on the filter paper.

Let sample remain heated or store in a desiccator until ready for sieving.

Separation by Decantation

25. Pour the dispersed sample into one or more test tubes, centrifuge tubes, high-form beakers, or small graduated cylinders. Stir well and let settle for the time required for a 1/32 mm particle to settle to the bottom (see Table 3).

TABLE 3   Time for quartz particle to settle 10 cm in water[a]

Diameter			Temperature							
Phi	mm		15°C		20°C		25°C		30°C	
4	1/16	0.062	0 min	32 sec	0 min	29 sec	0 min	25 sec	0 min	23 sec
4.5		0.044	1	5	0	57	0	51	0	45
5	1/32	0.031	2	10	1	55	1	42	1	31

[a] For a distance of settling(s) other than 10 cm, multiply above times by $s/10$.

26. Carefully decant liquid above sediment cake. Collect liquid in a 1000-ml cylinder if the fines are to be analyzed.

27. Transfer all sediment to one container. Add distilled or demineralized water (usually 10 cm above the sediment cake). Stir well. Let settle for required time (see Table 3) and decant. Repeat until decanted liquid is clear, usually about five times. After the first three settlings, dispersal is often helped by using water adjusted to pH 10 with the dispersing agent (about 1 ml of 10% dispersing solution per liter of water to give a 0.01% solution).

28. Dry sample in a 110°C oven, over a hot plate or under an infrared drying lamp. The drying time will be decreased greatly if the

sediment is transferred to a large funnel lined with rapid-filtering filter paper and the sediment dried on the filter paper. Let sediment remain heated or store in a desiccator until ready for sieving.

SIEVE ANALYSIS

For efficient sieving the sample must be composed of individual dry grains. According to Bartel (1960, in Müller, 1967, p. 65) with a surface moisture as little as 1%, adhesion forces exist that can overcome the weight of grains smaller than 1 mm.

□ △ 29. Build up a nest of clean screens for subdivisions desired with the coarsest screen on top. Half-height screens will allow a larger number of screens to be used at one time. A lid should be put on the top and a pan at the bottom of the nest of sieves.

□ △ 30. Pour dry sediment onto top screen in nest. Make certain that all sediment passes the top screen.

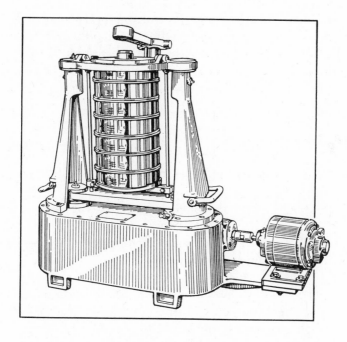

Fig. 7   Ro-Tap mechanical shaker. (From W. S. Tyler Co., 1967, p. 15)

□ △ 31. Place in Ro-Tap mechanical shaker (Fig. 7) and shake for 10 minutes.

Most workers have accepted 10 minutes of shaking in a Ro-Tap as an arbitrary standard, although some use 15 minutes. A long shaking time will result in more material passing through each screen (Whitby, 1958, p. 4); but because of inaccuracies in sieves (Table 1), long shaking times result primarily in the near mesh-size particles passing through the too-large holes (Müller, 1967, p. 75). The best sieving is one in which the sum total of the inaccuracies caused by the fines not passing through a sieve and by the coarse material passing through oversize holes is the smallest.

□ △ 32. Empty each sieve onto a large (15 x 15 in.) sheet of paper. The removal of sand is helped by striking the rim of the sieve with either the palm of the hand or the wooden handle of a screen brush along the general direction of the *diagonals* of the wire mesh and brushing the bottom of the sieve with a sieve brush. Use a soft brass wire sieve brush on sieves coarser than 100 mesh. For sieves finer than 100 mesh, use only a nylon bristle sieve brush. Be careful not to push wires apart.

□ 33. Add the fines passing the bottom (1/16 mm or smaller) screen to the cylinder containing the fines in step 22 or 26.

□ △ 34. Weigh each fraction to the nearest 0.01 gm. Make calculations as shown on Fig. 3. If determined, the weight shown on line 7 will be used as the sample weight in determining percentages.

CARE OF SIEVES

□ △ 35. After each use all sieves should be carefully cleaned (see step 32) and stored.

36. Occasionally, more thorough cleaning of the screens may be needed (W. S. Tyler Co., 1967, p. 19).

    (a) Wash sieves in warm soapy water using the special sieve nylon and brass sieve brushes.

    (b) If this treatment fails to remove most of the lodged particles, dip sieves in a boiling 5% solution of acetic acid and then use sieve brushes on sieves. Wash sieves thoroughly to remove the acid.

ACCURACY OF SIEVES

Three different types of sieves may be purchased. Most commercially available sieves are manufactured to meet the tolerances established under ASTM Specifications Ell-61. (See Table 1.) The National Bureau of Standards will, for a fee, check a set of sieves and will certify them if they meet ASTM specifications. The manufacturer selects matched sieves to give results for a given sample that are comparable to those obtained from the manufacturer's Master Sieves. Matched sieves are the most accurate available.

TESTING SIEVES

Sieves may be checked for accuracy in several different ways (ASTM, 1966, and W. S. Tyler Co., 1967, p. 39).

Use of Standard Samples

The use of calibrated glass spheres is recommended for checking and determining the effective sieve openings. Calibrated glass spheres may be obtained from the Supply Division, National Bureau of Standards, Washington, D. C. Three standard samples are now available at $9.50 each: No. 1017, 0.050 to 0.230 mm; No. 1018, 0.210 to 0.980 mm; and No. 1019, 0.90 to 2.55 mm. Instructions are provided for using the glass spheres in calibrating sieves.

For routine checking of sieves each laboratory should maintain its own standard sample. A set of sieves should be checked periodically with a standard size split of the standard sample to see if the set continues to give the same results. A new set of sieves can also be checked against the standard to see if the sets give comparable results. If they do not, calibration factors can be calculated for each sieve that will make the results comparable.

Measurement of Openings

Several methods of measuring openings are given in ASTM Specification Ell-61. One method is to use a microscope and measure the openings. Six nonoverlapping fields of view are selected. In each field measure at least 50 openings perpendicular to the wires, with the openings being located in a diagonal direction across the field (Fig. 8). The openings in three of the fields should be measured at right angles to those in the other three fields. Tabulate the results and check against Table 1.

REFERENCES

American Society for Testing Materials, 1963, Grain size analysis of soils, D422–63, pp. 203–214, *in* 1967 Book of ASTM Standards, Pt. 11, Philadelphia.

_____, 1966, Sieves for testing purposes, Ell-61, pp. 446–452, *in* 1966 Book of ASTM Standards, Pt. 30, Philadelphia.

Folk, R. L., 1968, Petrology of sedimentary rocks, Hemphills, Austin, Texas, 170 pp.

Jackson, M. L., L. D. Whitting, and R. P. Pennington, 1949,

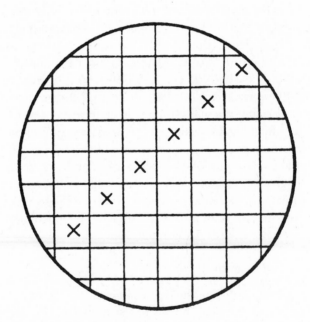

Fig. 8 Testing sieves by microscopic measurement of openings located along diagonals of openings.

Segregation procedures for mineralogical analysis of soils, *Soil Sci. Soc. Am. Proc.*, 14, 77—81.

Krumbein, W. C. and F. J. Pettijohn, 1938, Manual of sedimentary petrology, Appleton-Century-Crofts, 549 pp.

Leith, C. J., 1950, Removal of iron oxide coatings from mineral grains, *J. Sed. Pet.*, 20, 174—179.

McManus, D. A., 1965, A study of maximum load for small diameter screens, *J. Sed. Pet.*, 35, 792—796.

Müller, German, 1967, Sedimentary petrology, Part I, Methods in sedimentary petrology, translated by Hans-Ulrich Schmincke, Hafner Publishing Co., 283 pp.

Rolfe, B. N., R. F. Miller, and I. S. McQueen, 1960, Dispersion characteristics of montmorillonite, kaolinite, and illite clays in waters of varying quality and their control with phosphate dispersants, *U.S. Geol. Surv. P.P. 334—G*, pp. 229—273.

Shergold, F. A., 1946, The effect of sieving loading on the results of sieve analysis of natural sands, *Soc. Chemical Industry Trans.*, 65, 245—249.

Whitby, K. T., 1958, The mechanics of fine sieving, pp. 3—24, *in* Symposium on particle size measurement, ASTM Spec. Publ. No. 235, Philadelphia, 303 pp.

W. S. Tyler Co., 1967, Testing sieves and their uses (Handbook 53), W. S. Tyler Co., Mentor, Ohio, 48 pp.

CHAPTER 4

SEDIMENTATION ANALYSIS

JON S. GALEHOUSE

San Francisco State College, San Francisco, California

The three objectives in determining the size distribution of a sediment or a sedimentary rock are description, comparison, and interpretation. The first is to add to the overall physical description of the sediment or rock. The second is to compare the size distribution of the sediment or rock with that of others. The third is to make interpretations concerning the sedimentological history of the deposit. Sieve analysis, sedimentation analysis, and size measurement in thin section all fulfill the first two objectives about equally well, with one or another of the methods being more appropriate for particular samples (see Chapters 3 and 5). Many investigators, however, are convinced that sedimentation analysis is superior to the other two methods for interpreting sedimentological history. Sieve analysis and microscopic measurements determine the physical size of the grains comprising the rock or sediment. Most deposits, however, are laid down at a particular time and a particular place because of the settling velocities of their components, which are not only a function of physical size but also a function of grain shape and grain density. Sedimentation analysis is the only sizing technique that utilizes settling velocity; and, consequently, interpretations concerning sedimentological history based on it tend to be more valid than interpretations based on physical size alone.

DISPERSAL

In order for a sedimentation analysis of the size distribution of a sample to be valid, the different grains comprising the sample must be physical entities; that is, they must not be attached to one another. This is true also for a sieve analysis of sand-sized particles, and the techniques discussed in Chapter 3 apply also to sand being prepared for sedimentation analysis.

It cannot be overemphasized that great care must be taken in preparing any sample for size analysis. The objective of the investigation is to determine the size distribution of the sample. What is often determined, however, is the effectiveness or destructiveness of the disaggregating or dispersing technique.

Fine-grained sediments, especially those containing clay, usually require considerably more treatment to prepare them for sedimentation analysis than sand. Clay suspended in water tends to have a negative ionic charge. Clay particles attract either positively charged ions in the solution or the positive ends of water molecules. Positively charged solubles attracted to the clay, in turn, attract negatively charged solubles that surround them. The overall effect is a negatively charged mass that repels all the other negatively charged masses in the solution. This is the ideal condition that must be maintained for a valid sedimentation analysis of the clay.

To determine whether a dispersal method need be attempted, the following procedure should be followed. Place the sample for sedimentation analysis in a 1000-ml cylinder, add distilled water (distilled water should be used throughout the entire sedimentation analysis), stir for 5 or 10 minutes, and allow to stand overnight. Then check for any signs of flocculation (aggregation into small lumps). This can be done by withdrawing a drop, placing it on a slide, and observing it under the microscope. If there is no flocculation, individual particles showing Brownian movement will be present. If it has flocculated, the sample will show aggregates of particles. Flocculation can also be detected by the presence of a "mushy-looking" layer near the bottom of the cylinder with very clear water overlying it. If there are signs of flocculation, the sample must be dispersed into its individual components before attempting a sedimentation analysis.

The author has found that a primary cause of flocculation is too great a concentration of sediment in the cylinder. About 10 gm of clay per liter is a good value (see p. 76), although satisfactory results have often been obtained on samples as small as 2 gm or as large as 25 gm.

Investigators have used many different kinds and concentrations of chemical substances (called peptizers) for dispersing clays. This is because no single peptizer or method works for all samples. It is often simply a matter of trial and error to find a particular peptizer or particular method that works for the particular sample or group of samples in question. Nevertheless, certain substances and procedures have found more general applicability. A discussion of some of these follows.

Some investigators (e.g., Griffiths, 1967, pp. 50-51) emphasize that the combination of a strong base and a weak acid makes the best peptizer. The following method using sodium carbonate and sodium oxalate has been used quite successfully:

Bring the clay-rich suspension to 0.005 M sodium carbonate and 0.005 M sodium oxalate. To do this, make a 0.1 M solution of the two by adding 10.6 gm of sodium carbonate and 13.4 gm of sodium oxalate to one liter of distilled water. Place the clay-rich suspension to be dispersed in a 1000-ml cylinder, add 50 ml of the carbonate-oxalate solution, and bring the volume to one liter by adding distilled water. Stir the suspension for 5 or 10 minutes. Allow it to stand overnight and then check for flocculation as described above.

Some investigators have successfully used substances such as sodium hydroxide, ammonium hydroxide, or sodium hexametaphosphate as peptizers (see Rolfe et al., 1960). The author has obtained good results with sodium hexametaphosphate which is marketed as Calgon and is relatively inexpensive. One milliliter of 10% Calgon solution per estimated gram of clay is a very successful concentration (see Chapter 3).

With the exception of ammonium hydroxide, which evaporates during drying of the sample, the investigator should be sure to note the type and concentration of peptizer used, for this must be taken into account during the subsequent calculation of the results of the pipette analysis (see p. 85).

Hydrogen peroxide is also used as a dispersant. Whereas the above-mentioned peptizers act to maintain the electrical repulsion among particles, H_2O_2 affects the substance to be dispersed in two different ways. It oxidizes organic material which often inhibits dispersion and must be removed. In addition, the H_2O_2 generates oxygen in the pore spaces of the substance, which in effect pushes individual particles away from one another. One does not need to keep track of the amount and concentration of H_2O_2 used because on heating it completely disassociates into H_2O and O_2.

Procedures using various concentrations of H_2O_2 have been used successfully. As mentioned previously, it is often a matter of trial and error to find the right combination for a particular type of sample. A general method, based in part on that used in the laboratory of Tj. H. van Andel, is to place the sample in a 1000-ml beaker and add a 10 to 15% solution of H_2O_2 until it covers the sample. The H_2O_2 will quickly start to bubble. Allow the beaker to stand overnight and then add about 50 ml of a 30% solution (the concentration that is commercially available) of H_2O_2 and allow to stand for a few hours. Continue this procedure until there is no more reaction with the organic material. There will always be a certain bubbling when H_2O_2 is added, which is simply due to the dissociation of the H_2O_2 itself. This bubbling is quite distinct from the more vigorous bubbling that results from the reaction with the organic material. An efficient procedure is to add H_2O_2 to the samples in the early morning and then again in the late afternoon or evening.

After all the organic material is oxidized, the sample is heated in boiling water for about 1 hour to make sure all the H_2O_2 has dissociated. The sample must now be washed to remove the solubles. Filter candles are the most efficient equipment to use for washing clay-sized material (Fig. 1). These are available commercially[1] and are connected to a suction or vacuum pump. About 500 ml of distilled water is mixed with the sample and then all the liquid is drawn out through the filter candle, leaving the insoluble portion of the sample behind. This procedure is repeated six to eight times.

[1] Selas Corporation, Dresher, Pennsylvania. #FP-126 Microporous filter, about 1 1/2 by 8 in. with a porosity of 0.03. Cost is about $11 per filter.

Van Waters and Rogers, San Francisco, California. Mandler Filter Candle, Fine Porosity, #2, about 1 by 8 in. Cost is $6.50 per filter.

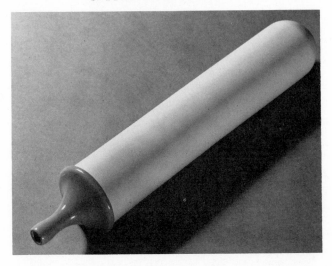

Fig. 1   Filter candle. (Courtesy of Selas Corporation)

Because the dispersal technique employed may affect the grain-size determination, one of the above procedures or a combination or modification of them should be standardized for a particular investigation dealing with many samples. A standardized procedure will save time and will at least make the comparisons among the samples much more valid.

SETTLING VELOCITY

Stokes' Law

Settling velocities used in the sedimentation analysis of silt and clay are usually computed from the now-famous settling law developed by G. G. Stokes in 1851. Stokes' law pertains to the terminal fall velocity of a sphere in a fluid and is explained as follows:

VRF (the viscous resistance to fall of a sphere in a fluid) = $6\pi r\mu v$
where r = radius of the sphere in cm,
μ = viscosity of the fluid in dyne-sec/cm^2 (poises),
v = fall velocity in cm/sec.

And *NDF* (the net downward force on a sphere in a fluid) = the force of gravity on the sphere minus the buoyant force of the fluid

$$NDF = \frac{4}{3}(\pi r^3 d_s g) - \frac{4}{3}(\pi r^3 d_f g)$$

where r = radius of the sphere in cm,
   d_s = density of the sphere in gm/cm^3,
   d_f = density of the fluid in gm/cm^3,
   g = acceleration due to gravity in cm/sec^2.
However, the terminal fall velocity is reached when *VRF* = *NDF*,

that is, when

$$6\pi r \mu v = \frac{4}{3}(\pi r^3 d_s g) - \frac{4}{3}(\pi r^3 d_f g)$$

or when

$$v = \frac{2(d_s - d_f)g r^2}{9\mu},$$

which is Stokes' law, where v is now the terminal fall velocity of the sphere.

Stokes' law as used in sedimentation analysis at a particular temperature is commonly simplified to

$$v = CD^2$$

where C is a constant equaling

$$\frac{(d_s - d_f)g}{18\mu}$$

and d_s = 2.65 gm/cm^3 (the density for quartz),
   d_f = the density of distilled water at the particular temperature,
   g = 980 cm/sec^2,
   μ = the viscosity of distilled water at the particular temperature,

and D = the *diameter* of the sphere in cm.

Table 1 lists the values of C at particular temperatures. These were calculated using values for density to six significant figures and values for viscosity to four significant figures as given in the *Handbook of Chemistry and Physics*, 49th edition. The acceleration due to gravity (g) was taken as 980 cm/sec^2, a value rounded to three significant figures, which is accurate for most places in the United States. Investigators working at latitudes less than 33° or greater than 43° or at altitudes above 1000 m should probably use a slightly different value for g (see *Handbook of Chemistry and Physics*, 49th edition, p. F-144).

Table 1 also lists values for D^2 so that velocities can be calculated quickly. If the settling velocity is known for a particular temperature, D can be calculated as

$$D = \frac{\sqrt{v}}{\sqrt{C}}$$

Values for $\sqrt{C}$ are also included in Table 1.

Stokes' law cannot be applied indiscriminately to all particles settling in a fluid. In the strictest theoretical sense, it is only valid under the following conditions and limitations.

1. *Particles must have reached terminal fall velocity.* For particles within the size range of applicability of Stokes' law, the terminal fall velocity is reached almost instantaneously. Weyssenhoff (1920) has shown that for a sphere with a diameter of 50 μ, the terminal fall velocity is reached in about 0.003 second. For smaller particles, the time is even less.

2. *Particles must be rigid.* All particles analyzed sedimentologically fulfill this condition.

3. *Particles must be smooth.* Most particles analyzed sedimentologically are not smooth. Arnold (1911) has shown that within the size range of applicability of Stokes' law, grains with irregular surfaces do not have any appreciable difference in settling velocity from smooth grains and the theoretical condition has no practical validity.

4. *No slippage or shear may take place between the particle and the fluid.* This depends on the wettability of the particle in the fluid, and the condition is fulfilled when water is used as the fluid.

5. *The fluid must be of infinite extent in relation to the particles.* A particle settling near the wall of a container will have its settling

TABLE 1   Values used in calculations involving Stokes' law
(see text for discussion)

Temperature in Degrees Centigrade	Constant (C)	$\sqrt{C}$
18	8,538	92.40
19	8,756	93.57
20	8,975	94.74
21	9,198	95.91
22	9,421	97.06
23	9,648	98.22
24	9,876	99.38
25	10,107	100.53
26	10,340	101.69
27	10,575	102.83

Diameter in ϕ	Diameter in cm		Diameter^2 in cm^2	
	Fractional	Decimal	Fractional	Decimal
4	1/160	6.250×10^{-2}	1/25,600	3.906×10^{-3}
5	1/320	3.125×10^{-2}	1/102,400	9.766×10^{-4}
6	1/640	1.562×10^{-2}	1/409,600	2.441×10^{-4}
7	1/1280	7.812×10^{-3}	1/1,638,400	6.104×10^{-5}
8	1/2560	3.906×10^{-3}	1/6,553,600	1.526×10^{-5}
9	1/5120	1.953×10^{-3}	1/26,214,400	3.815×10^{-6}
10	1/10,240	9.766×10^{-4}	1/104,857,600	9.537×10^{-7}
11	1/20,480	4.883×10^{-4}	1/419,430,400	2.384×10^{-7}

velocity decreased by an amount dependent on the nearness of the wall and the size of the particle (see Krumbein and Pettijohn, 1938, Fig. 21, p. 99). In the size range of Stokes' law, the wall effects are negligible if the sedimentation vessel is greater than 4 cm in diameter. Most 1000-ml graduated cylinders used in the pipette analyses of silt and clay are larger than this.

6. *Particle concentration must be less than 1%.* If particle concentration is high, the individual particles will interfere with one another during settling and the actual viscosity of the fluid will be different from that of the pure fluid. The maximum allowable concentration depends on the viscosity of the fluid, the particle size, the range in particle size, and the particle shape. For quartz particles which are within the range of applicability of Stokes' law and which

are settling in distilled water, the particle concentration should not exceed about 1% by volume (Irani and Callis, 1963, p. 60). This means a maximum of about 25 gm of sample can be used in a 1000-ml cyclinder for a pipette analysis. Better results are more consistently obtained if the sample is about 10 gm (a concentration of about 0.4%).

7. *Particles must be greater than 0.5μ in diameter.* Very small particles are affected by the Brownian movement of the molecules of the fluid. This keeps the particles from falling in a straight line, and consequently the resistance to fall is no longer only a function of the particle size and the viscosity of the fluid. Krumbein and Pettijohn (1938, p. 101) report that Stokes' law holds down to 0.1μ. Later reports (e.g., Irani and Callis, 1963, p.60) suggest that the law is valid only down to 0.5μ under ordinary settling conditions. If, however, centrifugal sedimentation is used, a modification of the law is valid down to 0.1μ.

8. *Particles must not be greater than 50μ in diameter.* The upper limit to the size of particles settling according to Stokes' law is a function of the temperature and Reynold's Number of the fluid and the density of the particles. Above this limit there is turbulence during settling. Oseen (1913) determined that Stokes' law is theoretically valid only up to 50μ. However, Rubey (1933) shows that observed settling velocity differs little from the theoretically determined Stokes' values up to about 140μ. Most investigators use Stokes' law up to the lower size limit of sand, 62.5μ, realizing that there may be a slight error in the 50 to 62.5μ fraction.

9. *Particles must be spheres.* In nature, practically no particles are perfect spheres. Wadell (1934a), using the same basic assumptions as Stokes, developed a settling velocity formula that takes particle shape into consideration. Wadell used a particle shape between that of a sphere and a disk which is much closer to the average shape in nature than is a sphere. In essence, Wadell's formula reduces the settling velocity determined by Stokes' law. To convert any Stokes value to a Wadell value, multiply by 0.64.

In summary, Stokes' law or Wadell's formula are essentially valid in the range of silt- and clay-sized particles. Sedimentation analyses run by pipette or by sedimentation balance are based mainly on these theoretical considerations.

Particles above the Range of Stokes' Law

Rubey (1933) devised a general formula for the settling velocity of a sphere that applies theoretically to particles within the size range of Stokes' law as well as to coarser particles. These theoretical calculations are valid as long as the net downward force acting on a particle in a fluid is the force of gravity on the particle minus the buoyant force of the fluid (see p. 74). For sand-sized and coarser particles, however, there are additional factors which must be taken into consideration that tend to invalidate the theoretical calculations.

During the settling of larger particles above the range of Stokes' law, a drag force develops that tends to decrease the settling velocity. The drag force originates because the larger particles falling at a greater velocity develop a wake or a low pressure zone behind (above) them as they settle (Zeigler and Gill, 1959, p. 4). Turbulence forms in these wakes that in essence takes energy from the system and decreases the settling velocity. The nonspherical shape of the particles produces additional turbulence during settling.

On the other hand, two factors tend to increase the settling velocity of coarser particles. Sedimentation analysis of sand is done with a settling tube (see p. 89) in which the sand grains are introduced into the top of the tube and allowed to settle to the bottom. Cook (1969, p. 781) found that groups of sand grains tend to fall as a unit until the grains become dispersed, which requires settling through a distance of several centimeters. In addition, smaller grains tend to be entrained by larger grains and accelerated. Cook (1969, p. 782) states that the net result of these interactions is that "...a sample settles at a faster rate than that predicted by the hydraulic characteristics of the grains. It is apparent that interaction effects are a function of sample sorting and skewness, and as such their magnitude is practically impossible to predict."

Fortunately, the settling velocities of particles above the range of Stokes' law can be determined experimentally because the individual particles are large enough to work with. Zeigler and Gill (1959) have compiled very useful tables and graphs for the settling velocity of quartz above the range of Stokes' law. Variables considered are the settling velocity, size, and shape of the quartz particles in addition to the temperature of the water in which the settling takes place.

Sedimentation Diameter

Sedimentation analysis actually determines the settling velocities of particles with different sizes, shapes, and densities. In practice, these settling velocities are expressed in terms of the diameters of quartz spheres that will settle at the same velocities. These diameters are usually calculated using either Stokes' law or Wadell's formula for silt and clay, and experimental results for sand. The actual physical size or diameter of the particles is *not* determined, only the "sedimentation diameter" which is a function of particle size, shape, and density. However, the sedimentation diameter is of much more significance than the mean physical diameter (see p. 69). The sedimentation diameter is analogous to the "norm" of petrology; that is, it is somewhat artificial, but is worthwhile because of the interpretations that can be made from it.

PIPETTE ANALYSIS OF SILT AND CLAY

By far the pipette method is the one most widely used for analyzing the grain size distribution of silt and clay. Usually the settling velocities used in pipette analyses are calculated from Stokes' law, although the following discussion can also apply to settling based on Wadell's modification of Stokes' law. The only difference is in the withdrawal times listed in Table 2 and the diameters calculated on p. 86 and p. 87. To convert any of these numbers to correspond to Wadell's formula, multiply by 0.64.

The first sample taken during a pipette analysis determines the total amount of silt and clay in the cylinder. It is taken at such a time and depth that no particular size fraction has completely settled past the sampling point. The subsequent samples, however, are taken at such a time and depth that a particular size fraction has settled past the sampling point. For example, as listed in Table 2, at 20°C the pipette sample taken at 7 minutes, 36 seconds after restirring the cylinder and at a depth of 10 cm will contain only particles with a sedimentation diameter finer than 6.0ϕ (15.6μ). All the coarser particles will have settled past the 10-cm depth.

A pipette analysis takes considerable time (see Table 2); however, 6 to 10 analyses run concurrently can be done in slightly more time

TABLE 2 Pipette withdrawal times calculated from Stokes' law

Diameter in ϕ Finer Than	Diameter in Microns Finer Than	Withdrawal Depth in cm	Elapsed Time for Withdrawal of Sample in Hours (h), Minutes (m), and Seconds (s)									
			18°	19°	20°	21°	22°	23°	24°	25°	26°	27°
4.0	62.5	20	20s	20s	20s	20s	20s	20s	20s	20s	20s	20s
4.5	44.2	20	2m 0s	1m 57s	1m 54s	1m 51s	1m 49s	1m 46s	1m 44s	1m 41s	1m 39s	1m 37s
			Restir	Restir	Restir	Restir	Restir	Restir	Restir	Restir	Restir	Restir
5.0	31.2	10	2m 0s	1m 57s	1m 54s	1m 51s	1m 49s	1m 46s	1m 44s	1m 41s	1m 39s	1m 37s
5.5	22.1	10	4m 0s	3m 54s	3m 48s	3m 42s	3m 37s	3m 32s	3m 27s	3m 22s	3m 18s	3m 13s
6.0	15.6	10	8m 0s	7m 48s	7m 36s	7m 25s	7m 15s	7m 5s	6m 55s	6m 45s	6m 36s	6m 27s
7.0	7.8	10	31m 59s	31m 11s	30m 26s	29m 41s	28m 59s	28m 18s	27m 39s	27m 1s	26m 25s	25m 49s
8.0	3.9	5	63m 58s	62m 22s	60m 51s	59m 23s	57m 58s	56m 36s	55m 18s	54m 2s	52m 49s	51m 39s
9.0	1.95	5	4h 16m	4h 9m	4h 3m	3h 58m	3h 52m	3h 46m	3h 41m	3h 36m	3h 31m	3h 27m
10.0	0.98	5	17h 3m	16h 38m	16h 14m	15h 50m	15h 28m	15h 6m	14h 45m	14h 25m	14h 5m	13h 46m
11.0	0.49	5	68h 14m	66h 32m	64h 54m	63h 20m	61h 50m	60h 23m	58h 59m	57h 38m	56h 20m	55h 5m

than one analysis run separately. Therefore most investigators set up some kind of an assembly line for dealing with large numbers of samples and start processing a new group of samples each day.

A considerable amount of equipment is also needed for pipette analyses (see Table 3). Most of this equipment, however, has multipurpose utility and should be present in any well-equipped sedimentological laboratory.

TABLE 3.   Equipment needed for pipette analyses

1.	1000-ml graduated cylinders (Plastic is more practical in the long run than glass.)
2.	50-ml evaporating dishes (About ten evaporating dishes are needed for each graduated cylinder. Each one should be numbered and weighed. These weights should be permanently recorded for future reference.)
3.	Stirring rod (see Fig. 2)
4.	20-ml pipette with rubber tube attached to top
5.	Distilled water
6.	Electric clock with large second hand
7.	Thermometer
8.	Drying oven
9.	Accurate balance
10.	Portable interval timer (optional)
11.	Desiccator (optional)

The sample to be pipetted must be composed only of silt- and clay-sized particles. Coarser material is removed by washing the sample through a 62μ stainless steel sieve. Silt and clay which have passed through the sieve are placed in a 1000-ml graduated cylinder and checked for flocculation (see p. 70). If there is no flocculation, or after dispersal has been completed; distilled water is added until the volume is exactly 1000 ml. The temperature is now read on a thermometer that has been standing in a cylinder of distilled water. This thermometer should be checked periodically during the entire analysis. There probably will be little or no variation in temperature for pipette withdrawals down to 8ϕ (a time interval of about 1 hour). Before smaller sizes can be sampled, however, the temperature may have varied considerably. If the temperature in the laboratory cannot be stabilized by some method of air conditioning or if the cylinders cannot be placed in a constant temperature bath, the sampling should probably be terminated at 8 or 9ϕ (see p. 87).

After determining the temperature, the appropriate withdrawal times are read from Table 2. The sample is then stirred (see Fig. 2) for 2 or 3 minutes in order to insure an even distribution of sediment throughout the cylinder. The last minute of stirring should consist of smooth, long strokes with the stirrer traversing the entire length of the cylinder. As the stirrer is withdrawn from the cylinder, timing begins. Twenty seconds after the stirring is completed, the first pipette sample is withdrawn. An efficient procedure for this and

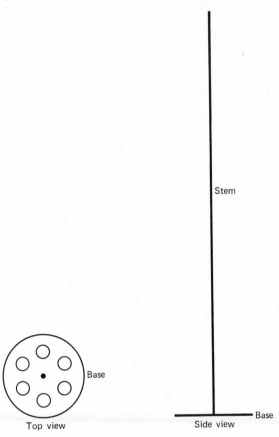

Top view   Side view

Fig. 2 Stirring rod. The stirring rod must be constructed to fit the particular 1000-ml graduated cylinders that are being used. The stem of the stirring rod is attached to the center of the base. The base can be made of metal, plastic, or rubber and should have a diameter about 5 mm less than the inside diameter of the 1000-ml graduated cylinders. Circles represent holes cut in the base.

subsequent withdrawals is to place the pipette into the cylinder at the proper depth 10 seconds before the withdrawal time (the pipette stem should have previously been marked at 5, 10, and 20 cm from the tip end). At the exact withdrawal time, begin sucking through the rubber tube until exactly 20 ml is in the pipette. This will take about 5 to 10 seconds. As the liquid reaches the 20-ml mark, pinch the rubber tube and remove the pipette from the cylinder. Let the pipette drain into an evaporating dish. Then, in order to rinse the pipette, suck up 20 ml of pure distilled water and allow it to drain into the same evaporating dish. Then prepare for the second pipette withdrawal. If samples are to be taken only at even phi intervals, follow the same procedure as for the first withdrawal. If additional samples are to be taken at 4.5ϕ and 5.5ϕ, restir the cylinder after completing the 4.5ϕ withdrawal and start the timing as before.

For samples of material finer than 8ϕ, withdrawal times (see Table 2) are rounded off to the nearest minute. Be sure to note withdrawal times and the evaporating dish numbers on the data sheet (see Fig. 3). It is quite useful to keep a portable interval timer in the laboratory. Then during the last few pipette samplings when there is considerable time between withdrawals, one can go about other laboratory work with the timer set to ring shortly before the next withdrawal is to be made. After the sample is taken, the evaporating dish is placed in a drying oven set at about $100°C$ and allowed to dry overnight.

At this point in the analysis of silt and clay by the pipette method, opinion differs among sedimentologists as to the best procedure to follow. One method (e.g., Folk, 1965, p. 38) is to remove the samples from the oven and have them stand in the laboratory for about an hour before weighing them. This allows the samples to absorb water from the atmosphere. The amount absorbed is a function of the type of clay minerals present and the humidity of the laboratory air. After the sample has reached equilibrium with the atmosphere, it is weighed. The other popular method among sedimentologists is to take the samples from the oven and place them in a desiccator until weighing is begun. They are then removed, one by one, from the desiccator, quickly placed in the balance containing a drying agent such as calcium chloride, and weighed.

Most investigators admit that the desiccator method is the more theoretically sound. For example, if the relative humidity varies from

Sample Number _____   Date _____

Temperature _____   Operator _____

Type and Amount   Comments:
of Dispersant _____

Diameter in ϕ	Depth of Withdrawal in cm	Time of Withdrawal	Evaporating Dish Number
4.0	20		
4.5	20		
		Restir	
5.0	10		
5.5	10		
6.0	10		
7.0	10		
8.0	5		
9.0	5		
10.0	5		
11.0	5		

ED = Evaporating Dish S = Sample D = Dispersant = _____ gm

4ϕ	4.5ϕ	5ϕ	5.5ϕ	6ϕ
$ED+S+D =$ ____ gm	$ED+S+D =$ ____ gm	$ED+S+D =$ ____ gm	$ED+S+D =$ ____ gm	$ED+S+D =$ ____ gm
minus ED ____	minus ED ____	minus ED ____	minus ED ____	minus ED ____
$S + D$ ____	$S + D$ ____	$S + D$ ____	$S + D$ ____	$S + D$ ____
minus D ____	minus D ____	minus D ____	minus D ____	minus D ____
S ____	S ____	S ____	S ____	S ____
x50	x50	x50	x50	x50
Total wt. S in 1000 ml	Total wt. S in 1000 ml	Total wt. S in 1000 ml	Total wt. S in 1000 ml	Total wt. S in 1000 ml

7ϕ	8ϕ	9ϕ	10ϕ	11ϕ
$ED+S+D =$ ____ gm	$ED+S+D =$ ____ gm	$ED+S+D =$ ____ gm	$ED+S+D =$ ____ gm	$ED+S+D =$ ____ gm
minus ED ____	minus ED ____	minus ED ____	minus ED ____	minus ED ____
$S + D$ ____	$S + D$ ____	$S + D$ ____	$S + D$ ____	$S + D$ ____
minus D ____	minus D ____	minus D ____	minus D ____	minus D ____
S ____	S ____	S ____	S ____	S ____
x50	x50	x50	x50	x50
Total wt. S in 1000 ml	Total wt. S in 1000 ml	Total wt. S in 1000 ml	Total wt. S in 1000 ml	Total wt. S in 1000 ml

Fig. 3   Pipette data sheet.

hour to hour or day to day in the laboratory, the sample will weigh more on a humid day and less on a dry day. In addition, a particular kind of clay mineral in one sample may absorb more moisture than a different kind of clay mineral in another sample. Finally, although the clay being analyzed most certainly did contain absorbed moisture during its deposition, the basic theory concerning sedimentation analysis treats the sedimentation fluid as a constant in which the silt and clay are completely submerged.

On the other hand, the "equilibrium" method which does not use the desiccator is much more practical. Although it is not as theoretically sound as the desiccator method, the equilibrium method has been extremely successful in the past. For one thing, it is almost impossible to keep all moisture from the sample. In the 1 or 2 seconds it takes to move the sample from the desiccator to the balance, some weight gain will occur. Furthermore, it is extremely difficult to get the balance completely moisture-free. Therefore the amount of moisture absorbed may be a function of the length of time it takes to weigh the sample. Because one cannot weigh all the samples with the same speed, an additional error is introduced.

In view of this discussion, the author recommends that the equilibrium method be used and the laboratory kept at a constant temperature and humidity.

After the total weight of the evaporating dish plus its contents is recorded on the data sheet (see Fig. 3), the weight of the clean evaporating dish (ED) and the weight of the dispersant (D) are subtracted from the total. The weight of the dispersant (D) in the evaporating dish is calculated as follows:

$$D = \frac{\text{(molecular weight of dispersant) (molarity of dispersant in 1000-ml cylinder)}}{50}$$

After subtracting ED and D, the remainder is S, the weight of the silt and clay in the evaporating dish. Because S is from 20 ml or 1/50 of the original volume of the 1000-ml cylinder, it must be multiplied by 50 to get its total weight in the cylinder.

If a pipette withdrawal is not taken at the exact stipulated time, a new withdrawal depth can be calculated at which the desired size of

material can be sampled at a new time. The following reasoning and
equations apply.

$$v = CD^2$$

where v = terminal fall velocity in cm/sec
       C = constant calculated from Stokes' law (see Table 1),
       D = sedimentation diameter in cm (see Table 1 for values of D^2).

$$v = \frac{X}{t}$$

where v = terminal fall velocity in cm/sec,
       X = depth to which the particles have settled in cm,
       t = amount of time the particles have settled in seconds.

$$\text{Therefore} \quad CD^2 = \frac{X}{t}$$

$$X = CD^2 t$$

where X is the new depth at which the sample is to be taken at the
new time t.

An even simpler method of calculating the new depth is as follows:

$$Vn = \frac{Xn}{tn}$$

where Vn = terminal fall velocity in cm/sec,
       Xn = new depth at which the sample is to be taken,
       tn = new time at which the sample is to be taken.

$$Vs = \frac{Xs}{ts}$$

where Vs = terminal fall velocity in cm/sec,
       Xs = standard depth at which the sample should have
             been taken,
       ts = standard time at which the sample should have
             been taken.

However, for the same size particles,

$$Vn = Vs$$

$$\text{and} \frac{Xn}{tn} = \frac{Xs}{ts}$$

$$\text{Therefore } Xn = \frac{tnXs}{ts}$$

where Xn is the new depth at which the sample is to be taken at the new time tn.

Using the weights obtained from a pipette analysis of silt and clay, cumulative percentages *coarser* than a particular grain size are calculated as follows:

$$\text{Cumulative \%} = \frac{100(G + S + SC - P)}{G + S + SC}$$

where G = weight of the gravel fraction,
S = weight of the sand fraction,
SC = weight of the silt and clay fraction,
P = weight of a particular pipette sample times 50.

Cumulative percents are calculated for each pipette withdrawal, and these results together with those from the size analysis of the sand and gravel fractions are plotted to form the cumulative curve for the entire sample (see Chapter 6).

It may not be feasible to run a complete pipette analysis of the silt and clay fraction. Because of the effects of Brownian movement (see p. 77), or because of temperature considerations (see p. 81), or because of a lack of time, the analysis may be terminated before the finest fractions are sampled. Folk (1965, p. 39) suggests that the unsampled distribution be interpolated. This can be done on ordinary arithmetic graph paper by extending the cumulative curve in a straight line from the last data point to 14 ϕ (0.06μ) at 100%. This manipulation assumes that all the clay particles are larger then 14 ϕ. Because the interpolation assumes the point in question (i.e., the size distribution), it is questionable whether this interpolation adds anything to the size analysis (see Chapter 6).

Hydrometer, manometer, centrifuge, photometric, elutriation, and decantation methods are also used occasionally for sedimentation analysis of silt and clay. These methods are not nearly so widely used in sedimentology as the pipette method and no attempt will be made to describe them. Those interested in details of these methods are referred to Krumbein and Pettijohn, 1938, p. 119-124; Irani and Callis, 1963, p. 71-89; ASTM, 1963, p. 203-214; and Müller, 1967, p. 87-93.

SEDIMENTATION BALANCES FOR SILT AND CLAY

Recently, the Cahn[1] and Sartorius[2] companies have placed on the market sedimentation balances that semiautomatically determine the size distribution of particles within the range of Stokes' law. Although the two differ in detail, they consist mainly of three components:

1. An automatic balance that continuously weighs the amount of sample that has settled.
2. A strip chart recorder that keeps a continuous record of the weight of sample that has settled.
3. A sedimentation cylinder in which the sample settles.

The main advantage of these devices is that one can place the sample in them and perform other activities while the sample is being run. The main disadvantages are that only one sample per day can be run and that the complete setup costs about $4000. For those interested in additional details and information concerning these sedimentation balances, contact the manufacturer or distributor in the footnotes.

---

[1]Cahn RG Electrobalance with #2800 Particle Sedimentation Accessory, manufactured by Cahn Instrument Co., 7500 Jefferson Street, Paramount, California.

[2]Sartorius Sedibal Sedimentation Balance, distributed by Brinkmann Instruments, Inc., 115 Cutter Mill Road, Great Neck, New York.

SETTLING TUBES FOR SAND

Whereas sedimentation analysis is commonly applied to particles finer than 4ϕ (62.5μ), sieve analysis is presently the most popular method for analyzing sand (see Chapter 3). The author thinks, however, that sedimentation analysis of sand in settling tubes will become so increasingly popular that in the near future (10 to 20 years) it will replace sieve analysis as the most common method.

Reasons for the increased popularity of sedimentation analysis of sand include the following:

1. A grain of sand is usually deposited at a particular time and place because of its settling velocity. Settling velocity of a sand grain is a function of its volume, density, roundness, and sphericity. Sedimentation analysis is the only "sizing" technique that takes all four of these factors into consideration. Sieving merely separates grains according to their least cross-sectional area.

2. Sedimentation analysis of sand is much faster than sieve analysis. A complete sedimentation analysis takes about 10 to 20 minutes, whereas a comparable sieve analysis may take several hours.

   (a) Only one weighing of the total sample is necessary for sedimentation analysis. One weighing of each sieve fraction is necessary for sieve analysis.

   (b) Each sieve must be cleaned after each sieve analysis.

   (c) The settling tube can be constructed so that the output from the sedimentation analysis is printed automatically on magnetic tape that can be fed into a computer which has been programmed to calculate all the size and sorting parameters.

3. Sedimentation analysis requires only 2 to 10 gm of sample, whereas sieve analysis usually requires about 100 gm of sample. This fact is extremely important for analyzing the coarse fraction of marine cores which often contain only a small amount of sand.

4. Sieve openings commonly change size with use.

5. Sedimentation analysis of sand produces a continuous cumulative curve with each sand grain being recorded. Sieve analysis produces a curve drawn from a very limited number of points. To study in more detail a particularly interesting or unusual size parameter, one need only read additional points off the original

cumulative curve of a sedimentation analysis. With the sieve analysis technique, one must run another analysis using a smaller phi interval (additional sieves).

One of the first devices used for the sedimentation analysis of sand was the Emery Tube (Emery, 1938). It was used to determine the size distribution of sand by volume percent. This was accomplished by introducing a sample at the top of the water column in the Emery Tube and then measuring the height of sediment that accumulates at the bottom of the tube through time. It was assumed that the height was proportional to volume; thus volume percents were determined. Any conversion of volume percent directly to weight percent may introduce a significant error due to density and shape considerations.

Another of the early settling tubes was developed by Dutch sedimentologists at the Hague Sedimentological Laboratory and was used to determine size distribution by weight percent (Tj. H. van Andel, personal communication). Subsequently, numerous investigators have built their own settling tubes for analyzing sand by modifying and improving these early devices (e.g., Tj. H. van Andel at Scripps Institution of Oceanography and now at Oregon State University, F.-C. Kogler at Der Universitat Kiel, and C. Phipps at the University of Sydney). Unfortunately, most of these improved settling tubes for the analysis of sand are not described in the literature and are not commercially available.

On the other hand, a few of the improved settling tubes have been described in detail. One is the visual accumulation tube developed by the Inter-Agency Committee on Water Resources at St. Anthony Falls Hydraulic Laboratory (Report #11, 1957). The visual accumulation tube is essentially a modification of the Emery Tube in which the sample is introduced mechanically rather than manually, an automatic tapper is used to jar the tube throughout the analysis, and a manually operated recorder traces the volume of accumulated sediment on a motor-driven time chart (Report #11, 1957, p. 18). The settling velocities used to calibrate the tube were determined experimentally. The visual accumulation tube has limited scientific value because, like the Emery Tube, it only determines volume percent.

Another device described in detail is the settling tube used at the Woods Hole Oceanographic Institution (Whitney, 1960; Zeigler, et al., 1960; Schlee, 1966). This tube ". . . measures size by a pressure differential between two columns of water having a common head. The change caused by introduction of sediment within one of the columns is measured by a water pressure transducer, and the output is fed to a recorder. As the sand grains settle past the pressure tap to the transducer, the pressure they caused is no longer detected by the transducer, and the pressure differential becomes less and less with the passage of time." (Schlee, 1966, p. 403). Benthos, Inc.[1] produces a Rapid Sediment Analyzer, Type 341, which is commercially available and is patterned after the settling tube at Woods Hole. Five to ten grams of sand are used for each analysis and this can be recovered if desired. The Rapid Sediment Analyzer can handle material in the 50μ to 4-mm size range. The main disadvantage of this device is that it costs between $5000 and $6000 (1969 prices).

Felix (1969) has recently described in detail the settling tube used at the University of Southern California for rapid size analysis of sand. This device which records weight percent consists of a Plexiglas tube about 2 m long with an inner diameter of 11.2 cm, a thin plastic disk on which the sediment settles, a monofiliment line which connects the disk near the bottom of the tube with a transducing cell (strain gauge) mounted over the top of the tube, a battery-powered amplifier to boost the output from the transducing cell, a paper chart recorder, and a wire mesh device for introducing the sediment into the water column.

"As the sediment collects on the pan at the bottom of the tube, the transducing cell converts strain into an electrical signal of a few millivolts. The signal is fed to the amplifier and thence to the chart recorder, which plots a curve of strain versus time." (Felix, 1969, p. 777).

The main advantage of a USC-type settling tube is the cost. Felix (1969) lists model numbers and addresses at which all the components can be purchased and states that the total cost is under $1100 (1968 prices). Cook (1969) in a companion paper to Felix's paper discusses in detail the calibration of the USC settling tube as well as theoretical and practical considerations that must be taken into account in calibrating any settling tube.

For the investigator interested in details on the availability, construction, and calibration of settling tubes for sand, the above

[1] For more information, write to Benthos, Inc., North Falmouth, Massachusetts, and ask for Data Sheet #3-02-681.

references as well as the following ones should be consulted: Camp (1946); Report #4 (1941); Rouse (1937); Sahu (1964); Schulz, Wilde and Albertson (1954); Schiller (1932); Wadell (1934b); and Zeigler and Gill (1959).

REFERENCES

American Society for Testing Materials, 1963, Grain size analysis of soils, D422–63, 203–214, *in* 1967 A.S.T.M. Standards, Part 11, Philadelphia.

Arnold, H. D., 1911, Limitations imposed by slip and inertia terms upon Stokes' law for the motion of spheres through liquids, *Philosophical Mag.*, 22, 755–775.

Camp, T. R., 1946, Sedimentation and the design of settling tanks, *Trans. Am Soc. of Civil Eng.*, 111, paper 2285, 895–958.

Cook, D. O., 1969, Calibration of the University of Southern California automatically recording settling tube, *J. Sed. Pet.*, 39, 781–786.

Emery, K. O., 1938, Rapid method of mechanical analysis of sands, *J. Sed. Pet.*, 8, 105–110.

Felix, D. W., 1969, An inexpensive recording settling tube for analysis of sands, *J. Sed. Pet.*, 39, 777–780.

Folk, R. L., 1965, Petrology of sedimentary rocks, Hemphill's, 159 pp.

Griffiths, J. C., 1967, Scientific method in analysis of sediments, McGraw-Hill Book Co., 508 pp.

Handbook of chemistry and physics, Forty-Ninth Edition, 1968–69, Chemical Rubber Publ. Co.

Irani, R. R., and C. F., Callis, 1963, Particle size: measurement, interpretation, and application, John Wiley and Sons, 165 pp.

Krumbein, W. C., and F. J., Pettijohn, 1938, Manual of sedimentary petrography, Appleton-Century-Crofts, 549 pp.

Müller, German, 1967, Sedimentary petrology, Part I, Methods in sed. pet., translated by Hans-Ulrich Schmincke, Hafner Publishing Co., 283 pp.

Oseen, C. W., 1913, Über den Gultigkeitsbereich der Stokes'schen

Widerstandformel, *Arkiv for Matematik, Astronomi Fysik,* 6, 1910; 7, 1911; 9, 1913.

Report #4, 1941, Methods of analyzing sediment samples, Subcommittee on Sedimentation, Inter-Agency Committee on Water Resources, St. Anthony Falls Hydraulic Laboratory, Minneapolis, Minnesota, 203 pp.

Report #11, 1957, The development and calibration of the visual-accumulation tube, Subcommittee on Sedimentation, Inter-Agency Committee on Water Resources, St. Anthony Falls Hydraulic Laboratory, Minneapolis, Minnesota, 109 pp.

Rolfe, B. N., R. F. Miller, and I. S. McQueen, 1960, Dispersion characteristics of montmorillonite, kaolinite, and illite clays in waters of varying quality, and their control with phosphate dispersants, U. S. Geol. Surv. P.P. 334-G, 229–273.

Rouse, H., 1937, Nomogram for the settling velocity of spheres, Nat. Res. Council, Rept. of the Committee on Sedimentation, May 1, 1937, Exhibit D, 57–64.

Rubey, W. W., 1933, Settling velocities of gravel, sand, and silt particles, *Am. J. Sci.,* 225, 325–338.

Sahu, B. K., 1964, Depositional mechanisms from the size analysis of clastic sediments, *J. Sed. Pet.,* 34, 73–83.

Schiller, L., 1932, Fallversuche mit Kugeln und Scheiben, *Handbuch der Experimentalphysik, Leipzig,* 4, No. 2.

Schlee, J., 1966, A modified Woods Hole rapid sediment analyzer, *J. Sed. Pet.,* 36, 403–413.

Schulz, E. F., R. H. Wilde, and M. L. Albertson, 1954, Influence of shape on the fall velocity of sedimentary particles, Colorado A & M Research Foundation report to the Missouri River Division of the Corps of Engineers, U. S. Army, Omaha, Nebraska, MD Sediment Series No. 5, 161 pp.

Stokes, G. G., 1851, On the effect of the internal friction of fluids on the motion of pendulums, *Trans. Cambridge Philosophical Soc.,* 9, Pt. 2, 8–106.

Wadell, H., 1934a, Some new sedimentation formulas, *Physics,* 5, 281–291.

_____, 1934b, The coefficient of resistance as a function of Reynolds number for solids of various shapes, *J. Franklin Inst.,* 217, 459–490.

Weyssenhoff, J., 1920, Betrachtungen über den Gültigkeitsbereich der Stokesschen und der Stokes-Cunninghamschen Formel: *Annalen der Physik*, 62, 1—45.

Whitney, G. G., 1960, The Woods Hold rapid analyzer, Woods Hole Oceanog. Inst., unpublished ms., Ref. No. 60-36, 19 pp.

Zeigler, J. M., and B. Gill, 1959, Tables and graphs for the settling velocity of quartz in water, above the range of Stokes' law, Woods Hole Oceanog. Inst., unpublished ms., Ref. No. 59—36, 13 pp.

Zeigler, J. M., G. G. Whitney, Jr., and C. R. Hayes, 1960, Woods-Hole rapid sediment analyzer, *J. Sed. Pet.*, 30, 490—495.

CHAPTER 5

GRAIN-SIZE MEASUREMENT IN THIN-SECTION

DANIEL A. TEXTORIS

University of North Carolina, Chapel Hill, North Carolina

One of the most commonly measured parameters in sediments and weakly consolidated rocks is grain size. This is accomplished normally by direct measurement, sieve analysis, or some sedimentation technique. Major reasons for studying size, or more commonly size distribution, in Recent sediments are to determine physical effects of the environment on this parameter and then to use this knowledge in interpreting older sediments or rocks. To do this, size analyses must be performed also on consolidated rocks so the investigator can compare data. Since most older rocks cannot be disaggregated without affecting the original grain size, thin-sections must be used to obtain size data.

Graton and Fraser (1935) have graphically shown the distribution of voids and uniform spheres on a random plane section (Fig. 1). It is obvious that true sphere diameter is seldom shown in section; indeed, many spheres do not appear to be in contact, when in reality they are.

Krumbein (1935) performed a similar experiment by embedding lead shot, all of the same diameter, in sealing wax. This was ground

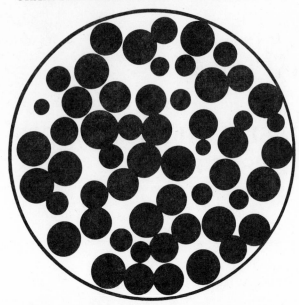

Fig. 1 Distribution of voids and spheres on a random plane section of an aggregate of uniform spheres chance-packed in a container. (After Graton and Fraser, 1935. By permission of University of Chicago Press)

to a polished surface. The average radius computed from the observed data was 0.763 of the actual radius. Thus a correction factor must be used in order to convert grain size in thin-section to that obtained by sieving. Another problem that may arise when a comparison is made between Recent sediments and rocks is that sediments are usually collected as grab samples. Hence the investigator measures size distribution for the entire homogenized sample. Fine bedding or laminae, which could give a bimodal plot, cannot be taken into consideration. In rock studies by thin-section, non-homogeneity can be observed and taken into account when performing size analyses. For this purpose thin-sections or polished sections should always be cut perpendicular to bedding.

Thin-section analyses allow observation of overgrowths on individual grains, and permit compensation for this diagenetic effect if so desired. Partial replacement of grains by various cements may have taken place. In this case, sieve analysis would give a finer than true size distribution. Compensation can also be made in this example if original detrital grain outlines are discernable.

Use of thin-sections allows the investigator to identify minerals with ease. This is particularly important if certain minerals are to be selected for measuring. Particle shape and degree of cementation may be determined. Original textures are preserved in thin-sections so that grain orientation, grain support versus matrix support, and so on, can be ascertained and used for final evaluation of whole-rock parameters.

SELECTION OF ROCK FOR ANALYSIS

If a rock contains grains with no overgrowths, if cement or matrix replacement of grains is minor, and if the rock is poorly indurated or cemented with calcite so that disaggregation of individual grains can be made, then sieve analysis will produce the best analytical results. However, if thin-section analysis is chosen, detrital rocks with 70% or more quartz (Friedman, 1958) are most amenable to study. Included are quartz arenites, subarkoses, and subgraywackes, which tend to be well sorted and consist of grains with high sphericity. Fine-grained detrital rocks tend to be polymineralic, poorly sorted, have a considerable amount of matrix, and consist of grains with low sphericities. These cannot be studied in thin-section, with results comparable to sieve-size data, to any degree of accuracy. Allochemical rocks, such as oosparites, some biosparites, and other rocks with well-sorted and highly spherical grains, even if replaced by silica or dolomite, are conveniently studied in thin-section.

SELECTION OF A MINERAL FOR ANALYSIS

If the rock is monomineralic, there is no question concerning grain selection for measurement. Since quartz is the most common clastic mineral in detrital rocks, selection of this mineral is considered important (van der Plas, 1962; Griffiths, 1967). Griffiths (1952) deduced theoretically that it would not be possible to find a simple, overall constant (correction factor for changing apparent size to true size) that could be used for all sandstones. Griffiths (1967, pp. 64-65) indicated that the independent variable, apparent long dimension, in thin-section may be defined as $P = f(s, sh, o, p)$ for a monomineralic rock or one in which one major mineral is selected for measurement. This measure apparently contains "confounded" information from variation in size (s), shape (sh), orientation (o), and packing (p) as well as from their interactions.

METHODS OF GRAIN SELECTION

Point Counting

Point counting provides the fastest and most accurate method of grain selection that are in common use. It requires using a micrometer ocular with cross-hairs for a petrographic microscope, or a similar arrangement with a binocular microscope for polished surfaces or large thin-sections of coarse rocks. A grid system is set up so the entire slide is traversed. If grains to be measured are fairly well sorted, the traverses should be made at right angles to bedding. If not, thin beds or laminae should be traversed parallel with bedding. A mechanical stage is used to move the thin-section in a systematic way on the microscope stage at a controlled spacing by using gears (Chayes, 1949). Every grain which comes under the cross-hair is measured, both in monomineralic or quartz-rich rocks, until some predetermined number has been reached. A laboratory counter may be used to keep continuous tabulation.

In a series of papers, Friedman (1958, 1965a, 1965b) discussed the use of this technique.

Ribbon Counting

Van der Plas (1962, 1965) urged the use of the ribbon-counting method for selecting grains. Equipment and attachments needed are the same as those for point counting, except for the gears. An arbitrary starting point is chosen, and the slide is moved perpendicular to the scale in the micrometer ocular. One may use, for example, the 20 and 80 lines on the scale as boundaries of the ribbon. The width of this ribbon should be as large as, or larger than, the largest diameter of grains to be measured in thin-section. As the slide is moved, all grains encountered in the path of this ribbon are measured until some predetermined number of grains has been measured. The pattern should be laid out systematically across the entire slide. Grains with midpoints within the band are counted. The pattern of movement of the mechanical stage is governed here, as above, by the presence or absence of bedding or laminae.

MEASUREMENT OF INDIVIDUAL GRAINS

Usually, investigators measure the largest apparent dimension of the grain (Krumbein, 1935; Rosenfeld et al., 1953; Friedman, 1958), or at least they choose a unique axis so that different rocks or sediments can be compared (Griffiths, 1967, pp. 64–65). The most common method is to use an ocular micrometer, which has been calibrated to the metric system with a stage micrometer. One could measure the largest apparent dimension by projecting the thin-section on a screen or from photomicrographs, but these methods involve one or more extra preparatory steps.

If point counting is done, the spacing of points and traverses should be equal to, or greater than, the mean grain size (Dennison and Shea, 1966). Point counting results are unbiased regardless of the closeness of spacing, but the standard deviation of the unbiased procedure is minimized with the spacing suggested.

Measurements are usually recorded in millimeters or microns and then converted to phi (ϕ) units and plotted on graphs. Data may be recorded on paper with preset divisions, for example, 0.25ϕ, or on mechanical laboratory counters with similar divisions. The size frequency distribution obtained in this way is a number frequency, and not a weight frequency as is obtained in sieved samples.

Problems involving measured axes are discussed in detail by Smith (1966, 1968) and Sahu (1968a).

NUMBER OF GRAINS TO BE COUNTED PER THIN-SECTION

Most investigators measure between 200 and 500 grains per thin-section (Krumbein, 1935; Krumbein and Pettijohn, 1938; Friedman, 1958; van der Plas, 1962; Griffiths, 1967). Possibly as few as 100 to 200 measurements are necessary for a single analysis to satisfy specified confidence limits of the mean (Rosenfeld, et.al., 1953).

WEIGHT PERCENT VERSUS NUMBER PERCENT

A great amount has been written about the significance of the weight percent obtained by sieving compared to the number percent obtained by point and ribbon counting. In particular, the problem revolves around conversion of the two sets of data to a comparable base. Actually, if a study remains in the framework of the suite of samples, there is no need for this consideration. The problem is critical if attempts are made to compare Recent sediments with well-cemented rocks. The problem is thoroughly discussed in theoretical and empirical terms by Krumbein and Pettijohn (1938, pp. 126–134), Greenman (1951), Pelto (1952), Rosenfeld, Jacobsen, and Ferm (1953), Roethlisberger (1955), Packham (1955), Friedman (1958), van der Plas (1962), and Sahu (1964, 1966, 1967, 1968b).

INSTRUMENTS FOR GRAIN SIZE ANALYSIS

Semiautomatic

The Zeiss TGZ 3 Particle-Size Analyzer was designed for use with a suitable enlarged photomicrograph (Fig. 2). The photo is mounted on the instrument, and as each grain is chosen, the operator controls a handwheel that adjusts the diameter of a light beam from below the photograph in the center of the instrument. The diameter of the beam may be adjusted so that it approximates the area of the grain being measured. If the grain is spherical, the beam boundary will coincide with the grain boundary. The beam can be adjusted to measure largest apparent diameter, thus corresponding to measurements commonly done with the microscope.

The iris diaphragm may be adjusted from 1.2 through 27.7 mm or 0.4 through 9.2 mm. After the grain is measured, the operator controls a footswitch that causes a punch to mark the grain image measured and to activate a bank of counters which has 48 continuous categories. After some experience, one may count and classify up to 1000 grains in 15 minutes.

Schubel and Schiemer (1967) constructed an inexpensive instrument using a Vickers A.E.I. Image Splitting Eyepiece coupled to a microscope for direct observation. Each particle image is split into

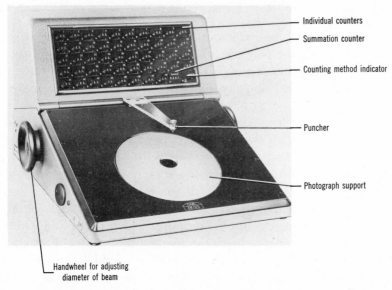

Individual counters
Summation counter
Counting method indicator
Puncher
Photograph support
Handwheel for adjusting
diameter of beam

Fig. 2 Zeiss Electronic Particle-Size Analyzer TGZ 3. (Courtesy of Carl Zeiss, Inc.)

two identical coplanar particle images that can be superimposed or displaced with respect to each other. The diameter of a spherical grain is found by shearing the images until they are tangent, then reading the micrometer displacement and converting the reading into units of length with a constant determined by the optical setup. Largest apparent dimensions may also be measured by shearing the images parallel to the long axis of the superimposed grain images to obtain the length.

The micrometer may be coupled with a bridge to 16 electric impulse counters, each corresponding to a certain displacement interval of the micrometer and to a certain grain-size interval. One-half of the bridge is a high resolution potentiometer which, coupled to the micrometer, generates a voltage analog of the micrometer setting. The other half of the bridge is a logarithmic voltage divider. When a grain image is "split," the operator depresses a footswitch which activates the stepping switch and the micrometer generated analog voltage is compared with the increasing voltage scale which represents the desired size intervals. With some practice, 1000 particles per hour can be measured.

Automatic

The Quantimet Image Computer, made by Metals Research Ltd., Cambridge, England (Fig. 3), is a device for measuring grain size automatically. A thin-section is placed on the microscope stage. The beam image is split; one ray goes into the binocular eyepiece for viewing and the other into a television camera. A television monitor shows the field covered and indicates which parameters the instrument is measuring on what features. The camera output is also fed into an electric detector that responds to the changes in output voltage as the scanning spot in the camera tube passes over grain boundaries in the field. The discriminator can be set to respond to areas darker or lighter than a selected threshold. Signals obtained from these areas are fed into a computer which derives the required information and presents it on the meter, for example, particle size by number of grains. Different size settings would be used to obtain the size distribution. The instrument does not convert number frequency to weight frequency nor does it add a correction factor for the effect of sectioning.

Fig. 3 Quantimet Image Analyzing Computer. (Metals Research Ltd. Courtesy of Jarrell-Ash Co.)

CORRELATION OF THIN-SECTION DATA WITH SIEVE DATA

A number of investigations have been conducted on the problem of conversion of thin-section size frequency data to sieve-size frequency data once the actual measurements have been made. Both the conversion from number to weight frequency and the correction for the sectioning effect must be considered. Rosenfeld, Jacobsen, and Ferm (1953) determined the necessary conversion factors for several different rock types and concluded there is no generally applicable correction factor for converting thin-section measurement to equivalent sieve-size, but found that empirically derived conversions with statistically determined confidence limits may be used within the framework of a single study. Except for the simplest possible idealized cases, they reported no success. Their analyses, measuring apparent long dimensions, showed that uncorrected thin-section frequencies can yield coarser sizes than sieve frequencies, and cumulative curves by both methods are generally parallel to one another. Friedman (1958) obtained similar results.

Friedman made grain-size analyses on 38 rock samples by both sieving and thin-section measurement methods. He worked mainly with quartz arenites and included some subgraywackes and graywackes, all rocks which fit ideal conditions mentioned earlier. He measured apparent long dimensions of up to 500 grains per thin-section.

A plot of thin-section quartile parameters against sieve quartile parameters indicates a linear relationship between the two techniques. For all practical purposes, the same linear function expresses the relationship between quartile parameters as determined by the two methods so that all three quartile measures (Q1, Md, Q3) can be combined into one plot. A graph based on 114 quartile measures for sieve and thin-section analyses forms the basis for the derivation of sieve-size distribution from thin-section data.

Fig. 4 demonstrates the linear relationship. It can be used to plot a cumulative frequency curve for sieve-size distribution from thin-section data as follows (Friedman, 1958):

The 5th, 16th, 25th, 50th, 75th, 84th, and 95th percentiles for a given thin-section cumulative frequency curve are selected. These parameters are derived by interpolation from a straight-line plot on probability paper.

The equivalent sieve values for the above percentiles are determined with the aid of Fig. 4.

The sieve-size cumulative frequency curve is constructed with the new values.

To simplify conversion, Friedman (1958) developed a special graph paper (Fig. 5) that permits conversion without mathematical computations. The correlation diagram (Fig. 4) served as the basis for the construction of this graph. Scale A is subdivided into equal intervals of 0.25φ. Scale B is divided into 0.1φ intervals. Thin-section cumulative frequency values are plotted according to scale A, and the converted sieve-size distribution parameters are read on scale B. It should be noted that in Fig. 4 thin-section parameters are considered the independent variable and therefore neither that diagram nor the graph are adapted for converting sieve-size to thin-section parameters.

Friedman (1962) later extended this study of frequency value conversion to moment measures. He found that a linear 1:1 relationship exists between mean grain size for sieve and thin-section analyses (Fig. 6). The same conversion equation used for quartile conversions can be used again. For mean deviation and standard deviation a linear relationship holds also, but the conversion equation does not work well. Indeed, parameters compiled from observed thin-section values correlate better with sieve data than converted values. For higher moments (skewness and kurtosis) the correlation between sieve and thin-section data is apparently not significant.

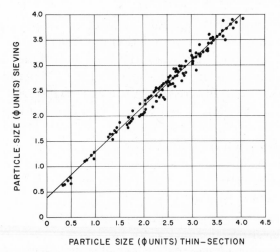

PARTICLE SIZE (φ UNITS) THIN−SECTION

Fig. 4   Single overall correlation line for determining approximate sieve-size distribution from thin-section analyses. Statistical parameters were derived by plotting cumulative curves on probability paper. The equation which combines all three quartile parameters is

Q (calculated) = 0.3815 + 0.9027 Q (observed in thin-section). (After Friedman, 1958. By permission of University of Chicago Press)

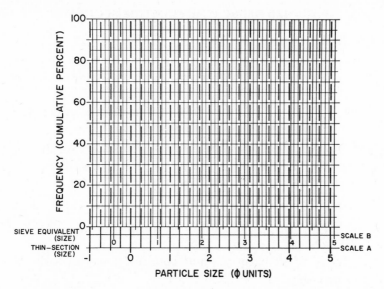

Fig. 5 Graph paper for converting thin-section frequency values to sieve equivalents. (After Friedman, 1958. By permission of University of Chicago Press)

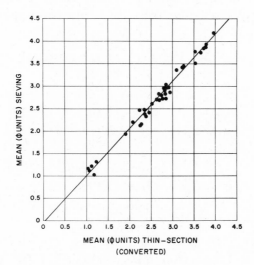

Fig. 6 Regression line for converted thin-section mean and sieve mean. The equation used is

mean (converted thin-section) = 0.3815 + 0.9027 Mean (observed in thin-section).

(After Friedman, 1962. By permission of Society of Economic Paleontologists and Mineralogists)

REFERENCES

Chayes, F., 1949, A simple point counter for thin-section analysis, *Am. Mineralogist*, **34**, 1–11.

Dennison, J. M., and J. H. Shea, 1966, Reliability of visual estimates of grain abundance, *J. Sed. Pet.*, **36**, 81–89.

Friedman, G. M., 1958, Determination of sieve-size distribution from thin-section data for sedimentary petrological studies, *J. Geol.*, **66**, 394–416.

———, 1962, Comparison of moment measures for sieving and thin-section data in sedimentary petrological studies, *J. Sed. Pet.*, **32**, 15–25.

———, 1965a, In defense of point counting analysis, a discussion, *Sedimentology*, **4**, 247–249.

———, 1965b, In defense of point counting analysis: hypothetical experiments versus real rocks, *Sedimentology*, **4**, 252–253.

Graton, L. C., and H. J. Fraser, 1935, Systematic packing of spheres with particular relation to porosity and permeability, *J. Geol.*, **43**, 785–909.

Greenman, N. N., 1951, The mechanical analysis of sediments from thin-section data, *J. Geol.*, **59**, 447–462.

Griffiths, J. C., 1952, Measurement of the properties of sediments, *Bull. Geol. Soc. Am.*, **63**, 1256–1257.

———, 1967, Scientific method in analysis of sediments, McGraw-Hill Book Co., 508 pp.

Krumbein, W. C., 1935, Thin-section mechanical analysis of indurated sediments, *J. Geol.*, **43**, 482–496.

———, and F. J. Pettijohn, 1938, Manual of sedimentary petrography, Appleton-Century-Crofts, 549 pp.

Packham, G. H., 1955, Volume-, weight-, and number-frequency analysis of sediments from thin-section data, *J. Geol.*, **63**, 50–58.

Pelto, C. R., 1952, The mechanical analysis of sediments from thin-section data: a discussion, *J. Geol.* **60**, 402–406.

Roethlisberger, H., 1955, An adequate method of grain-size determination in sections, *J. Geol.*, **63**, 479–584.

Rosenfeld, M. A., L. Jacobsen, and J. C. Ferm, 1953, A comparison of sieve and thin-section technique for size analysis, *J. Geol.,* **61,** 114–132.

Sahu, B. K., 1964, Transformation of weight frequency and number frequency data in size distribution studies of clastic sediments: *J. Sed. Pet.,* **34,** 768–773.

_____, 1966, Thin-section analysis of sandstones on weight-frequency basis, *Sedimentology,* **7,** 255–259.

_____, 1967, Generation of cumulative frequencies from the corrected phi size moments of random thin-section size analysis data, *Sedimentology,* **8,** 329–335.

_____, 1968a, Discussion of the article "Grain size measurement in thin-section and in grain mount," by R. E. Smith, *J. Sed. Pet.,* **38,** 266–268.

_____, 1968b, Thin-section size analysis and the moment problem, *Sedimentology,* **10,** 147–151.

Schubel, J. R., and E. W. Schiemer, 1967, A semiautomatic microscopic particle size analyzer utilizing the Vickers image splitting eyepiece, *Sedimentology,* **9,** 319–326.

Smith, R. E., 1966, Grain size measurement in thin-section and in grain mount, *J. Sed. Pet.,* **36,** 841–843.

_____, 1968, Grain size measurement in thin-section and in grain mount: reply to comment by Sahu, *J. Sed. Pet.,* **38,** 268–271.

van der Plas, L., 1962, Preliminary note on the granulometric analysis of sedimentary rocks, *Sedimentology,* **1,** 145–157.

_____, 1965, In defense of point counting analysis, a reply, *Sedimentology,* **4,** 249–251.

CHAPTER 6

MATHEMATICAL TREATMENT OF SIZE DISTRIBUTION DATA

EARLE F. McBRIDE

University of Texas at Austin, Austin, Texas

Grain-size analyses determined by grain measurement, microscopic measurement in thin-section, sieving, or sedimentation methods yield raw data on percent of grains in a sample that occur in each size class. How grain-size data from a single sample or from a larger number of samples are treated depends on the goal of the project. Grain-size analyses are made for one or more of the following reasons:

1. To describe samples in terms of statistical measures.
2. To correlate samples from similar depositional environments or stratigraphic units.
3. To determine the agent (wind, river, turbidity current, etc.) of transportation and deposition.
4. To determine the process (suspension, traction, saltation, etc.) of final deposition.
5. To determine the environment of deposition (channel, flood plain, beach, dune, neritic marine, etc.).

Since most work in this field has been based on the weight percent of each size class, data produced by measurement of grains in thin-section, which provides percent by the number of grains of each class, are generally converted to weight percent using Friedman's (1958, Fig. 8) graph. Greenman (1951) suggests a different approach to this problem.

For reconnaissance work in sediment distribution patterns, a scheme as simple as mapping of percent sand in samples may show gradients that can be adequately interpreted. However, for most purposes a more detailed treatment of size data is required. The three most commonly used manipulations are (a) plotting data as histograms, frequency distribution curves, or other type of graph, with visual interpretation of plots; (b) computation of moment statistics (mean, standard deviation, skewness, and kurtosis) from raw size-percentage data; and (c) computation of descriptive statistics from intercepts taken visually from graphs (graphic-computational technique). Each technique has its advantages and disadvantages, and considerable controversy exists over the most suitable way to handle data. More than 12 basically different schemes to handle grain-size data have been proposed (see references at end of chapter), and typically each new worker introduces a slight modification of a previous scheme. The logical procedure is to use whatever scheme works best for a particular goal; however, there is little agreement among specialists as to what is "best." This chapter describes the schemes of data treatment most commonly used in the United States.

SIZE SCALES

Most workers in the United States use the geometric grain-size scale devised by Udden (1898, 1914), starting at 1 mm with a constant ratio of 2 (or 1/2) between classes; and most workers use the class name modifications proposed subsequently by Wentworth (1922). Krumbein (1934) introduced the Phi (ϕ) scale as a log transformation of millimeters in order to simplify computation of statistical parameters. Because the Phi scale is now used almost exclusively for computation and is gradually replacing the millimeter scale it is used in this chapter. The graphic and mathematical distinctions among the arithmetic, geometric, and logarithmic grain-size parameters are treated by Krumbein and Pettijohn (1938, pp. 228–267).

GRAPHICAL PRESENTATION

Histograms

Histograms are simple bar graphs showing the percent of grains in each size class. They are easy to construct, but their shape depends in part of the size of the class interval used (1 ϕ, 1/2 ϕ, or 1/4 ϕ interval) and the arbitrary choice of class limits (see Fig. 1 and Table 1). There is no quantitative way of comparing histograms of different samples; therefore histograms are inadequate for most grain-size studies.

Triangular Diagrams

Triangular diagrams are used to show the relative amounts of three components: sand-silt-clay, gravel-sand-mud, fine-medium-coarse sand, and so on (Fig. 2). Marine geologists commonly use the scheme in reconnaissance studies involving large numbers of samples. If each field of a triangle is given a different pattern or color, the data can be presented in the form of a facies map for areal interpretation. This is

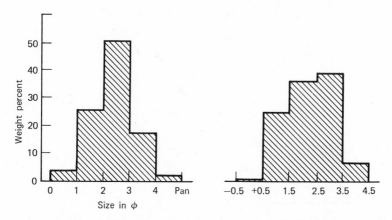

Fig. 1   Histograms showing the same grain-size distribution but constructed by using different class boundaries. Cumulative curves of the distribution are shown in Figs. 3 and 4.

done by giving each datum point on the map the same pattern or color as the grain-size field in which the sample plots on the triangle, then drawing boundaries between areas of different pattern on the map.

TABLE 1   Grain-size distribution plotted in Figs. 1, 3, and 4

Class Limits in ϕ	Frequency Weight %	Frequency in 1 ϕ Classes		Cumulative Frequency, Weight %
		Boundary at Even ϕ Classes	Boundary at 1/2—ϕ Classes	
0— .5	0.9		0.9	0.9
.5—1.0	2.9	3.8		3.8
1.0—1.5	12.2		25.1	26.0
1.5—2.0	13.7	25.9		39.7
2.0—2.5	23.7		37.4	53.4
2.5—3.0	26.8	50.5		80.2
3.0—3.5	12.2		39.0	92.4
3.5—4.0	5.6	17.8		98.0
> 4.0	2.0		7.6	100.0

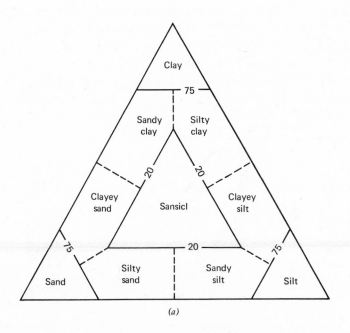

(a)

Fig. 2   Three sediment nomenclature schemes for sand-silt-clay mixtures: (a) Shepard, 1954; (b) Folk, 1954; (c) Gorsline, 1960.

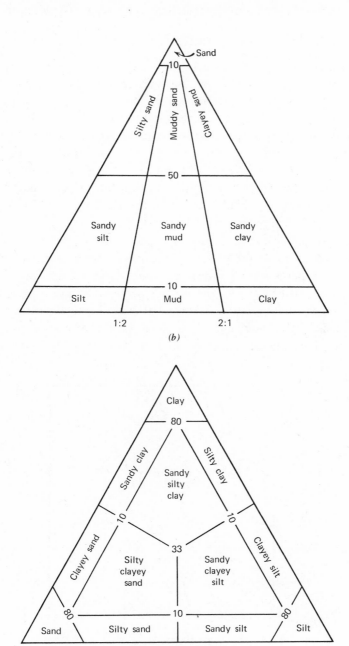

Fig. 2 (Continued)

Cumulative Frequency Curves

Frequency curves are constructed to show the percent of grains coarser or finer than any given grain size for an individual sample. In the United States it is conventional to plot gain size on the abscissa with smaller grain sizes to the right and to plot cumulative percentage on the ordinate. Data are cumulated beginning from the origin, or coarse end of the abscissa, and the resulting curve is a "coarser-than" curve, a curve showing percent of grains coarser than a given size. However, all possible variations of ordinate and abscissa have been used, especially among European workers, who have shown little tendency toward uniformity.

Early workers plotted cumulative percentages on an arithmetic scale, on which most samples plot as an S-shaped curve. However, a probability percentage ordinate scale, on which log-normal distributions plot as a straight line, is preferable; and a probability percentage ordinate scale is essential if data are to be extrapolated from a size-distribution curve (see p. 115). Because most unimodal sediments have nearly à normal Gaussian distribution when a

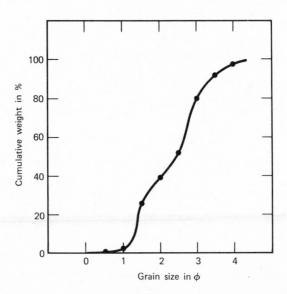

Fig. 3 Cumulative curve drawn on arithmetic percent ordinate. The curve has two conspicuous inflection points, indicating the sample is bimodal. The cumulative curve must be fitted through all data points.

geometrical grain-size scale (ϕ) is used on the abscissa, they approach a straight line when plotted on probability percentage paper. Differences between samples are more conspicuous when dealing with nearly straight lines than with S-shaped curves.

The data to be plotted must be first cumulated beginning at the coarse end of the scale. The cumulative weight percent is plotted at corresponding size class boundaries (not the midpoint of class fields).

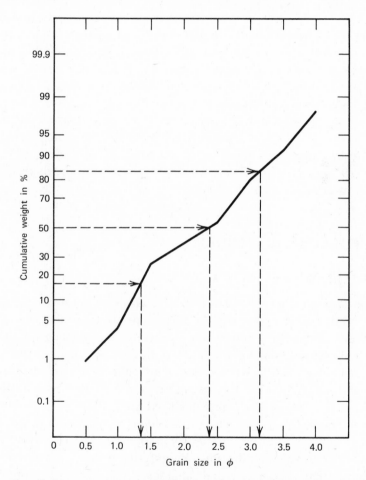

Fig. 4 Cumulative curve drawn on probability percent ordinate. Same sample as shown in Fig. 3. The dashed lines show how the phi percentiles are interpolated: The 16th, 50th, and 84th percentiles are the grain sizes at which the 16th, 50th, and 84th percentage lines intersect the cumulative curve.

On arithmetic percentage paper the data points must be connected using a French curve, whereas on probability percentage paper the points may be connected either by a curve or by straight-line segments. On probability paper the use of straight-line segments assumes that samples are normally distributed between size classes, but both techniques described are equally interpretive.

Frequency Curves

The first derivative of a cumulative frequency curve is a bell-shaped curve known as the frequency (distribution) curve. The bell-shaped curve is generated from a histogram by theoretically increasing the number of size classes to infinity, which correspondingly reduces the class interval to zero. Frequency curves have no graphic advantage over the corresponding cumulative curves, and are difficult to construct accurately. Techniques for graphic construction of frequency curves are given by Krumbein and Pettijohn (1938, pp. 190–195), and mathematical solutions are given by Brotherhood and Griffiths (1947) and Bush (1951).

COMPUTATIONAL TECHNIQUE

Moment Statistics

One approach to the quantitative analysis of grain-size data is to characterize each size analysis by a derived number or set of numbers, and then compare and contrast samples using the derived numbers. Because most grain-size distributions approach a normal (or Gaussian) distribution when ϕ size is plotted on an arithmetic scale, conventional moment statistics can be used to characterize individual samples. The descriptive statistics used are *mean size* (a measure of central tendency of the curve), *standard deviation* (a measure of spread around the mean, or sorting), *skewness* (a measure of symmetry of the distribution around the mean), and *kurtosis* (a measure that has no visual analogy, but which is commonly inappropriately related to peakedness (Baker, 1968)).

The moment statistics applied to grain size analysis are defined as follows:

Mean $\bar{x}_\phi = \dfrac{\Sigma fm}{n}$ (1)

Standard deviation $\sigma_\phi = \sqrt{\dfrac{\Sigma f(m - \bar{x}_\phi)^2}{100}}$ (2)

Skewness $Sk_\phi = \dfrac{\Sigma f(m - \bar{x}_\phi)^3}{100\,\sigma_\phi{}^3}$ (3)

Kurtosis $K_\phi = \dfrac{\Sigma f(m - \bar{x}_\phi)^4}{100\,\sigma_\phi{}^4}$ (4)

where f = weight percent (frequency) in each grain-size grade present,
m = midpoint of each grain-size grade in phi values
n = total number in sample which is 100 when f is in percent.

Table 2 shows the way calculations are made for the four moment statistics. The technique of computing the parameters is commonly called the method of moments, because, as Table 2 shows, computations involve multiplying a weight (frequency, in percent, by a distance (from the midpoint of each size grade to the arbitrary origin of the abscissa); this is analogous to computing moments of inertia for the grains in each size grade in the distribution. For a rigorous treatment of statistical theory of moment parameters the reader should consult a textbook on statistics.

Computations (Table 2) are made in a similar manner regardless of whether 1ϕ, $1/2\phi$, or $1/4\phi$ grain-size classes are used, although time required for the computation is longer when smaller size classes are used. The work is simplified by putting raw data on punched cards and having the computations made by a computer. Several programs for grain-size analyses have been described (Kane and Hubert, 1963; Collias et al., 1963; Schlee and Webster, 1967).

Computations made from formulas (2), (3), and (4) for standard deviation, skewness, and kurtosis (Table 2) are subject to successive roundoff errors. An alternate method of computation is to expand and simplify to the following formulas:

$$\sigma_\phi = \frac{1}{100} \Sigma fm^2 - \bar{x}^2_\phi \tag{5}$$

$$Sk_\phi = \frac{\dfrac{1}{100} \Sigma fm^3 - \dfrac{3}{100} x_\phi \Sigma fm^2 + 2\bar{x}^3_\phi}{\sigma^3_\phi} \tag{6}$$

$$K_\phi = \frac{\dfrac{1}{100} \Sigma fm^4 - \dfrac{4x_\phi}{100} \Sigma fm^3 + \dfrac{6}{100} \bar{x}^2_\phi \Sigma fm^2 - 3\bar{x}^4_\phi}{\sigma^4_\phi} \tag{7}$$

Computations are made as shown in Table 3. This procedure has the advantage of having less roundoff error and involves simpler calculations that can be made conveniently on a desk calculator. The numbers in the various powers of m columns are the same for all computations, and these numbers can be incorporated into data sheets.

A useful scheme for simplifying hand calculation of grain-size moments by converting the midpoints of size classes to integers (the coded technique) is described by Krumbein and Pettijohn (1938, pp. 249–253).

One of the advantages of the method of moments is that the entire distribution is used in computing the summary statistics rather than only a part of the distribution as in the graphic-computational technique described below. A problem arises, however, with an "open-ended" distribution that results from a sieve analysis of a sand sample that has a large pan fraction. In this instance, either the distribution of "fines" must be determined by a sedimentation technique (by pipetting or by using a hydrometer) or a distribution must be assumed for the unanalyzed fraction. For sand samples with pan fractions less than about 8% of the total weight, if fines are chiefly silt-size grains, an assumed size of 4.25 or 4.5ϕ should be used; if chiefly clay, an assumed average size of 8 or 10ϕ should be used.

TABLE 2 Form for computing moment statistics using ½ φ classes

Class Interval, φ	m Midpoint, φ	f Weight %	fm Product	m −x̄ Deviation	(m −x̄)² Deviation Squared	f(m −x̄)² Product	(m −x̄)³ Deviation Cubed	f(m −x̄)³ Product	(m −x̄)⁴ Deviation Quadrupled	f(m −x̄)⁴ Product
0−0.5	.25	.9	.2	−2.13	4.54	4.09	−9.67	−8.70	20.60	18.54
0.5−1.0	.75	2.9	2.2	−1.63	2.66	7.71	−4.34	−12.59	7.07	20.50
1.0−1.5	1.25	12.2	15.3	−1.13	1.28	15.62	−1.45	−17.69	1.63	19.89
1.5−2.0	1.75	13.7	24.0	−0.63	0.40	5.48	−0.25	−3.43	0.16	2.19
2.0−2.5	2.25	23.7	53.3	−.13	.02	0.47	.00	0.00	0.00	0.00
2.5−3.0	2.75	26.8	73.7	0.37	.13	3.48	.05	1.34	0.02	0.54
3.0−3.5	3.25	12.2	39.7	.87	.76	9.27	.66	8.05	0.57	6.95
3.5−4.0	3.75	5.6	21.0	1.37	1.88	10.53	2.57	14.39	3.52	19.71
>4.0	4.25	2.0	8.5	1.87	3.50	7.00	6.55	13.10	12.25	24.50
Total		100.0	237.9			63.65		−5.53		112.82

119

TABLE 3 Alternate form for computing moment statistics using ½φ classes

Class Interval φ	m Midpoint, φ	f Weight %	fm Product	m^2 Midpoint Squared	fm^2 Product	m^3 Midpoint Cubed	fm^3 Product	m^4 Midpoint Quadrupled	fm^4 Product
0–0.5	.25	.9	.2	.063	.05	.015	0.01	.0039	0.00
0.5–1.0	.75	2.9	2.2	.56	1.62	.42	1.22	.316	0.92
1.0–1.5	1.25	12.2	15.3	1.56	19.03	1.95	23.79	2.43	29.65
1.5–2.0	1.75	13.7	24.0	3.06	41.92	5.36	73.43	9.36	128.23
2.0–2.5	2.25	23.7	53.0	5.06	119.92	11.39	269.94	25.60	606.72
2.5–3.0	2.75	26.8	73.7	7.56	202.61	20.80	557.44	57.15	1531.62
3.0–3.5	3.25	12.2	39.7	10.56	128.83	34.33	418.83	111.5	1360.30
3.5–4.0	3.75	5.6	21.0	14.06	78.74	52.73	295.29	197.6	1106.56
> 4.0	4.25	2.0	8.5	18.06	36.12	76.77	153.54	246.2	492.40
Total		100.0	237.6		628.84		1793.49		5256.40

An additional problem arises through an erroneous assumption built into moment computations. For computation all grains within a size class are treated as if they are in a size equal to the midpoint of the size grade; that is, the inertia of each size grade is assumed to be concentrated at the midpoint, whereas it is actually nearer to the mean than the midpoint, so that the moment arm is shorter than the mean-to-midpoint distance. The midpoint assumption is not strictly accurate for most distributions. A formula (Sheppard's correction) for correcting the error in computing standard deviation has not been used much by sedimentologists because errors in technique (incomplete disaggregation of grains, faulty sieves, etc.) commonly overshadow this computational error.

GRAPHIC COMPUTATIONAL TECHNIQUE

Graphic computational techniques for characterizing grain-size distributions were introduced to avoid the lengthy calculations required by moment statistics before development of the electronic computer. Most graphically derived statistics in current use are approximations of moment statistics $(\bar{x}, \sigma, Sk, K)$, although some workers still use the now outmoded quartile statistics (so-called because they are based on grain-size values at the 25th, 50th, and 75th percentiles on the cumulative curve). Although graphic-computational statistics describe a slightly different property of sample distributions than do moment statistics (Friedman, 1962b; Middleton, 1962; Folk, 1966, p. 80), they have been equally as successful as moment statistics in distinguishing environments and processes of deposition, are easily computed, and have the advantage that the cumulative curve that must be drawn to determine the percentiles may show traits of the distribution that would not otherwise be obvious (Folk, 1966, p. 77).

The first step in the graphic-computational technique is to draw a cumulative percentage curve on probability paper in the manner described on page 115. S-shaped cumulative curves drawn on arithmetic paper are not suitable because of the difficulty of extrapolation of percentiles in the nearly horizontal tails of the curves. The curve must be drawn through all data points, not between points, as in a smoothed curve of best fit. (See Fig. 3.) The grain-size value needed to compute each graphic statistic (Table 4) is then read from the graph by finding the grain-size value in ϕ that corresponds with each percentile value to be used in the computation (Fig. 4 and Table 5). The notations $\phi16$, $\phi50$, and $\phi84$ (verbally described as the 16th, 50th, and 84th percentiles) refer to the grain-size in ϕ at the 16th, 84th, and 50th percent values on the cumulative curve.

If the grain-size distribution is open-ended and percentiles in the tails of the distribution are needed for computation, the curve must be arbitrarily extended. One way is to extend straight lines from the last two data points on each end of the curve. For mud samples, Folk (1966, p. 75) recommends that grain sizes be determined to 10ϕ, and that a straight-line distribution be assumed beyond 10ϕ to 100% at 14ϕ.

TABLE 4   Formulas for computing "graphical" parameters

Mean		Efficiency, %
Trask (1930)	Median, $\phi50$	64
Otto (1939), Inman (1952)	$M\phi = (\phi16 + \phi84)/2$	74
Folk and Ward (1957)	$Mz = (\phi16 + \phi50 + \phi84)/3$	88
McCammon (1962)	$(\phi10 + \phi30 + \phi50 + \phi70 + \phi90)/5$	93
McCammon (1962)	$(\phi5 + \phi15 + \phi25 \ldots + \phi85 + \phi95)/10$	97
Sorting		
Krumbein (1934)	$QD\phi = (\phi75 - \phi25)/1.35$	37
Otto (1939), Inman (1952)	$(\phi84 - \phi16)/2$	54
Folk and Ward (1957)	$(\phi84 - \phi16)/4 + (\phi95 - \phi5)/6.6$	79
McCammon (1962)	$(\phi85 + \phi95 - \phi5 - \phi15)/5.4$	79
McCammon (1962)	$(\phi70 + \phi80 + \phi90 + \phi97 - \phi3 - \phi10 - \phi20 - \phi30)/9.1$	87
Skewness		
Krumbein and Pettijohn (1938)	$Skq\phi = [\phi25 + \phi75 - 2(\phi50)]/2$	
Inman (1952)	$\alpha_{1\phi} = \dfrac{\phi16 + \phi84 - 2(\phi50)}{\phi84 - \phi16}$	
	$\alpha_{2\phi} = \dfrac{\phi5 + \phi95 - 2(\phi50)}{\phi84 - \phi16}$	

TABLE 4   Formulas for computing "graphical" parameters (Continued)

Mean	
Folk and Ward (1957)	$Sk_I = \dfrac{\phi84 + \phi16 - 2\phi50}{2\,(\phi84 - \phi16)} +$
	$\dfrac{\phi95 + \phi5 - 2\phi50}{2\,(\phi95 - \phi5)}$

Kurtosis	
Krumbein and Pettijohn (1938)	$Kqa = \dfrac{\phi75 - \phi25}{2\,(\phi90 - \phi10)}$
Inman (1952)	$\phi = \dfrac{(\phi95 - \phi5) - (\phi84 - \phi16)}{\phi84 - \phi16}$
Folk and Ward (1957)	$K_G = \dfrac{\phi95 - \phi5}{2.44(\phi75 - \phi25)}$

TABLE 5 ϕ Percentiles interpolated from Fig. 4

Percentile	ϕ Size
3	0.92
5	1.07
10	1.22
15	1.33
16	1.35
20	1.42
25	1.46
30	1.66
35	1.86
45	2.21
50	2.38
55	2.54
65	2.71
70	2.80
75	2.90
80	3.00
84	3.15
85	3.20
90	3.42
95	3.71
97	3.88

Workers who use the graphical-computational technique to characterize grain-size distributions have suggested a number of formulas to be used. Most formulas yield approximations of moment statistics, although some have no analog in moment statistics. Table 4 lists the most commonly used formulas. The table also gives the efficiency of individual formulas in approximating the computed moment mean in samples drawn from normal distributions (data from McCammon, 1962). A comparison of the advantages of formulas proposed by different workers, as well as the geologic meaning of commonly used statistical parameters, was presented by Folk (1966).

Table 6 compares statistical values determined by moment and graphical techniques for the sediment sample given in Table 1.

TABLE 6   Comparison of moment and "graphical" parameters for sample given in Table 1

Mean	
Moment (first method)	2.38ϕ
Moment (second method)	2.38ϕ
Trask	2.38ϕ
Otto-Inman	2.25ϕ
Folk and Ward	2.29ϕ
McCammon (first method)	2.30ϕ
McCammon (second method)	2.30ϕ
Sorting	
Moment (first method)	0.64ϕ units
Moment (second method)	0.63ϕ units
Otto-Inman	0.90ϕ units
Folk and Ward	0.85ϕ units
McCammon (first method)	0.84ϕ units
McCammon (second method)	0.87ϕ units
Skewness	
Moment (first method)	-0.21
Moment (second method)	-0.16
Inman	-0.14
Folk and Ward	-0.13
Kurtosis	
Moment (first method)	6.63
Moment (second method)	4.93
Inman	0.47
Folk and Ward	0.75

REFERENCES

Baker, R. A., 1968, Kurtosis and peakedness, *J. Sed. Pet.*, **38**, 679–680.

Brotherhood, G. R., and J. C. Griffiths 1947, Mathematical derivation of the unique frequency curve, *J. Sed. Pet.*, **17**, 77–82.

Bush, J., 1951, Derivation of a size-frequency curve from the cumulative curve, *J. Sed. Pet.*, **21**, 178–182.

Collias, E. E., M. R. Rona, D. A. McManus, and J. S. Creager, 1963, Machine processing of geological data, Univ. Washington Tech. Rept. 87, 119 pp.

Folk, R. L., 1966, A review of grain-size parameters, *Sedimentology*, **6**, 73–93.

_____, 1954, The distinction between grain size and mineral composition in sedimentary rock nomenclature, *J. Geol.*, **62**, 334–359.

_____, and W. C. Ward, 1957, Brazos River bar, a study in the significance of grain-size parameters, *J. Sed. Pet.*, **27**, 3–27.

Friedman, G. M., 1958, Determination of sieve-size distribution from thin-section data for sedimentary petrological studies, *J. Geol.*, **66**, 394–416.

_____, 1962, On sorting coefficients, and the log-normality of the grain-size distribution of sandstones, *J. Geol.*, **70**, 737–756.

Gorsline, D.S., 1960, Lecture, Univ. of Texas at Austin.

Greenman, N. M., 1951, The mechanical analysis of sediments from thin-section data, *J. Geol.*, **59**, 447–462.

Inman, D. L., 1962, Measures for describing the size distribution of sediments, *J. Sed. Pet.*, **22**, 125–145.

Kane, W. T., and J. F. Hubert, 1963, Fortran program for calculation of grain size textural parameters on the IBM 1620 computer, *Sedimentology*, **2**, 87–90.

Krumbein, W. C., 1934, Size frequency distribution of sediments, *J. Sed. Pet.*, **4**, 65–77.

_____, and F. J. Pettijohn, 1938, Manual of sedimentary petrology, Appleton-Century-Crofts, 549 pp.

McCammon, R. B., 1962, Efficiencies of percentile measures for describing the mean size and sorting of sedimentary particles, *J. Geol.*, **70**, 453–465.

Middleton, G. V., 1962, On sorting, sorting coefficients, and the lognormality of the grain-size distribution of sandstones—a discussion, *J. Geol.*, **70**, 754–756.

Otto, G. H., 1939, A modified logarithmic probability graph for the interpretation of mechanical analyses of sediments, *J. Sed. Pet.*, **9**, 62–76.

Schlee, John, and Jacqueline Webster, 1967, A computer program for grain-size data, *Sedimentology*, **8**, 45–54.

Shepard, F.P., 1954, Nomenclature based on sand-silt-clay ratios, *J. Sed. Pet.*, **24**, 151–158.

Trask, P. D., 1930, Mechanical analysis of sediments by centrifuge, *Econ. Geol.*, **25**, 581–599.

Udden, J. A., 1898, Mechanical composition of wind deposits, Augustana Library Publ. 1, 69 pp.

———, 1914, Mechanical composition of clastic sediments, *Bull. Geol. Soc. America.*, **25**, 655–744.

Wentworth, C. K., 1922, A scale of grade and class terms for clastic sediments, *J. Geol.*, **30**, 377–392.

Selected References on the Use of Grain Size to Interpret Processes and Environments of Deposition

Bull, W. B., 1962, Relation of textural (CM) patterns to depositional environment of alluvial-fan deposits, *J. Sed. Pet.*, **32**, 211–217.

Curray, J. R., 1960, Tracing sediment masses by grain-size modes, 21st Inter. Geol. Congr., Copenhagen, Rept. Session, Norden, 119–130.

Doeglas, D. J., 1946, Interpretation of the results of mechanical analysis, *J. Sed. Pet.*, **16**, 19–40.

———, 1955, A rectangular diagram for comparison of size-frequency distributions, *Geologie Mijnbouw*, **17**, 129–136.

Folk, R. L., 1962, Sorting in some carbonate beaches of Mexico, *Trans. New York Acad. Sci., Series* II, **25**, 222–244.

Friedman, G. M., 1961, Distinction between dune, beach, and river sands from their textural characteristics, *J. Sed. Pet.*, **28**, 151–163.

Harris, S. A., 1958, Probability curves and the recognition of adjustment to depositional environment, *J. Sed. Pet.,* **28,** 151–163.

Inman, D. L., 1949, Sorting of sediments in the light of fluid mechanics, *J. Sed. Pet.,* **19,** 51–70.

Mason, C. C., and R. L. Folk, 1958, Differentiation of beach, dune, and aeolian flat environments by size analysis, Mustang Islands, Texas, *J. Sed. Pet.,* **28,** 211–226.

Moss, A. J., 1963, The physical nature of common sandy and pebbly deposits, Part 2, *Am. J. Sci.,* **261,** 297–343.

Passega, R., 1957, Texture as characteristic of clastic deposition; *Bull. Am. Assoc. Petrol. Geologists,* **41,** 1952–1984.

———, 1964, Grain size representation by CM patterns as a geological tool, *J. Sed. Pet.,* **34,** 830–847.

Rogers, J. J. W., and C. Strong, 1959, Textural differences between two types of shoestring sands, *Trans. Gulf Coast Assoc. Geol. Soc.,* **10,** 168–170.

Sahu, B. K., 1964, Depositional mechanisms from the size analysis of clastic sediments, *J. Sed. Pet.,* **34,** 73–83.

Tanner, W. F., 1964, Modification of sediment size distributions, *J. Sed. Pet.,* **34,** 156–164.

SECTION III

GRAIN ATTRIBUTES

CHAPTER 7

GRAIN SHAPE

WAYNE A. PRYOR

University of Cincinnati, Cincinnati, Ohio

Analysis of grain morphology is generally restricted to form descriptions of detrital particles. Three principal grain-form parameters are in common usage today: *shape, roundness,* and *pivotability*. Grain shape is defined in terms of the described spatial geometric form of grains; whereas grain roundness describes the relative sharpness (or lack of sharpness) of grain corners and edges. Pivotability (or rollability) is a relatively novel measurement based on the movement of grains, and is a multivariant measure of grain morphology.

There is an extensive and long-term literature on the subject of grain-form origin and interpretation and granular morphometry. A bibliography accompanying this chapter includes selected references pertaining to morphometric technology and grain-morphology theory and interpretation. For detailed historical insights into morphometric grain analysis, the reader is referred to Krumbein and Pettijohn (1938) and especially to the comprehensive book, *Granulometrische und morphometrische Messmethoden an Mineralkörnen, Steinen, und sonstigen Stoffen* (Köster, 1964), which has one of the most complete sets of references on grain morphology and morphometry available in the modern literature.

Grain morphometry is accomplished by four principal methods.

1. *Visual description.* The operator verbally describes grain morphology from visual observation of the grain or grain facsimile.

2. *Visual comparison.* The operator views the actual grain or grain facsimile and compares it to a standard reference.

3. *Direct measurement.* The operator makes dimensional measurements of the actual grain or grain facsimile.

4. *Response measurement.* The operator observes and measures the response of the actual grain to a set of standard physical conditions.

SHAPE

Grain shape is defined as the spatial geometric form of a grain. The shapes of grains can be described verbally by an operator in strictly qualitative terms. These terms depend on the observational ability of the operator in relation to operational definitions and consequently may vary from one operator to another. Descriptions may be in the form of geometric identities such as cubic, spherical, elliptical, prolate, platy, prismatic, tabular, acicular, cylindrical, and conical, or they can be described in terms of crystal-form such as euhedral, anhedral, hexagonal, and tetragonal. Grain shape can also be described in free-form terms such as globular, round, vermiform, reniform, shard-like, disk-shaped, needle-shaped, fusiform, irregular, lozenge, and rodlike. Although these descriptions may be extremely informative, they lack consistency in application and exactness of description. For this reason, grain shape is normally described quantitatively and expressed numerically by means of measurements. The most commonly used quantitative grain shape parameters are *sphericity, flatness ratio, roundness ratio,* and *elongation index.*

Sphericity

Since Wadell's introduction of the sphericity concept (Wadell, 1932), which entailed the measurement of grain surface area and volume, numerous attempts have been made to render sphericity a practical parameter to measure. This has been done by substituting a variety of measures for the difficult-to-obtain surface area dimensions.

Presently used sphericity calculations, based on Wadell's original sphericity, utilize measurements of actual grains, grains in thin-section, or grain projections.

ψ_w = Wadell's Working Sphericity (Wadell, 1935)

$$\psi_w = \sqrt{\frac{\frac{4\,A_p}{II}}{d_p}}$$

where A_p = projected area of grain,
d_p = diameter of smallest circumscribed circle around grain projection.

ψ_i = Krumbein's Intercept Sphericity (Krumbein, 1941)

$$\psi_i = \sqrt[3]{\frac{L \cdot I \cdot S}{L^2}}$$

where L = longest axis,
I = intermediate axis,
S = short axis.

ψ_p = Maximum Projection Sphericity (Sneed and Folk, 1958)

$$\psi_p = \sqrt[3]{\frac{S^2}{L \cdot I}}$$

where L = longest axis,
I = intermediate axis,
S = short axis.

The projection sphericity (ψp), as suggested by Sneed and Folk (1958), takes into consideration the hydraulic behavior of the particle and therefore is a more actualistic concept of sphericity and the preferred measure.

For sphericity measurements of grains in thin-section and grain projections, where only two dimensional data are available, Riley (1953) proposed a projection sphericity, ψr.

ψr = Riley Projection Sphericity (Riley, 1953)

$$\psi r = \frac{d_i}{D_c}$$

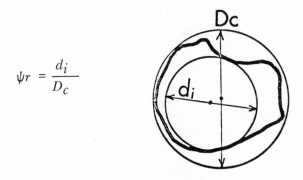

where D_c = diameter of smallest circumscribing circle,
      d_i = diameter of largest inscribing circle.

Zingg (1935) proposed a shape classification similar in concept to sphericity. Zingg's classification is based on ratios of the three principal intercept axes; long, short, and intermediate, I/L and S/I. Zingg devised four shape classes: I, disc; II, spheroid; III, blade; and IV, rod. Zingg's classification scheme, including modifications by Krumbein (1941) and Brewer's added classes of planar and acicular (Brewer, 1964), is shown in Fig. 1.

Sneed and Folk (1958) combined maximum projection sphericity (ψp) with the intercept ratios of S/L and $L\text{-}I/L\text{-}S$ into a form triangle (Fig. 2) with four major classes: I, compact, II, platy, III, bladed, and IV, elongated, which can be subdivided into a total of ten subclasses.

Perhaps the most widely used method of determining sphericity is the *visual comparator*. The ease of operation enables large numbers of grains to be classified, which accounts for its popularity. A comparator can be constructed by an operator by simply collecting a series of grains of graded sphericity, determined by any one of the

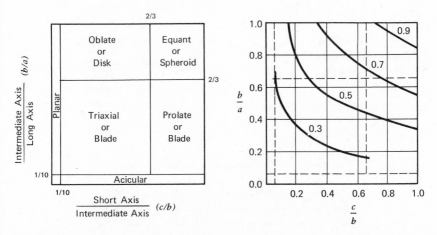

Fig. 1 Shape classes based on ratios of specific axes and relationship of sphericity (ψ_I) to the shape classes. (After Krumbein, 1941, and Brewer, 1964)

preceding techniques, and mounting them on slides and/or photographing them. Grains to be classified are compared to the standards, and a decision is made relative to the sphericity classification of the grains. Powers (1953) constructed such a comparator utilizing photographs of grains. Powers' chart presents two classes ot sphericity, *high sphericity* and *low sphericity*, in combination with six classes of grain roundness. Krumbein and Sloss (1955) constructed a grain silhouette comparator (Fig. 3) illustrating four classes of sphericity in combination with five classes of roundness.

Flatness Ratio and Roundness Ratio

These two grain shape measures were introduced by Wentworth (1919) in one of the first attempts to quantify grain shape. They were initially the ratios of surface curvature radii, but were later modified to use the more easily derived major geometric diameters of length (A), breadth (B), and thickness (C).

$$\text{Flatness ratio} = \frac{A + B + C}{3}$$

$$\text{Roundness ratio} = \frac{A + B}{2C}$$

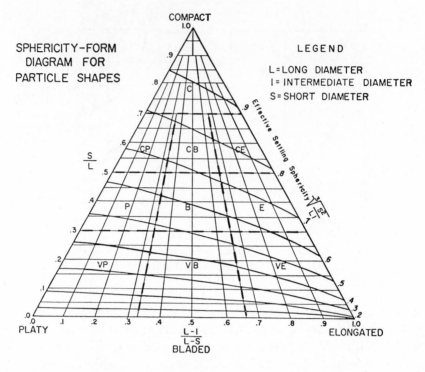

Fig. 2 Form triangle showing form classes and relationship of sphericity (ψ_p). (After Sneed and Folk, 1958)

Cailleux (1947, 1952) adopted the roundness ratio, but required that the three principal dimensions meet a right angles, and designated this as a flatness ratio. Thus the Cailleux flatness ratio is given as

$$F_r = \frac{L + \ell}{2E}$$

where L = greatest length,
ℓ = greatest width,
E = greatest thickness measured normal to L and ℓ.

Elongation Index

The elongation index is described by Schneiderhöhn (1954) as the ratio of the greatest width to the greatest length.

$$\text{Elongation index} = \frac{W}{L}$$

Another method of measuring this parameter is suggested by Folk (1965, p. 9) utilizing the least projection widths and lengths. Table 1 gives the classes and indices derived by Folk.

$$\text{Elongation index} = \frac{W_p}{L_p}$$

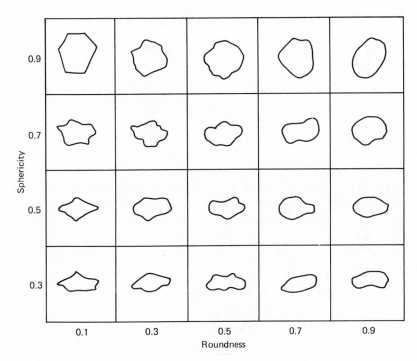

Fig. 3   Chart for visual estimation of roundness and sphericity. (From Krumbein and Sloss, 1955)

TABLE 1   Elongation classes and indices (after Folk, 1965)

Elongation Classes	Elongation Indices
Very equant	> 0.75
Equant	0.75-0.72
Subequant	0.72-0.69
Intermediate	0.69-0.66
Subelongate	0.66-0.63
Elongate	0.63-0.60
Very elongate	< 0.75

Oblateness—Prolateness

A measure of oblateness and prolateness recently has been introduced by Dobkins and Folk (1968):

$$OP = S \ \frac{L/ (L - I)}{(L - S - 0.50)}$$

where L = long axis length
  I = intermediate axis length
  S = short axis length

Effective Settling Sphericity

Grain shape may be classified in terms of a response measurement. The relationship between the fall velocity of a spherical grain and grain size is well known; however, the velocity change as a grain varies from a sphere is less well known. Herein is a potential shape measurement given in terms of fall velocity. Briggs, McCulloch, and Moser (1962) discuss the sphericity-fall velocity relationship utilizing the concept of dynamic shape factor (DSF). Sneed and Folk (1958) discuss this relationship as effective settling sphericity and relate it to their form triangle (Fig. 2).

ROUNDNESS

Grain roundness is defined as the description of the relative sharpness of grain corner, or more simply as a description of grain surface curvature.

The verbal description of grain roundness or angularity includes all the uncertainty of any qualitative description. However, Russell and Taylor (1937), Pettijohn (1949), and later Schneiderhöhn (1954) set limits to a verbal classification (Table 2) and accompanied this with a visual silhouette comparator (Fig. 4), in which five classes of roundness are shown and described.

TABLE 2   Roundness classes (after Schneiderhöhn, 1954)

Classes	Definition
I. Angular	Strongly developed faces with sharp corners. Sharply defined, large reentrants with numerous small reentrants
II. Subangular	Strongly developed flat faces with incipient rounding of corners. Small reentrants subdued and large reentrants preserved
III. Subrounded	Poorly developed flat faces with corners well rounded. Few small and gently rounded reentrants, and large reentrants weakly defined
IV. Rounded	Flat faces nearly absent with corners all gently rounded. Small reentrants absent and large reentrants only suggested
V. Well rounded	No flat faces, corners, or reentrants discernible, and a uniform convex grain outline

As with shape parameters, many quantitative measures for roundness have been proposed. Köster (1964) lists six major methods of calculation, of which the following three seem to be the most widely used.

P_d = Degree of Roundness (Wadell, 1933)

$$P_d = \frac{\Sigma\left(\frac{r}{R}\right)}{N}$$

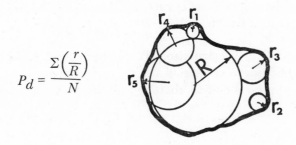

where r = curvature radius of individual corners,

N = number of corner radii *including corners whose radii are zero,*

R = radius of maximum inscribed circle.

P_r = Roundness Ratio (Wentworth, 1933)

$$P_r = \frac{r_1}{R}$$

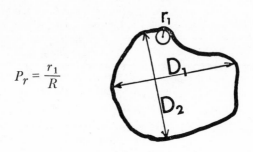

where r_1 = radius of smallest corner,

$R = \dfrac{D_1 + D_2}{4}$ = mean grain radius,

D_1 = longest dimension of grain,

D_2 = greatest width of grain normal to D_1.

P_i = Rounding Index (Cailleux, 1947)

$$P_i = \frac{2\,r_1}{L}$$

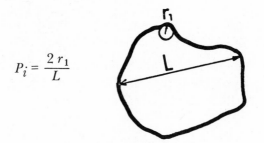

where r_1 = radius of smallest corner,

L = greatest length.

Utilizing the Wadell calculation, Pettijohn (1949), Powers (1953), and Folk (1955) have combined the degrees of roundness (P_d) with descriptive roundness classes, thereby unifying the two classifications (Table 3).

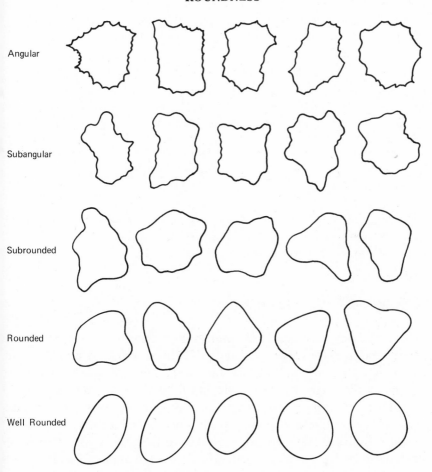

Fig. 4 Roundness chart of Russell and Taylor. (1937, after Schneiderhöhn, 1954)

The visual comparator presented by Powers (1953) is one of the most widely used devices for grain roundness studies. Powers presented the six classes of roundness and two classes of sphericity as grain photographs. The operator is faced with comparing a grain with the photographs and deciding the relative similarity of the two. In practice, many operators decide whether the grain is round or angular and include grains in one of these two major classes. The ratio of round to angular grains is then expressed for the sample as a whole, in percent of angular grains or percent of round grains. The

TABLE 3   Degree of roundness classes and indices

Class Name	Roundness Indices Pettijohn (1949)	Powers (1953)	Folk (1955) Rho scale
Very angular		0.12–0.17	0.00–1.00
Angular	0.0–0.15	0.17–0.25	1.00–2.00
Subangular	0.15–0.25	0.25–0.35	2.00–3.00
Subrounded	0.25–0.40	0.35–0.49	3.00–4.00
Rounded	0.40–0.60	0.49–0.70	4.00–5.00
Well rounded	0.60–1.00	0.70–1.00	5.00–6.00

boundary between round and angular is placed between the round and subround classes, or 0.5 degrees of roundness on the Powers scale (1953), or 4.00 on Folk's (1955) rho scale.

MORPHOMETRIC TECHNIQUES

Quantitative analysis of grain morphology requires measurement of particle radii, diameters, and lengths. The measurements are made on either actual grains or on grain facsimile. The operator's choice of facsimile type is largely dependent on the particle size, and to a lesser extent on the type of parameter required. It is obvious that direct measurements can be made only on gravel or larger sized particles, whereas indirect methods are required to measure sand or smaller sized particles which cannot be held and manipulated by hand.

Direct Measurement

The various length measurements can by made on large particles with a simple scale or ruler (the metric system is recommended). The "shapometer" (Tester and Bay, 1931) or its modification may be also applied to length measurements. Measurements of curved surfaces (e.g., radii and diameters) are more difficult. Wentworth (1919) devised a convexity gauge and a set of curvature templates to measure surface curvatures in the field. However, a transparent "Wadell circular scale" (Fig. 5) can be constructed for use with large particles. In practice, the transparent circular scale is placed over the corner to be measured and the radius or diameter is determined.

Indirect Measurement

This kind of measurement can be made on either small or large particles and involves the use of projected images or photographs of particles. Large particles can be manually oriented for the best desired two-dimensional aspect, but for small particles a grain mount is desirable because the largest and intermediate lengths or diameters lie normally in the plane of projection.

One method of indirect measurement utilizes particle images from a grain mount or thin-section projected on the ocular lens of a microscope. The desired lengths are measured from the images with a calibrated ocular micrometer, and radii of curvature can be measured with ocular diaphragm devices described by Alling (1941) and Hörnsten (1959).

The most common method of radii of curvature measurement does not require specialized microscopic accessories. Grain images are projected with a microprojector, or an enlarged photograph of the grains is used. In this instance the enlarged image is measured with a scale or ruler and the radii are measured with a transparent "Wadell circular scale" (Fig. 5).

Photomicrographs are by far the best technique; they are easily and cheaply produced, easily measured, and constitute a permanent record. The grain mount or thin-section may be placed in the film-holder of an enlarger and normal photographic printing techniques employed. The use of a high contrast paper, such as Kodabromide F3, insures sharp, clear images.

Boggs (1967) describes the use of grain photographs and the Zeiss TGZ 3 Particle Size Analyzer in analysis of grain sphericity and roundness. In this technique the photographic print is mounted on the analyzer and the radius measurements are rapidly and automatically tabulated for each grain image.

PIVOTABILITY

Pivotability or rollability is a measure of the motion response of a grain to a set of standard physical conditions in a gravity-driven system. The term pivotability, as related to grain morphology, was introduced by Shepard and Young (1961). However, Glezen and

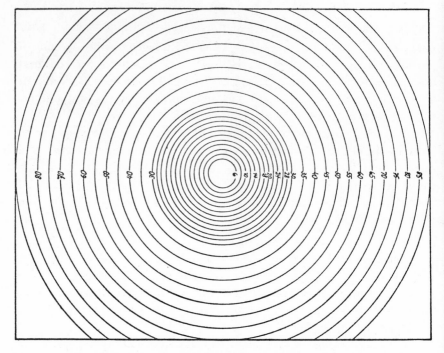

Fig. 5 Circular scale for measurement of radii from grain drawings, projections, or photographs. (From Köster, 1964)

Ludwick (1963) point out that the concept has been used by numerous industries to shape-sort particles for many years. In 1963, Glezen and Ludwick, and Kuenen, simultaneously presented the results of many years of development of devices to measure the shape of grains by an "actualistic" movement of grains down a slope. In both devices the velocities of grains, of narrow specific size ranges, were measured and classified into velocity groups.

The Automated Grain-Shape Classifier of Glezen and Ludwick (1963) is a highly sophisticated instrument that measures the velocity of small numbers of grains as they roll down a roughened and inclined series of chutes. Initial results were confined to grains in the diameter size range of 1.00 to 0.42 mm. By utilizing multiple chutes it is possible to analyze automatically six samples of 1500 grains each in 2.5 hours. Three shape classes, based on velocity groups, were distinguished: (a) fast grains, spherical; (b) slow grains, tabular; and (c) intermediate grains, intermediate. Glezen and Ludwick determined that the percentage of tabular grains in a sample (tabularity) is the most geologically significant measure of pivotability.

A less expensive and less complicated device called the Rock and Roll Shape Sorter is described by Kuenen (1963). This is a single, hemicylindrical trough, slightly inclined along its long axis, and rotated back and forth, and normal to its long axis by a motor drive. Only the size grades between 1.00 and 0.15 mm were used in Kuenen's experiments. The fractional samples, several hundred milligrams each, are introduced at the upper end of the "rock and roll" chute and the grains migrate to the bottom of the chute in a zigzag path. At the lower end of the chute the grains are funneled into vials that are replaced at given time intervals, so that 12 fractions of successively lower velocity grains are separated from each other. The velocity grades are equated to relative pivotability grades.

In both the Automated Grain-Shape Classifier of Glezen and Ludwick (1963) and the Rock and Roll Shape Sorter of Kuenen (1963), the shape, roundness, center of gravity, specific gravity, moment arms, impingement points, and intergrain interference factors determine grain velocities and rollability or pivotability characteristics. Pivotability is therefore an actualistic and multi-variant measure of grain morphology.

SUMMARY

In utilizing morphometric analysis in geologic problems, investigators are frequently confronted with the need to describe large numbers of sediment samples and even larger numbers of individual detrital grains. In addition, grain morphology is usually not the only parameter to be utilized, since multivariant solutions to geologic problems are generally more reliable than those based on single parameters. Hence investigators are faced with ever-expanding numbers of analyses, for many of which the value is not initially known, so that for routine morphometric analysis it is usually advisable to select relatively rapid techniques.

The ultimate choice of methodology to be employed in describing the morphology of detrital grains, however, depends on the consideration of a number of factors, especially the following:

1. Size of particles to be described
2. Amount of available sample
3. Induration-state of sample material
4. Parameters desired for particular problem
5. Parameters obtainable from sample material
6. Desired sophistication of parameters
7. Availability of special equipment
8. Expense in time and money
9. Investment-reward ratio

Each problem therefore presents a somewhat unique situation or set of factors, so that no one technique can be said to be absolutely better than another, because this is a problem-dependent decision.

REFERENCES

Alimen, H., and D. Fenet, 1954, Granulométrie de Sables d'erg aux environs de la Saoura, *Soc. Geol. France, Compte Rendus*, No. 9—10, 183—185.

Alling, H. L., 1941, A diaphragm method for grain size analysis, *J. Sed. Pet.*, 11, 28—31.

Anderson, G. E., 1926, Experiments on the rate of wear of sand grains, *J. Geol.*, 34, 144—158.

Aschenbrenner, B. C., 1955, A photogrammetric method for the tri-dimensional measurement of sand grains, *Photogrammetric Eng.*, 21, 376—382.

Atherton, E., et al., 1960, Differentiation of Caseyville (Pennsylvanian) and Chester (Mississippian) sediments in the Illinois Basin, Illinois Geol. Surv. Circ. 306, 36 pp.

Boggs, Sam, Jr., 1967, Measurement of roundness and sphericity parameters using an electronic particle size analyzer, *J. Sed. Pet.*, 37, 908—913.

Brewer, Roy, 1964, Fabric and mineral analysis of soils, John Wiley and Sons, 470 pp.

Briggs, L. I., D. S. McCulloch, and F. Moser, 1962, The hydraulic shape of sand particles, *J. Sed. Pet.*, 32, 645–656.

Cailleux, A., 1945, Distinction des galets marins et fluviatiles, *Bull. Soc. Geol. France*, 15, 375–404.

———, 1947, L'indice d'émoussé des grains de sable et grès, *Rev. Geomorphologic Dynamique*, 3, 78–87.

———, 1952, Morphoskopische Analyse der Geschiebe und Sandkörner und ihre Bedeutung fur die Paläoklimatologie, *Geol. Rundschau*, 40, 11–19.

Cox, E. P., 1927, A method of assigning numerical and percentage values to the degree of roundness, *J. Paleon.*, 1, 179–183.

Curray, J. R., and J. C. Griffiths, 1955, Sphericity and roundness of quartz grains in sediments, *Bull. Geol. Soc. Am.*, 66, 1075–1096.

Dobkins, J. E., Jr., and R. L. Folk, 1968, Pebble shape development on Tahiti-Nui, *Bull. Am. Assoc. Petrol. Geologists*, 52, 525.

Eissele, K., 1957, Kritische Betrachtung einer Methode zur Bestimmung des Rundungsgrades von Sandkornern, *Naues Jahrb. Geologic. u. Paläontologie*, 9, 410–419.

Emrich, G. H., and F. J. Wobber, 1963, A rapid method for estimating sedimentary parameters, *J. Sed. Pet.*, 33, 831–843.

Fischer, Georg, 1933, Die Petrographie der Grauwacken, *Jahr. der Preussischen Geol. Landesanstalt.*, 54, 322–323.

Folk, R. L., 1955, Student operator error in determination of roundness, sphericity and grain size, *J. Sed. Pet.*, 25, 297–301.

———, 1965, *Petrology of sedimentary rocks*, Hemphill's, 159 pp.

Glezen, W. H., and J. C. Ludwick, 1963, An automated grain-shape classifier, *J. Sed. Pet.*, 33, 32–40.

Gregory, H. E., 1915, Note on the shape of pebbles, *Am. J. Sci.*, 39, 300–304.

Griffiths, J. C., 1967, Scientific method in analysis of sediments, McGraw-Hill Book Co., 508 pp.

Guggenmoos, Theresia, 1934, Über Korngroben und Kornformenverteilung von Sanden verschiedener geologischer Entstehung, *Neues Jahrb. fur Min., Geologie u. Paläontologie*, 72, Abt. B, 429–487.

Hoernes, H., 1911, Gerölle und Geschiebe, *Verh. der Kaiserlich-Königlichen Geol. Reichsanstalt,* 12, 267–274.

Hörnsten, A., 1959, A method and set of apparatus for mineralogic-granulometric analysis with a microscope, *Bull. Geol. Inst. Univ. Uppsala,* 38, 105–136.

Hövermann, J., and H. Poser, 1951, Morphometrische und morphologische Schotter-analysen, Proc. 3rd Inter. Congr. of Sed., Groningen-Wageningen, The Netherlands, pp. 135–156.

Huitt, J. L., 1954, Three-dimensional measurement of sand grains, *Bull. Am. Assoc. Petrol. Geologists,* 38, 159–160.

Köster, Erhard, 1964, Granulometrische und morphometrische Messmethoden an Mineralkornern, Steinen und sonstigen Stoffen, F. Enke Verlag, Stuttgart, 336 pp.

Krumbein, W. C., 1941, Measurement and geologic significance of shape and roundness of sedimentary particles, *J. Sed. Pet.,* 11, 64–72.

———, 1942, The effects of abrasion in the size, shape, and roundness of rock fragments, *J. Geol.,* 49, 482–520.

Krumbein, W. C., and F. J. Pettijohn, 1938, Manual of sedimentary petrography, Appleton-Century-Crofts, 549 pp.

Krumbein, W. C., and L. L. Sloss, 1955, Stratigraphy and sedimentation, W. H. Freeman and Co., 660 pp.

Kuenen, Ph. H., 1956, Experimental abrasion of pebbles; rolling by current, *J. Geol.,* 64, 336–368.

———, 1963, Pivotability studies of sand by a shape-sorter, pp. 208-215, *in* Developments in sedimentology, 1, Elsevier Publishing Co., 464 pp.

Lamar, J. E., 1927, Geology and economic resources of the St. Peter Sandstone in Illinois, *Bull. Illinois Geol. Surv.,* 53, 148–151.

Mackie, W., 1897, On the laws that govern the rounding of particles of sand, *Trans. Edinburgh Geol. Soc.,* 7, 298–311.

Müller, German, 1967, Sedimentary petrology, Part 1, Methods in sed. pet., Hafner Publishing Co., 283 pp.

Pentland, A., 1927, A method of measuring the angularity of sands, *Royal Soc. Canada, Proc. and Trans. Series. 3,* 21, Appendix C, Titles, and Abstracts, p. XCIII.

Pettijohn, F. J., 1949, Sedimentary rocks, 1st ed., Harper and Bros., 526 pp.

———, 1957, Sedimentary rocks, 2nd ed., Harper and Bros., 718 pp.

Poser, H., and J. Hövermann, 1952, Beitrage zur morphometrischen und morphologischen Schotteranalyse, *Braunschweigische Wiss. Ges.*, Abh. 4, 12–36.

Powers, M. C., 1953, A new roundness scale for sedimentary particles, *J. Sed. Pet.*, 23, 117–119.

Reichelt, G., 1961, Uber Schotterformen und Rundungsgradanalyse als Feldmethode, *Petermanns Geographische Mitt.*, 105, 15–24.

Riley, M. C., 1953, A new roundness scale for sedimentary particles, *J. Sed. Pet.*, 11, 94–97.

Robson, D. A., 1958, New technique for measuring roundness of sand grains, *J. Sed. Pet.*, 28, 108–110.

Rosefelder, A., 1961, Contribution a l'analyse tectural des sediments, *Bull. Service carte geol. Algerie*, 29, 310 pp.

Russell, R. D., and R. E. Taylor, 1937, Roundness and shape of Mississippi River sands, *J. Geol.*, 45, 225–267.

———, 1937, Bibliography on roundness and shape of sedimentary particles, Rept. Comm. Sed., 1936–1937, National Res. Council, pp. 65–80.

Schneiderhöhn, P., 1954, Eine vergleichende Studie uber Methoden zur quantitativen Bestimmung von Abrundung und Form an Sandkornern, *Heidelberger Beitrage zur Min. u. Pet.*, 4, 172–191.

Shepard, F. P., and R. Young, 1961, Distinguishing between beach and dune sands, *J. Sed. Pet.*, 31, 196–214.

Sneed, E. D., and R. L. Folk, 1958, Pebbles in the lower Colorado River, Texas: a study in particle morphogenesis, *J. Geol.*, 66, 114–150.

Sorby, H. C., 1880, On the structure and origin of non-calcareous stratified rocks, *Quart. J. Geol. Soc. London*, 36, 46–91.

Szadeczky-Kardoss, E. V., 1933, Die Bestimmung des Abrollungs-grades, *Zentralblatt fur Min., Geologic u. Palaontologie, Abt. B*, 389–401.

Tester, A. C., 1931, The measurement of the shapes of rock particles, *J. Sed. Pet.*, 1, 3—11.

——, and H. X. Bay, 1931, The shapometer: a device for measuring the shapes of pebbles, *Science*, 73, 565—566.

Tricart, J., 1951, Études sur le faconnement des galets marins, Proc. 3rd Inter. Congr. Sed., Groningen-Wageningen, The Netherlands, pp. 245—255.

Valeton, I., 1955, Beziehungen zwischen petrographischer Baschaffenheit, Gestalt und Rundungsgrad einiger Flussgerölle, *Petermanns Geographische Mitt.*, 99, 13—17.

Wadell, H. A., 1932, Volume, shape and roundness of rock particles, *J. Geol.*, 40, 443—451.

——, 1933, Sphericity and roundness of rock particles, *J. Geol.*, 41, 310—331.

——, 1934, Shape determinations of large sedimental rock-fragments, *Pan-Am. Geologist*, 61, 187—220.

——, 1935, Volume, shape, and roundness of quartz particles, *J. Geol.*, 43, 250—280.

Wentworth, C. K., 1919, A laboratory and field study of cobble abrasion, *J. Geol.*, 27, 507—521.

——, 1922, The shapes of pebbles, *Bull. U.S. Geol. Surv. 730-C*, pp. 91—114.

——, 1922, The shapes of beach pebbles, *U.S. Geol. Surv., P.P. 131-C*, pp. 75—83.

——, 1933, The shapes of rock particles: a discussion, *J. Geol.*, 41, 306—309.

——, 1936, An analysis of the shapes of glacial cobbles, *J. Sed. Pet.*, 6, 85—96.

Wright, A. E., 1957, Three dimensional shape analysis of fine-grained sediments, *J. Sed. Pet.*, 27, 306—312.

Ziegler, Victor, 1911, Factors influencing the rounding of sand grains, *J. Geol.*, 19, 645—654.

Zingg, Th., 1935, Beitrage zur Schotteranalyse, *Schweizerische Mineralogische u. Petrographische Mitt.*, 15, 39—140.

CHAPTER 8

GRAIN SURFACE TEXTURE

DAVID H. KRINSLEY

Queens College, City University of New York, Flushing, N.Y.

STANLEY V. MARGOLIS

University of California at Riverside, Riverside, California

Electron microscopy has been applied to sediments by many investigators (Folk and Weaver, 1952; Folk, 1965; Harvey, 1966; Krinsley and Donahue, 1968d; Oldershaw, 1968), and evidence regarding grain size, grain surface characters, grain-to-grain relationships, and even slight differences in composition has been obtained. Biederman (1962), Krinsley and Takahashi (1962a, 1962b, 1962c, 1964), Kuenen and Perdok (1962), Porter (1962), Krinsley et al. (1964), Krinsley and Schneck (1964, p. 277), Bramer (1965), Krinsley and Newman (1965), Krinsley and Funnell (1965), Waugh (1965), Swift, Byers, and Krinsley (1966), Angelucci (1966a, 1966b), Margolis (1966a, 1966b), Hamilton, and Krinsley (1967), Soutendam (1967), Wolfe (1967), Geitznauer et al. (1968), Krinsley and Donahue (1968a, 1968b, 1968c), and Krinsley and Margolis (1969) have examined the surfaces of quartz sand grains with the electron microscope; it is possible to distinguish littoral (beach), eolian (dune), glacial and postdepositional environments on the basis of surface textures. Frequently, textures representing one episode may be superimposed on another; in addition, some of the observed textures have been reproduced experimentally (Krinsley and Takahashi, 1962c).

To avoid unnecessary complications in interpreting surface textures, examined first were grains from modern unconsolidated deposits of known origin that had been subjected to mechanical and chemical processes during erosion, transportation, and deposition. Samples representing known eolian, littoral, and glacial environments were collected from many parts of the world to establish characteristic surface textures; the textures observed were duplicated experimentally. After criteria for distinguishing various environments had been established through examining recent deposits, the same criteria were applied to the study of older sands, that is, ancient quartz sand grains from known environments, distinguished by stratigraphic and paleoecological criteria, were examined to check the procedure. Finally, sand-grain surfaces from unknown ancient environments were studied. This chapter will describe briefly the techniques used in preparing surface textures of quartz sand grains for electron microscopic study, the characteristic surface textures for the various environments, and the experimental work used to duplicate these textures.

TECHNIQUES

The preparation of replicas has been described in some detail in the literature; see, for instance, Rochow and others (1960), Haine (1961), and Kay (1961). The following method described (Krinsley and Takahashi, 1964) is a modification of these techniques.

Preparation

The sample is first examined under a binocular microscope, then described, and the grains selected for study. Polycrystalline grains are rejected to avoid the complexities of grain boundaries. About 25 grains between 0.5 and 1.5 mm in diameter are chosen from the original sample; the maximum grain axis is recorded for future reference. The selected grains are then placed in concentrated nitric acid, boiled for ½ hour, washed with distilled water, boiled in a stannous chloride solution for ½ hour, and finally washed with distilled water. Organic debris may be removed in a strong oxidizing solution of 1.5 gm each of potassium dichromate and potassium permanganate dissolved in 15 ml of concentrated sulfuric acid (McIntyre and Be, 1967).[1] The grains should be washed again and finally dried. This procedure is necessary because the sand grains must be absolutely clean before replication; obscuring material may be transferred from the grain to the film replica if these precautions are not taken. Another technique for cleaning is simply to replicate individual grains as many as six or seven times and use the last replica for study. Loose debris is removed from grain surfaces quite effectively this way. Ultrasonic vibration should *not* be used; we have found that abrasion will occur under some circumstances (Porter, 1962).

Replication

After thorough cleaning, the grains are ready to be replicated. One or two drops of acetone is placed on some cellulose acetate film (Faxfilm[2]) taped to a glass slide. The acetone softens the cellulose film, and the grain is pressed down on the soft film with a probe; care must be taken that the grain does not break through the acetate and contact the glass. Grains must not be dropped in a pool of liquid Faxfilm, because the liquid will then flow completely around the grain and coat the upper surface. After several minutes the acetate is sufficiently hardened so that the grain may be removed either with a probe or fingernail (Fig. 1, step 1). Generally, grains larger than 0.25 to 0.50 mm in diameter must be used, for it is extremely difficult to embed and remove smaller samples. Some success with smaller material is possible if the grains are sprinkled on the softened acetate film the way salt is sprinkled on food. After hardening, it is possible to remove many of the small grains with a probe, but usually some grains cannot be removed. The presence of sand grains on the replica is not bothersome, but the unremoved grains cannot be examined. Another difficulty with small grains is that the films produced are usually so small that they are unstable and tend to tear because of their high curvature. However, if a large number of grain impressions

[1] There is no evidence that these treatments have any effect on quartz surface textures; experimental work has been done in this area.
[2] Ladd Research Industries, Box 901, Burlington, Vermont, is one source.

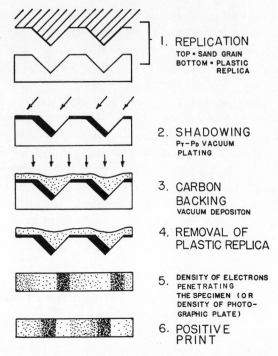

1. REPLICATION
TOP = SAND GRAIN
BOTTOM = PLASTIC
REPLICA

2. SHADOWING
Pт-Pᴅ VACUUM
PLATING

3. CARBON
BACKING
VACUUM DEPOSITON

4. REMOVAL OF
PLASTIC REPLICA

5. DENSITY OF ELECTRONS
PENETRATING
THE SPECIMEN (OR
DENSITY OF PHOTO-
GRAPHIC PLATE)

6. POSITIVE
PRINT

Fig. 1 Diagrammatic representation of technique for study of sand grain surface textures with electron microscopy. (Krinsley and Takahashi, 1964)

are made, some should be usable. In addition, grains from beach and dune sands less than 0.5-mm maximum diameter generally show only chemical textures. Grains larger than 1.5 mm do not usually include definitive markings.

After the grains have been removed from the plastic, bits of loose material frequently adhere to the edge of the grain replica. This material should be cut carefully with a scalpel and removed because it may shield the replica from metal and carbon coating.

Since the acetate film is too thick to be viewed directly through the electron microscope, it is necessary to prepare a much thinner metal-carbon film (less than 200 Å) for successful penetration of electrons at 40 to 100 kv.

Shadowing

The process of preparing a thin film is known as shadowing. It is accomplished by evaporating a metal on the acetate replica at an angle of about 45° (Fig. 1, step 2). Shadows of objects on photographs (45° shadow angle) are equal in length to the height of the object. A platinum-palladium alloy is used as the shadowing agent because of high scattering power for electrons, ready evaporation under vacuum, lack of granulation under the electron beam, relatively low melting temperature, and lack of reaction with the tungsten filament used to heat it. Chromium may be used in place of platinum-palladium, but the surface texture produced is somewhat coarser. Other heavy metals can be substituted, but the two mentioned are quite adequate.

Cellophane tape may be used to fasten the specimens to the glass slide before shadowing.

The center of a piece of tungsten wire is bent in the shape of a V, and four or five twists of shadowing metal are wrapped around the tip; the tungsten holder is then placed in the evaporation unit. The evaporator is evacuated to approximately 3×10 mm Hg, and a current is passed through the tungsten wire (the authors use approximately 23 amp for 9 seconds), vaporizing the platinum-palladium metal at a 45° angle to the specimen. A smaller angle (about 30°) may be used if the specimen is of low relief. Next, a film of carbon to support the extremely fragile metal film must be evaporated directly down on the specimen (Fig. 1, step 3). Carbon is used because it is chemically inert, amorphous, and very transparent to electrons. The amount and rate of carbon deposition (voltage governs the rate of carbon evaporation) are both critical; either too little or too much carbon causes the film to break when the plastic is dissolved, whereas too high a filament temperature causes the film in the evaporator to burn so that breakage occurs. If the specimen has a brownish tinge (which can be observed with the naked eye), the temperature used was probably too high and the specimens may be unusable.

If metal and carbon films are applied to the specimen without breaking vacuum, it may be necessary to protect the carbon electrodes from the metal coating or vice versa. This can be done by placing a metal shield with a high vaporization temperature between the platinum-palladium metal and the carbon rods. This is necessary because vaporization of carbon having a thin layer of previously deposited metal may coat the specimen film with relatively thick metal globules opaque to electrons. Either the metal coating or carbon coating may be applied first.

Removal of Acetate Peel

To remove the peel from its plastic backing, a 2% solution of formvar in chloroform is allowed to run down the shadowed side of the specimen (see Fig. 2 for all steps). The chloroform will evaporate readily, leaving a film of formvar that strengthens the metal-carbon film; this is done because otherwise the replica may tend to disintegrate when the acetate is removed. The peel containing the

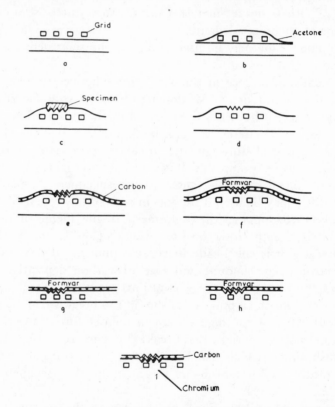

Fig. 2 (a) Grid resting on cellulose acetate film; (b) grid and acetate film with drop of acetone resting on top; (c) specimen embedded in acetate forming replica; (d) acetate film replica after removal of specimen using HCl; (e) acetate film replica after having been shadowed with carbon; (f) acetate film – carbon replica after flooding with Formvar; (g) acetate film after having been cut away from cellulose acetate sheet; (h) Formvar-carbon film after removal of cellulose acetate film with acetone; (i) final replica after removal of Formvar and shadowing through grid bars at an angle with a heavy metal. (After Kay, 1961; Hyde and Krinsley, 1964)

thin, shadowed film replica is cut with a sharp knife and placed (shadowed side up) upon a 3-mm diameter, 200 mesh screen that can be inserted in the electron microscope.

The acetate peel is now ready to be separated from the platinum-palladium-carbon-formvar replica. It may be immersed in acetone, but usually this causes the metal film replica to float off the screen on which it had been placed, curl up, and become useless. A more efficient technique is to use a piece of stainless steel wire gauze, about 75 mesh, bent into the shape of an M (Fig. 3) The gauze is placed in a dish, and the 200 mesh screen with the acetate-backed replica is positioned at the trough point of the M so that it is sitting horizontally. Acetone is added with an eyedropper at the side of the dish until the liquid level just reaches and wets the screen. The dish is covered and left for 5 minutes; the 200 mesh screen is then removed and its base wiped on filter paper. The screen is placed back in the acetone on the M-frame.

The process described is repeated after 5 minutes, and the dish is covered for about 1½ hours. The acetone should be changed after each application. After this time, the screen is removed and placed on another M-frame in an empty dish. Chloroform is added at the side of the dish with an eyedropper until the liquid level just reaches the screen's edge. The dish is covered and left for 10 minutes; the chloroform solution is changed with an eyedropper and the specimen left for an additional 10 minutes. The M-frame should then be lifted from the chloroform bath and the screen removed and permitted to dry. This process will remove both formvar and acetate, leaving a metal-carbon film which can be examined with the electron microscope.

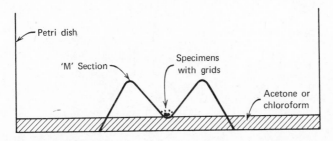

Fig. 3   Grid with cellulose acetate chromium-carbon film resting on 'M' section. (Hyde and Krinsley, 1964)

A somewhat different method for removing the cellulose acetate is to use an acetone reflux unit (Fig. 4) of the type sold by Ladd Research Industries (P. O. Box 901, Burlington, Vermont). After the replicas have been plated and shadowed with metal and carbon, they are cut to size and mounted on 200 mesh, 3-mm diameter grids and placed in position on the gauze finger of a reflux unit. The gauze is gently flooded with acetone, thus securing the specimens and preventing the loss of grids inside the unit when the finger is replaced.

The finger is then positioned inside the unit, and acetone in the bowl is heated to boiling point. Water hoses are connected so that acetone condenses in an upper condenser tube, and the acetone vapor inside the reflux unit washes away the bulk of the acetate sheet within 5 minutes. The hoses are reconnected so that water is permitted to flow through the finger condensing the acetone on the

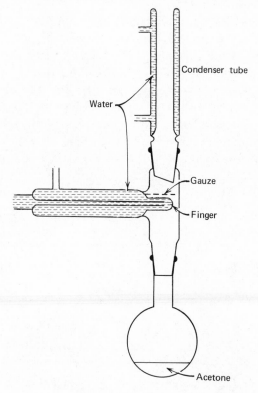

Fig. 4 Diagram of reflux unit in operating position. (Hyde and Krinsley, 1964)

grids; this process takes about 10 minutes. The system is then cooled and the grids are removed. Care must be taken that the grids are perfectly dry before placing them on filter paper, for otherwise the carbon film may be destroyed.

Although the carbon surrounding the impression may be destroyed, the actual impression remains intact because of its coarse structure. Films prepared in this manner rarely disintegrate and manipulation of the specimens is quite easy. It is important to wash the finger in acetone for about 30 seconds before reusing.

Another method of eliminating the acetate peel has been used by a number of workers, among them Hay et al. (1963). The metal and carbon are evaporated directly down on the specimen, which is replicated, dissolved in acid, and the specimen viewed directly without an acetate peel intermediary. However, the metal-carbon replica must be considerably stronger than in the techniques described above; it is also obvious that the specimen cannot be used again.

A method of examining the surfaces of sand grains with a light microscope is useful as an interim process. A carbon-metal replica is prepared as above and is placed on a 3-mm, 200 mesh grid; if the specimen is not needed for electron microscopy, it can be floated off and placed on a glass slide (Krinsley, et al., 1964; Soutendam, 1967).

The grid or slide is placed under a high-power light microscope; light scattering is reduced and somewhat better resolution than ordinarily obtained with a given light microscope can be attained. Contrast can be varied and improved by changing the thickness of the film, and surfaces of low relief can be highlighted by decreasing the shadowing angle. We examine all screens routinely before electron-microscopy; generally, poor specimens can be discarded immediately, thus saving time.

Electron Microscopy

After the acetate peel has been removed, the copper screen with the metal-carbon film is placed in the electron microscope and the image viewed on a fluorescent screen. The theory, operation, and photographic techniques of electron microscopy have been extensively published (e.g., see Haine, 1961 and Rochow et al., 1960) and thus will not be discussed here.

The entire process, starting with the original specimens and ending with photographs (the authors generally process about 20 specimens in a batch), may be accomplished with about 12 hours of work if the vacuum evaporator and electron microscope are in proper working condition.

Interpretation of Photographs

The image recorded on a photograph plate represents essentially the thickness of the platinum-palladium film deposited on the plastic replica. As shown in Fig. 1, step 2, shadowing with the metal at an angle results in heaviest deposition on the surfaces perpendicular to the direction of shadowing and lightest on the surfaces parallel to the direction of shadowing. After carbon backing and removal of the plastic replica, the cross section of the specimen to be examined with the electron microscope looks like the fourth diagram in Fig. 1. More electrons can penetrate the thinner section of the film more easily than the thicker portion: this means that areas covered with thicker films of metal appear less exposed on the photographic (negative) plate than thinner films (Fig. 1, step 5). On making a positive print, the darker areas represent a thicker metal film and the lighter areas the reverse (Fig. 1, step 6). The relation between the original surface topography and the final positive print becomes obvious when the first and the sixth diagrams in Fig. 1 are compared. The topographic expressions observed on the positive print appear as if the sand grain is observed under oblique light shining from the lower right-hand side of Fig. 1.

Other techniques in the literature (Rochow and others, 1960; Haine, 1961; Kay, 1961) might be modified to produce replicas of sand grains. The authors have tried several other methods, but the ones described in detail here are eminently satisfactory.

Interpretation of Surface Textures

The various surface textures recognized are enumerated below. (See Table 1 for summary.)

Textures of Glacial Origin

The following types of surface texture appear to be produced by glacial action and are characteristic of grains obtained from most

TABLE 1 Summary of surface texture characteristics (Krinsley and Donahue, 1968a)

Littoral.		Aeolian	
High Energy (Surf)	Medium- and Low-Energy Beach	Tropical Desert	Coastal
1. V-shaped patterns irregular orientation (a) $0.1\,\mu$ average depth (b) 2 Vs per square micron density 2. Straight or slightly curved grooves 3. Blocky conchoidal breakage-patterns	1. *En echelon* V-shaped indentations at low energy. As energy increases, randomly oriented Vs replace the *en echelon* features. A continuous gradation is present between high- and low-energy features.	1. Meandering ridges 2. Graded arcs 3. Chemical or mechanical action—regular pitted surfaces replacing the above features in many cases	1. Meandering ridges 2. Graded arcs

Glacial		Diagenetic	
Normal	Glacio-Fluvial	Wavy-Patterns	Worn (Soln.)
1. Large variation in size of conchoidal breakage-patterns 2. Very high relief (compared with grains from littoral and aeolian environments) 3. Semiparallel steps 4. Arc-shaped steps 5. Parallel striations of varying length 6. Imbricated breakage blocks which look like a series of steeply dipping hogback ridges. 7. Irregular small-scale indentations which are commonly associated with conchoidal breakage-patterns 8. Prismatic patterns, consisting of a series of elongated prisms and including a very fine-grained background	Rounding of glacial patterns 1-8	Curved branching irregular lines developed to varying degrees	Relatively flat, featureless surfaces

glacial tills and glacio-fluvial deposits. They are listed in approximate order of importance.

1. Large variation in size of conchoidal breakage-patterns; probably related to the large variation in the size of particles in glacial sediments (Plate 6, Krinsley and Donahue, 1968a).

2. Very high relief (compared with grains from littoral and aeolian environments); probably related to the relatively large size of particles and the large amount of energy available for grinding (Plate 6, Krinsley and Donahue, 1968a).

3. Semiparallel steps; probably caused by sheer stress (Plate 6, Krinsley and Donahue, 1968a).

4. Arc-shaped steps; probably representing percussion fractures (Plate 6, Krinsley and Donahue, 1968a).

5. Parallel striations of varying length; probably caused by the movement of sharp edges against the grains involved (Plate 41, Krinsley and Funnell, 1965).

6. Imbricated breakage blocks which look like a series of steeply dipping hog-back ridges (Plate 6, Krinsley and Donahue, 1968a).

7. Irregular small-scale indentations which are commonly associated with conchoidal breakage-patterns; probably caused by grinding (Plate 6, Krinsley and Donahue, 1968a).

In the authors' opinion, observation of four or more of these textures over large areas of a single grain may be taken as adequate evidence of a glacial origin. Textures 2-4 were first recognized and illustrated from glacial grains by Krinsley and Takahashi (1962b, 1962c; see also Krinsley and others, 1964), and they also reproduced them experimentally by grinding crushed quartz in ice against a quartz plate.

It is sometimes possible to tell from surface textures whether a grain has been embedded in a till with a more-or-less normal stone content or whether it was in one with relatively few stones. Breakage-patterns on the order of 30 or 40μ in diameter seem to imply a till with a normal stone content; when the maximum diameter of the breakage-patterns is about 20μ, a clayey till seems to be implied. Grinding experiments have shown that there is a direct relationship between the maximum size of the grinding material and the maximum size of breakage patterns.

Some glacial surfaces are observed that, even at magnifications of 1500 to 15,000X seem to have been worn down without any obvious mechanical or chemical markings having been impressed upon them. Krinsley and Takahashi (1952b) noted such a worn appearance, and it is thought to be produced by glacio-fluvial action; this type of surface is shown rather generally by late Wisconsin glacio-fluvial grains from the United States (see also Krinsley et al., 1964, Plate I,b).

Textures of Littoral Origin

The following types of surface texture appear to be produced by turbulent aqueous action and are characteristic of grains obtained from littoral (beach) environments.

8. Small, V-shaped indentations; probably caused by grain-to-grain collisions in an aqueous medium (Plates 1, 3, 8, Krinsley and Donahue, 1968a).

9. Straight, or slightly curved, grooves from 2 to 15μ long (which are usually much finer than those seen on glacial grains and occur singly); probably caused by a sharp edge of one sand grain being drawn lightly across another. (The process seems likely to be more rapid and to involve less energy than that producing its glacial counterpart.) (Plates 1, 3, Krinsley and Donahue, 1968a.)

10. Chatter marks, a series of subparallel indentations averaging about 0.5μ in length; probably produced by a portion of one grain skipping across another.

11. Blocky conchoidal breakage-patterns (these differ from their glacial counterparts in their smaller and more uniform size; they differ from those produced by aeolian action in having regular rather than curved sides, but may be confused with the curved sides; produced by grinding of pebbles and sand in an aqueous medium. Compare Plate 1 (Krinsley and Donahue, 1968a), beach, with Plate 4, dune, from the same publication and note the similarity. Generally, however, regularity is more pronounced in dune than in beach sand.

Textures 7 and 10 were first recognized and illustrated by Krinsley and Takahashi (1962a, 1962c; see also Krinsley et al., 1964), who reproduced them experimentally. V-shaped indentations, resembling those observed on nearly every modern beach specimen so far examined, were obtained by agitating crushed quartz in water on a high-speed shaking-table. The angle between the arms of the V-shaped patterns has subsequently been studied both on natural beach grains and on experimental grains. The skewness and kurtosis of the natural and artificial angle distributions do not differ significantly, suggesting that both were formed in the same way. Blocky conchoidal breakage-patterns were produced by agitating crushed quartz in water in a ball-mill. These patterns are not common on natural beach grains, and when they do occur they probably imply production in a pebbly environment.

V-shaped indentations on grains from a normal beach environment have a maximum density of about two V's per square micron (Krinsley and Takahashi 1962c; Krinsley et al., 1964) and average depths of about 0.1μ. In the course of present investigation, the authors have observed V-shaped indentations that are considerably shallower than 0.1μ and usually somewhat denser. They have not observed such shallow indentations on modern beach sands, but these were seen on modern shelf sands and, like normal beach V's, have been reproduced experimentally on a shaking-table. It appears that the shallower, denser pattern requires less energy for its formation; we think that it may be formed either in deeper water or under estuarine conditions.

Both textures 9 and 10 have been reproduced experimentally on a shaking-table. The straight or slightly curved grooves commonly exhibit several V-shaped indentations branching from them (Plate 1, Krinsley and Donahue, 1968); these may be formed by a slight rocking motion during the traverse of one grain across another, causing an adjacent point on the traversing grain to be forced down into the surface of the other. Chatter marks are only seen rarely.

With the using of an electron microscope crystallographically oriented etch pits have been observed on quartz sand grains from marine environments. Similar features have been produced by etching quartz in the laboratory with hydrofluoric acid and sodium hydroxide. Examination of quartz sand grain surfaces from beaches of different wave intensities along the Atlantic and Gulf coasts of the United States suggests a relationship between the effects of such chemical solution features and mechanical abrasion on sand grains (Plate 2, Krinsley and Donahue, 1968a).

Sand grains from beaches with low wave activity exhibit oriented etch pits readily attributed to the solution of quartz by sea water. These features are the surface expression of defects within the crystal. Sand grains from beaches with moderate wave energy show a combination of chemical etch features and those phenomena thought by Krinsley and Takahashi (1962a) to be caused by grain-to-grain impacts, whereas grains from high energy beaches predominantly show impact features (Margolis, 1968).

Textures of Aeolian Origin

The following types of surface texture appear to be produced by aeolian processes and are characteristic of both coastal and desert dunes.

12. Meandering ridges, apparently resulting from the intersection of slightly curved conchoidal breakage-patterns. These breakage-patterns differ from their glacial counterparts in their smaller size, never much larger than about 15μ in diameter, and greater uniformity; they differ from those produced by beach action in having curved sides rather than regular sides and, again, greater uniformity. These conchoidal breakage-patterns are probably caused by grain-to-grain collisions in an aeolian medium (see Krinsley and Takahashi 1962c, Fig. 1, etc.).

13. Graded arcs, in series, each with a somewhat smaller radius proceeding toward the center, and possibly having an even greater tendency to form meandering ridges than the conchoidal breakage-patterns previously mentioned and probably representing percussion fractures (they are more common than their glacial counterpart) (Plate 37, Krinsley and Funnell, 1965).

The meandering ridges and curved surfaces characteristic of dune-sand grains were first recognized and illustrated by Krinsley and Takahashi (1962a, 1962c; see also Krinsley et al., 1964), and were reproduced experimentally by agitating crushed quartz grains in a wind-circuit device. The conchoidal breakage-patterns and graded arcs recognized here are usually rapidly abraded and merge into meandering ridges under natural conditions, but these textures can be seen unabraded on experimental grains and occasionally on natural ones (Plate 37, Krinsley and Funnell, 1965).

It appears that coastal dunes contain only features 11 and 12

(Krinsley et al., 1964, Plate VI). Tropical-desert quartz sand grains, however, can be distinguished from coastal dune grains by the presence of relatively flat, somewhat irregular, and frequently pitted surfaces. These surfaces occur to a greater extent on desert sands than the meandering ridges and graded arcs mentioned. The meandering ridges and graded arcs probably are produced during periods of violent wind action; conversely, during times of lower wind velocity these features may be destroyed either by chemical action or abrasion by fines (see Soutendam, 1967, and Krinsley and Donahue, 1968a, for photographs).

Chemical Etching

This seems to produce curved, branching, somewhat irregular lines, and worn surfaces. The irregular lines appear to be initiated in surface depressions such as the V-shaped indentations, and may be either localized in depressions or spread over most of the grain surface (see especially Plate 3, Krinsley and Donahue, 1968a). Early in the process most of the preexisting mechanical textures can still be observed, but as etching proceeds, the V-shaped indentations tend to be obliterated and replaced by a series of jagged, connected branching patterns, commonly aligned *en echelon*. The large conchoidal breakage-patterns characteristic of glacial action seem to persist longer than most, but finally even they are lost, leaving a rather flat, monotonous surface covered by curved, branching, rather irregular lines. Similar textures have been produced experimentally by mechanically subjecting abraded quartz to high pressures and temperatures in a hydrothermal bomb (Krinsley and Donahue, Plate 7, 1968a).

Other portions of the same surface may be flat and monotonous with very few features. Both patterns have been found in great numbers on Paleozoic, Mesozoic, and Cenozoic grains. A very rough correlation appears to exist between the degree of diagenesis and the age of a particular group of sediments if those sediments are not too widespread and the samples are taken from comparable lithologies.

It has been found possible to distinguish evidence of as many as five sedimentary episodes on the surface of a single grain. During its sedimentary history a grain may be subjected to a succession of different mechanical and chemical processes, and the textures produced by later episodes may be superimposed on those produced by previous ones. Depressed areas below the general surface may be protected for a time from later mechanical action. Not every grain in a given sample will show every episode, but by comparison with the order of events shown by other grains in the same deposit it is frequently possible to build up a comprehensive picture of the history of the deposit as a whole.

Scanning Electron Microscopy

The scanning electron microscope (SEM) has been recently used to study a wide variety of objects (e.g., Smith and Oatley, 1955; Thornton, 1965; Hay and Sandburg, 1967; Sandburg and Hay, 1967, 1968; Krinsley and Margolis, 1969). Its wide range of magnification (about 15X to 50,000X), the ability to examine entire specimens rather than restricted portions of objects such as sand grains, and great depth of field (about equal to the transmission instrument) make it an extremely valuable tool for studying surface textures. Additionally, it is not necessary to construct a replica; surfaces can be examined directly, thus saving a great deal of preparation time and simplifying interpretation.

Although resolution of the SEM is usually no better than 250 A (versus 2000 A for the light microscope) as compared to 2.5 A for transmission electron microscope (TEM), the size of characteristic textural features on the surfaces of quartz grains is well within the resolution capabilities of SEM.

The construction of a sand grain replica is a time-consuming, rather complicated process, thus precluding the possibility of examining large numbers of grains, may introduce artifacts and distortion, and may not reproduce all of the original detail on the specimen. Replicas are notoriously fragile; 30 to 40% of specimens prepared may be unusable. It is usually difficult to determine what portion of the grain is being observed with the TEM, for low power magnification (about 200X) may not enable observation of the entire grain. Only half the grain is commonly replicated, and about 40% of that surface is supported on grid bars where it cannot be observed. Only one or, at the most, two grain surfaces can be mounted on one specimen grid and observed at a given time.

These problems can be eliminated with the scanning electron microscope. Grains are observed directly without the need for replication, thus avoiding most artifacts and distortion. Much more detail can be seen with the SEM, indicating that many small features are not observed or are somewhat distorted with the replica method. A number of grains can be directly observed at one time with the SEM, permitting selection of a particular grain to be zoomed in, on, and photographed at high magnification. Most microscopes are supplied with a tilting stage, so that considerably more than half a given grain can be observed. Finally, depth of focus and certain other SEM features give a simulated three-dimensional photograph that simplifies interpretation.

A metal specimen plug supplied by the instrument manufacturer is coated with a thin layer of silver paint and the specimens pressed into the paint. After the paint has dried for a few minutes, the specimen plug is placed in a vacuum evaporator on a rotating table. About 4 in. of thin gold wire is wrapped around a piece of tungsten wire approximately 4 in. above the specimen. Next a current is passed through the wire until the gold melts. Sufficient current is then applied to the wire so that it just glows, and this is continued until the gold has evaporated; this should take about 20 minutes. The specimens are then ready to be mounted for viewing in the SEM.

Textures of Littoral Origin

All the littoral features described in the previous section can be observed with the SEM. Figures 5 (TEM) and 6 (SEM) picture littoral grains as seen with both types of microscopes. Note the three-dimensional effect in Fig. 6; the V-shaped patterns are obviously indentations and the result of breakage. Additionally, breakage steps can be recognized on several of the V-shaped indentations. The straight scratches in Fig. 5 consist of a series of satellite blocks that have probably been formed during a single breakage episode. It is likely that the straight scratches grade into the large breakage blocks and finally into the V-shaped patterns as the amount of energy supplied to the grains decreases.

Figures 5 and 6 show the obvious differences in resolution of the two microscopes, but the two do supplement each other. V-shaped patterns, scratches, and blocks can be seen more clearly in Fig. 5 (TEM) because of the better resolution; however, the apparent depth is a function of the shadowing angle and may present an incorrect picture of the actual grain surface. Figure 6 (SEM) is not so sharply defined as Fig. 5 (TEM); however, an excellent three-dimensional

view is obtained from the photograph and certain kinds of features
are noted that are difficult to decipher in Fig. 5. For instance, the
indentations in Fig. 6 apparently consist of a group of small
rectangular breakage depressions gouged out most probably by a
sharp point on one grain skipping or saltating across the grain in
question. This cannot be determined from Fig. 5. However,
extremely fine detail can be seen in Fig. 5 that cannot be observed in
Fig. 6. Note particularly the fine, somewhat wavy lines running from
one black dot (depression) to another; this feature is diagenetic in
nature.

Fig. 5   Beach environment from late Pleistocene terrace deposits on the Georgia
coast. The large etch pit was probably formed under low energy
conditions. Mechanical V-shaped patterns, *b*, were probably formed after
the etch pit as were the features labeled *c* (straight or slightly curved
grooves). Diagenesis is indicated by *d*; note that it appears to be the final
feature (TEM). *Note:* All of these photographs were made of quartz
sand grains with either the transmission electron microscope or the
scanning electron microscope (TEM or SEM).

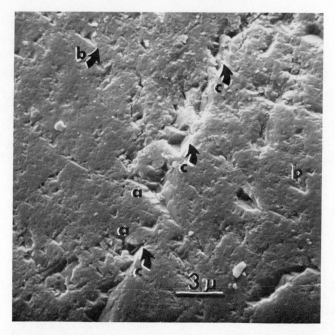

Fig. 6   Beach environment, surf zone in central California near Point Arguello. Features *b* and *c* are as indicated in Fig. 1. Feature *a* represents a blocky, conchoidal breakage pattern, worn down (SEM).

Eolian Action

All the eolian features mentioned, observed with the TEM (Fig. 7), can be seen on SEM photographs (Figs. 8 and 9). The meandering ridges are pictured in Figs. 8 and 9; they can be seen somewhat more clearly here than in TEM photographs (Fig. 7).

Graded arcs, as mentioned above, can also be observed on both kinds of photographs (Figs. 7, 8, and 9) but are more obvious in the SEM photographs, perhaps because of the simulated three-dimensional view.

Pitted surfaces can also be seen on SEM photographs (Figs. 8 and 9), but they may grade into the oriented fracture patterns as in the following discussion.

Figure 10 (SEM) is a typical surface found on ten samples of desert sands from various parts of the world. The features observed, which we call oriented fractures, although this does not necessarily imply that they are mechanical, are oriented ridges projecting from the

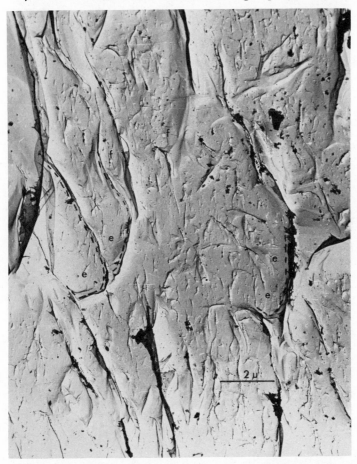

Fig. 7 Dune environment from late Pleistocene terrace deposits on the Georgia coast. Meandering ridges, *e*, enclose worn breakage blocks; note their general uniformity. Graded arcs, *f*, are not too common on this photograph although quite common on other dune sands. Note also the mechanical V-shaped patterns indicative of surf action. This grain was probably cycled back and forth between dune and beach (TEM).

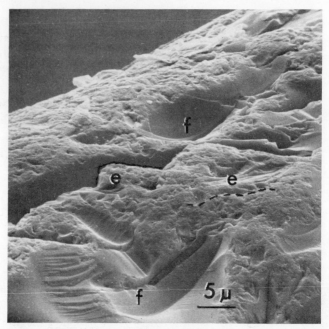

Fig. 8 Dune environment, pocket desert, New Mexico. Features *e* and *f* are as indicated in Fig. 3. Note the smooth, rounded, and slightly indented surface characteristic of eolian action (SEM).

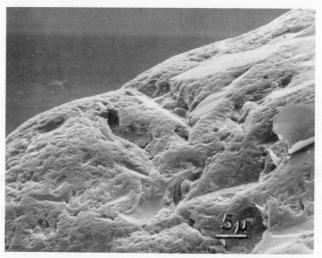

Fig. 9 Dune environment, pocket desert, New Mexico. This is another portion of the same grain as Fig. 4. Note how the breakage pattern scale may vary from one area to another. (SEM).

grain surface. These features are quite characteristic of hot, desert sands and are an additional criterion for eolian action not noted with the TEM.

Surface rounding is a feature of eolian grains that can be seen with the binocular microscope, but its relationship to other characteristic eolian features can be seen clearly in only Figs. 8 and 9 (SEM).

It has been difficult to separate desert from glacial sands with the TEM, for conchoidal and rectangular breakage appear on both. Desert pitting is not always easy to distinguish; however, with the SEM, the overall appearance of desert sands, the oriented fractures, and the relationship between the large breakage-patterns, rounding,

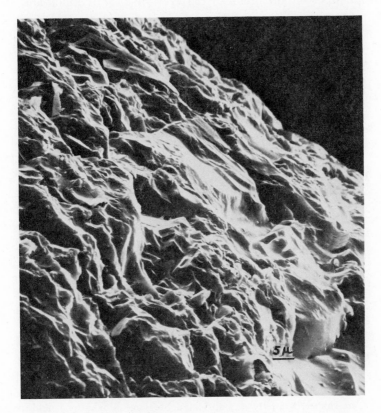

Fig. 10 Dune environment, Libyan desert. Oriented fracture patterns, a new feature seen only with the SEM, are very prominent (SEM).

and pitting are extremely characteristic. Desert sands can be therefore easily distinguished from the other categories described. No statements can as yet be made concerning coastal dune sands.

Glacial Textures

All the glacial features listed in the previous section as seen with TEM (Fig. 11) can be observed with SEM (Figs. 12 and 13). These figures indicate that the actual variety of features is considerably greater than that noted with the TEM. The amount of relief is extremely large as compared to other environments sampled; thus relief, by itself, can be used as a criterion for glacial action.

Sands that have been ice rafted and never exposed to terrestrial environments contain breakage-patterns of various kinds and sizes,

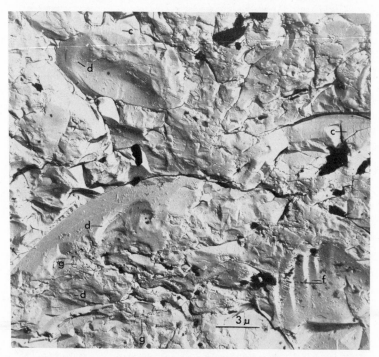

Fig. 11 Glacial environment, Late Wisconsin, Gardiners Island, Long Island, New York. Semiparallel steps are indicated on the photograph by c, arc-shaped steps by d, imbricated breakage blocks by f, and irregular small scale indentations by g. Note the tremendous variety and variable size of the features, important characteristics of glacial action (TEM).

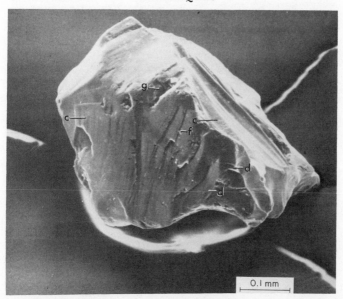

Fig. 12 Glacial environment, Miocene, North Pacific, Core CK13-3. The sand grain represented here was probably carried by floating ice to the place where it was dropped. Features *c, d, f* and *g* as indicated in Fig. 7. Note the tremendous depth of focus. The grain is characteristically jagged and irregular; the lack of weathering and diagenesis may be due to the fact that it probably was frozen into glacial ice from the time it was picked up (SEM).

but little or no evidence of solution (Fig. 13). On the other hand, grains deposited in terrestrial environments probably have been reworked somewhat by surface waters and certainly may have been subject to groundwater flow; these tend to show evidences of pitting and solution.

Parallel striations of varying length are more common than observed with the TEM; they are considerably more prominent and are one of the best criteria for glaciation (see Figs. 12 and 13).

Figure 14 (SEM) represents a common type of diagenetic pattern also observed in Fig. 1d which contains both beach and diagenetic patterns. The SEM pattern (which is probably closer to reality than the TEM photograph) is deeper and more prominent than those noted on TEM pictures. Otherwise previous interpretations seem to be correct.

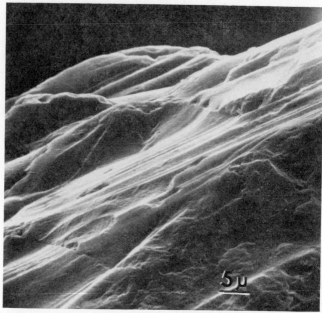

Fig. 13 Glacial environment, Miocene, North Pacific, Core CK13-3. Note the long, parallel ridges; these can be seen with the necessary degree of clarity only with the SEM and are one of the most important glacial characteristics. They may be the result of either cleavage or gouging (SEM).

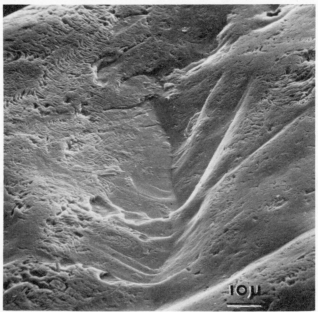

Fig. 14 Diagenetic environment, terrace deposits, Alligator Point, Florida. A typical, wavy, etched pattern characteristic of diagenesis is present; this texture has been duplicated experimentally (SEM).

REFERENCES

Angelucci, A., 1966a, Morfoscopical microscopic elettronico dei granuli di quarzo eolici e di spiaggo, *Accademico nazionale dei Lincie*, II, 1073–1077.

———,1966b, Preparazione di granuli di quarzoper lo studio morforcopico al microscopico elettronico; *Periodico di Mineralogia*, XXXV, 727–734.

Biederman, M. W., Jr., 1962, Distinction of shoreline environments in New Jersey; *J. Sed. Pet.*, 32, 181–200.

Bramer, H., Bestimmug der oberflachen beschaffenkeit von-Quartzkornern mit der Elektronenmikroskop, *Geologie*, 14, 1114–1117.

Folk, R. L., 1965, Some aspects of recrystallization in ancient limestones, in Dolomitization and limestone diagenesis; L. C. Pray and R. C. Murray, eds., S.E.P.M. Spec. Publ. 13, 180 pp.

———, and C. E. Weaver, 1955, A study of the texture and composition of chert, *Am. J. Sci.*, 250, 498–510.

Geitzenauer, K., S. Margolis and O. Edwards, 1968, Evidence consistent with Eocene glaciation in a South Pacific deep sea sedimentary core, *Earth and Planetary Sci. Letters*, 4, 173–177.

Haine, H. E., The electron microscope, John Wiley and Sons, 282 pp.

Hamilton, W., and D. Krinsley, 1967, Late Paleozoic glacial deposits of South Africa and southern Australia, *Bull., Geol. Soc. Am.* 78, 783–800.

Harvey, R. D., 1966, Electron microscope study of microtexture and grain surfaces in limestones, Illinois Geol. Surv. Circ. 404, 18 pp.

Hay, W., and P. Sandberg, 1967, The scanning electron microscope, a major breakthrough for micropaleontology, *Micropaleontology*, 13, 407–418.

———, K. Towe and C. Wright, 1963, Studies of the shell structure of Foraminifera via electron microscopy, *Micropaleontology*, 9, 171–195.

Hyde, P., and D. Krinsley, 1964, An improved technique for electron microscopic examination of Foraminifera, *Micropaleontology*, 10, 491–493.

Kay, D., 1961, Techniques for electron microscopy, Blackwell, Oxford, England, 331 pp.

Krinsley, D., and T. Takahashi, 1962a, The surface textures of sand grains, an application of electron microscopy, *Science,* **135,** 923–925.

———, 1962b, The surface textures of sand grains, an application of electron microscopy: glaciation, *Science,* **138,** 1262–1264.

———, 1962c, Applications of electron microscopy to geology, *Trans. New York Acad. Sci.,* **25,** 3–22.

———, 1964, A technique for the study of surface textures of sand grains with electron microscopy, *J. Sed. Pet.,* **34,** 423–426.

Krinsley, D., and M. Schneck, 1964, The paleoecology of a transition zone across an upper Cretaceous boundary in New Jersey, *Palaeon.,* **7,** 266–280.

Krinsley, D., T. Takahashi, M. Silberman, and W. Newman, 1964, Transportation of sand grains along the Atlantic shore of Long Island, New York: an application of electron microscopy, *Marine Geology,* **2,** 100–121.

———, N. Morton and J. Cutbill, 1964, The application of vacuum shadowing to the examination and photography of small objects, *Geol. Mag.,* **101,** 357–359.

———, and W. Newman, 1965, Pleistocene glaciation: a criterion for recognition of its onset, *Science,* **149,** 442–443.

———, and B. Funnell, 1965, Environmental history of sand grains from the Lower and Middle Pleistocene of Norfolk, England, *Quart. J. Geol. Soc. London,* **121,** 435–461.

———, and J. Donahue, 1968a, Environmental interpretation of sand grain surface textures by electron microscopy, *Bull. Geol. Soc. Am.,* **79,** 743–748.

———, 1968b, Diagenetic surface textures of quartz grains in limestone, *J. Sed. Pet.,* **38,** 859–863.

———, 1968c, Some methods to study surface textures of sand grains, a discussion, *Sedimentology,* **10,** 217–221.

———, 1968d, Pebble surface textures, *Geol. Mag.,* **105,** 521–525.

———, and S. Margolis, 1969, A study of quartz sand grain surfaces with the scanning electron microscope, *Trans. New York Acad. Sci.,* **31,** 457–477.

Kuenen, Ph. H., and W. B. Perdok, 1962, Experimental abrasion, 5: frosting and defrosting of quartz grains, *J. Geol.,* **70,** 648–658.

Margolis, S., 1966a, Electron micrography of modern and ancient quartz sand grains, M. S. Thesis, Florida State Univ., unpublished manuscript.

———, 1966b, Electron micrography of modern and ancient quartz sand grains, *Coastal Res. Notes,* **2**, 7–8.

———, 1968, Electron microscopy of chemical solution and mechanical abrasion features on quartz sand grains, *Sed. Geol.,* **2**, 243–256.

McIntyre, A., and A. Be, 1967, Modern Coccolithophoridae of the Atlantic Ocean: I. Placoliths and Cyrtoliths, *Deep-Sea Res.,* **14**, 561–597.

Oatley, C., 1966, The scanning electron microscope, *Sci. Progr.,* **54**, 483–495.

Oatley, C., W. Nixon, and R. Pease, 1965, Scanning electron microscopy, *Advances in Electronics and Electron Phys.,* **21**, 181–217.

Oldershaw, A. E., 1968, Electron microscopic examination of Namurian bedded cherts, North Wales (Great Britain), *Sedimentology,* **10**, 255–272.

Porter, J. J., 1962, Electron microscopy of sand surface textures, *J. Sed. Pet.,* **32**, 124–135.

Rochow, T. G., A. M. Thomas, and M. C. Botty, 1960, Electron microscopy, *Analytical Chem.,* **32**, 92R–103R.

Sandberg, P., and W. Hay, 1967, Study of microfossils by means of the scanning electron microscope, *J. Paleon.,* **41**, 999–1001.

Sandberg, P. A., and W. W. Hay, 1968, Application of scanning electron microscopy in paleontology and geology, Proc. Symp. Scanning Electron Microscope—The Instrument and Its Application, I.I.T. Res. Inst., Chicago, Illinois, 29–38.

Smith, and C. Oatley, 1955, The scanning electron microscope and its field of application, *Brit. J. Appl. Phys.,* **6**, 391–399.

Soutendam, C. J. A., 1967, Some methods to study surface textures of sand grains, *Sedimentology,* **8**, 281–290.

Swift, D. J. P., D. S. Byers, and D. Krinsley, 1966, A periglacial eolian sand at Debert, northern Nova Scotia: a preliminary report, *Maritime Sediments,* **2**, 25–30.

Thornton, P., 1965, The scanning electron microscope, *Sci. J.,* **1**, 66–71.

Waugh, B., 1965, Preliminary electron microscope study of the development of authigenic silica in the Penrith sandstone, *Yorkshire Geol. Soc. Proc.*, 35, 59—69.

Wolfe, M. J., 1967, An electron microscope study of the surface texture of sand grains from a basal conglomerate, *Sedimentology*, 8, 239—248.

SECTION
IV

TEXTURAL ANALYSIS

CHAPTER 9

SAMPLE IMPREGNATION

DANIEL J. STANLEY

Division of Sedimentology, Smithsonian Institution,
Washington, D.C.

Impregnation refers to the binding of friable or semiconsolidated materials through the introduction and penetration of cementing material into natural or artificially enlarged pore spaces. Embedment methods for encasing sedimentary specimens are also considered in this general discussion of impregnation. The purpose of this chapter is to outline procedures for the preservation of sediments that produce a minimum amount of grain displacement and disruption of fabric.

Impregnation techniques have been perfected over many years. Most earlier methods evolved from a need to facilitate the microscopic examination of texture, fabric, and mineralogy of friable sediments. Techniques for the consolidation of friable materials for thin-section preparation developed during the nineteenth century are outlined by, among others, Forbes (1869), Sorby (1882), and Johannsen (1918). Measurement of permeability and porosity and the detection of natural small fissures in sedimentary rocks have further stimulated the development of new techniques, particularly by the petroleum industry. Methods of embedment for mounting

183

and displaying specimens have also been refined and improved. During the last two decades, emphasis has been placed on developing methods that facilitate preservation and examination of sedimentary and biological structures, particularly in unconsolidated materials collected in outcrop exposures or in cores. Details of peel and impression techniques and methods for impregnating large surface areas are discussed in Chapter 10.

IMPREGNATION FOR THIN-SECTION PREPARATION

General

Numerous products have been used for artificially binding sedimentary materials for thin-section preparation, but only a few meet the necessary requirements. Ideally, the impregnating medium should (a) have a relatively low viscosity and surface tension for complete penetration during impregnation, (b) be stable at high temperatures, (c) harden at room temperatures, (d) be readily distinguishable from minerals in thin-section, that is, have an index of refraction near that of balsam or other known index, (e) be easily cut and ground during sectioning, and (f) be sufficiently hard to prevent enclosure of grains of abrasive, but sufficiently durable to eliminate chipping around pores during grinding. It is also important to avoid disruption of the grain-to-grain relationship during impregnation, especially when fabric and pore structures are to be examined optically and permeability is to be measured.

Early methods included boiling a sample in Canada balsam until it could absorb no more balsam, saturation with melted copal gum or Canada balsam-shellac mixtures (Johannsen, 1918), or saturation with various forms of wax, including paraffin, pure rubber dissolved in xylene, or gelatin solutions (Milner, 1962). Bakelite, a synthetic phenol resin, had some success (Ross, 1924; Leggette, 1928) and was in common use until World War II. Although this medium has great strength and resistance to solvents and penetrates dense, fine-grained materials with reasonable success, bakelite has the serious disadvantage of having a high index of refraction (1.634). In thin-section, bakelite appears as a pale yellow, isotropic substance with high relief.

The following sections outline techniques that fulfill different requirements. The details of actual thin-section preparation, however, are not treated in this discussion.

Canada Balsam Impregnation

An early and popular method for the preparation of soft, porous sediments, described by Forbes (1869), is to soak the sample in turpentine, then in Canada balsam, and heat the soaked sample until it hardens. A sample, after such impregnation, can be mounted on a glass slide in the usual way. The advantage of Canada balsam as a binding agent is its refractive index (cooked balsam: approximately 1.538) which approximates that of quartz.

More efficient impregnation of less permeable samples with balsam can be attained with vacuum (Schlossmaker, 1919). A specimen is placed on a shoulder of a container in a vacuum chamber and, after evacuation, the chip is caused to fall into the balsam at the bottom of the vessel. Waldo and Yuster (1937) modified this method by evacuating air (pressure to 1 mm) from a clean, dry specimen with an ordinary laboratory vacuum pump, immersing the sample in molten balsam, and later forcing balsam into the evacuated pores under pressure with compressed air. The pressure range employed is 1 mm Hg to about 3880 mm Hg (about 75 psi).

Sufficient balsam should be melted in order to cover completely a sample after it has been placed in the vacuum chamber. Temperatures of 66 to 76°C suffice to melt balsam; a porcelain crucible is used to hold the specimen immersed in molten balsam. After cessation of bubbling (indicating that most of the voids have been filled with molten balsam), air is allowed to reenter the evacuated chamber and pressure is applied to force balsam into the still unfilled pore system. The temperature is raised to about 120°C and the heating unit shut off. The temperature may continue to rise slowly to nearly 150°C, at which time the fluidity of the balsam produces good impregnation. After the heating unit has been turned off, the specimen is maintained under pressure for ½ to 1 hour to ensure good impregnation. After cooling to room temperature, the specimen can be ground. The disadvantages of this method are that the balsam becomes brittle and yellow.

Thermosetting Resin Impregnation

Impregnation of sediments with synthetic thermosetting resins is now common practice and presents several advantages. Polyester and other resins are excellent binders, are resistant to fracturing and chemical reaction, are not markedly affected by temperature change, and can be easily machined, ground, and polished (Cavanaugh and Knutson, 1960). A disadvantage of most resins, however, is their rather short shelf life and short pot life, which are the time intervals available for impregnation once the catalyst or hardener has been added. Furthermore, the index of refraction of most artificial resins (1.60 to 1.64) is higher than that of Canada balsam (approximately 1.54) which causes difficulty in making mineralogical determinations. The problem of distinguishing quartz from potash feldspar is an example (Müller, 1967).

Techniques and required equipment are relatively unsophisticated and inexpensive, and there is available a great variety of material including polyesters, styrenes, phenolics, polyurethanes, dialpthalates, epoxies, and combinations of these. Very many resins cure at elevated temperatures, whereas others cure at room temperature. Both types are discussed in this chapter. Numerous references such as the *Modern Plastics Encyclopedia* (McGraw-Hill, 1966, published annually) summarize the properties of the myriad of currently available products.

Castolite and Laminac Resins

Some thermosetting plastics of the polyester type produce transparent, colorless castings and can be used quite effectively to impregnate sandstones for thin-section preparation. Lockwood (1950) has used castolite and laminac resins, although many other resins can be substituted. In order to fill the smallest capillaries in the channel system of a specimen having a permeability as low as 1 millidarcy, it is necessary to use a viscosity-reducing agent. Monomeric styrene can be used for this purpose.

The following plastic-to-viscosity reducing agent ratios are recommended to obtain maximum penetration (percent by volume): 55% Castolite, 45% Monomeric styrene (viscosity, 44 centipoise at

TABLE 1   Materials and suppliers

Castolite	The Castolite Co., Woodstock, Illinois
Castolite hardener	The Castolite Co.
Laminac resin 4126	American Cyanamid Co., 30 Rockefeller Plaza, New York
Laminac catalyst 347	American Cyanamid Co.
Monomeric styrene	The Dow Chemical Co., Midland, Michigan

15.6°C); or 55% Laminac resin 4126, 45% Monomeric styrene (viscosity, 16 centipoise at 15.6°C).

Necessary equipment includes 30-ml tall-form beaker, aluminum sample holders, 500-ml wide-mouthed, thick-walled Erlenmyer flask, No. 10 neoprene stopper with a three-way stopcock, neoprene tubing with a pinch clamp for plastic introduction, shallow pan, funnel, and vacuum pump.

Dry specimens are mounted in aluminum holders and placed in a 30-ml beaker. Several samples can be treated in a single operation (Fig. 1). The beaker is placed in the Erlenmeyer flask and the flask closed with a No. 10 neoprene stopper. The flask is suspended in a water bath. The stopper with a three-way stopcock and a tube (for introduction of plastic) extends into the beaker. The upper end of this tube is fitted with neoprene tubing, a pinch clamp, and a funnel. One of the openings of the stopcock is connected to a rubber tube running to a vacuum pump. A mercury manometer with a range of 0 to 200 mm is connected between the flask and the vacuum pump.

Air is removed from pore spaces of the samples by reducing the pressure in the flask to approximately 1 mm Hg. Low pressure is maintained for 30 minutes, and the temperature of the water bath is maintained at about 15.6°C. A mixture consisting of thermosetting plastic (22 ml), monometric styrene (18 ml), and approximately 0.7 ml of catalyst (Laminac catalyst 347) is poured slowly into the funnel. The pinch clamp ensures the slow introduction of impregnating material. After a short time, pressure is increased to atmospheric by slowly opening the three-way stopcock. The beaker is removed from the flask, placed in a shallow pan of water, and

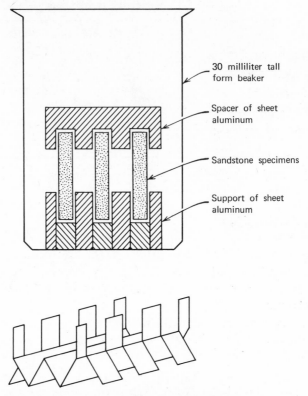

Oblique drawing of support made from aluminum sheet

Fig. 1 Three sandstone specimens, 3/4-in. square and 1/8-in. thick, on aluminum supports in 30-ml beaker for plastic impregnation (top) and shape of support (bottom). (After Lockwood, 1950)

heated to 60°C until initial hardening of the plastic takes place. Water must not be allowed to enter the beaker during this step. When the consistency of the binding material is that of stiffjelly, the beaker is placed in an 82°C oven for approximately 2 hours for final curing. After cooling, the excess-hardened plastic around the samples can be removed with a saw, and the impregnated specimens are ready for thin-sectioning.

Epoxy-Butyl Glycidyl Ether - Diethylenetriamine

A modification of Lockwood's (1950) technique which involves the use of a different group of thermosetting plastics has been outlined by Cavanaugh and Knutson (1960). One of the advantages of this method is the possibility of impregnating many samples (at least ten) during a single operation and completing saturation of unconsolidated sediments and sedimentary rocks with permeabilities greater than 0.1 millidarcy. This method, unlike the one described in the previous section, requires that samples be placed under considerable pressure during one phase of the operation. A pressure cell is thus required.

TABLE 2   Materials and suppliers

	Dow 331	The Dow Chemical Co., Midland, Michigan
Epoxy resin	Shell Epon 820	Shell Chemical Co., New York
	CIBA 6010	CIBA Chemical Co., Fairlawn, New Jersey
Butyl glycidyl ether	Diethylenetriamine	

Samples cut to about 2 X 2 X 1 cm are oven-dried at 125°C. Approximately 200 ml of plastic mixture is prepared for impregnation of ten samples. After cooling, labeled samples are placed in small, waxed paper cups and each covered by a mixture of 15 ml epoxy resin and 5 ml butyl glycidyl ether to which is added 20 ml of diethylenetriamine. The paper cups containing the samples and liquid mixture are placed in the pressure cell (Fig. 2), under as high a vacuum as possible, for 45 minutes. Pressure is increased to atmospheric, and 100 to 125 atm nitrogen pressure is applied for at least 1 hour. Samples are allowed to gel either at atmospheric pressure (pressure on the pressure cell must be released slowly to avoid foaming of the plastic) or under pressure for 10 to 12 hours. The equipment used to evacuate the samples and to supply fluid pressure could also be of the type recommended by Nuss and Whiting (1947) (Fig. 3) or by Ginsburg et al. (Figs. 8, 10). After samples have gelled, they are cured at about 125°C for a period of 2 hours. The plastic should harden on cooling. Excess plastic is cut away, and the impregnated samples are ready for thin-sectioning.

Rapid Thermosetting Method

A somewhat simpler and more rapid method of impregnating friable and semiconsolidated rocks has been developed by Mr. G.

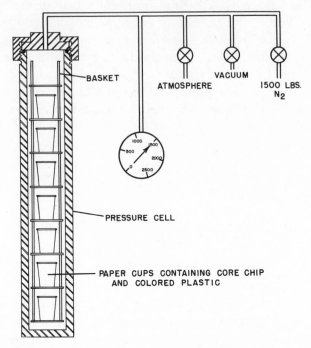

Fig. 2   Schematic of core saturation equipment. (From Cavanaugh and Knutson, 1960)

Moreland (1968) for use in the Department of Mineral Sciences, Smithsonian Institution.

TABLE 3   Materials and suppliers

Araldite plastic: AY-105   Chemical Coatings & Engineering Co., Media, Pennsylvania

Hardener: Hy-935A   Chemical Coatings and Engineering Co.

Toluene ($C_6H_3CH_3$, a viscosity-reducing agent)

The liquid resin can be prepared in advance for the preparation of numerous small chips by mixing thoroughly 3 gm of plastic hardener and 3 gm of Araldite plastic in a tall-form beaker. Five times this amount (by weight), or 30 gm, of toluene is added and mixed in order to obtain a satisfactory viscosity for complete penetration. The

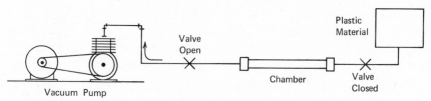

Valve
Open

Plastic
Material

Valve
Closed

Chamber

Vacuum Pump

Step 1
Chamber containing cores is evacuated

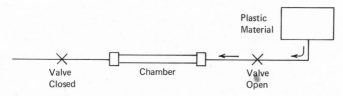

Plastic
Material

Valve
Closed

Chamber

Valve
Open

Step 2
Plastic material is forced into chamber
by atmospheric pressure

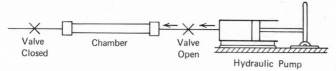

Valve
Closed

Chamber

Valve
Open

Hydraulic Pump

Step 3
Fluid pressure is applied on the chamber

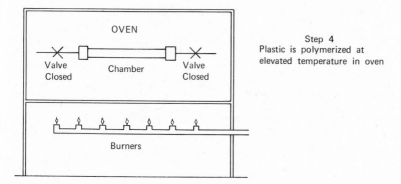

OVEN

Valve
Closed

Chamber

Valve
Closed

Burners

Step 4
Plastic is polymerized at
elevated temperature in oven

Fig. 3 Equipment and procedures for impregnating sediments under high pressures. (From Nuss and Whiting, 1947)

191

dried sample chips, placed in small, disposable plastic beakers, are completely covered with the liquid and placed in a vacuum for 15 to 30 minutes, or until bubbles are no longer visible and until such time as the plastic has penetrated even the smallest pores. Atmospheric pressure is then readmitted to the vacuum bell, and the sample chips are placed in a 200°C oven for 45 minutes to 1 hour. Samples are removed and allowed to cool to room temperature. After cooling, the impregnated sample chip is ground on a fine lap with fine abrasives to remove any scratches or saw marks before mounting on a glass slide. The hardened specimen and thin-section glass are placed on a hot plate heated to 200°C. A small amount of mixture consisting of one part plastic to one part hardener is added to the glass slide and the impregnated sample chip is placed on the glass. The specimen and glass slide must be left on the plate for 1 hour before allowing to cool to room temperature. The sample, cemented on the glass section, is then ready for thin-section preparation.

Artificial Plugs of Unconsolidated Sediment

A method for making cylinders of unconsolidated sediments, including carbonates, is outlined by Davies and Till (1968). Samples, thoroughly dried by adding acetone and warming, are placed in a 1-in. diameter glass tube.

Table 4   Materials and suppliers

Araldite CY-212	CIBA Chemical Co., Fairlawn, New Jersey
Hardener HY-951	CIBA Chemical Co.

A mixture consisting of ten parts Araldite plastic to one part hardener is added to the sample in the tube. As in previous methods, the tube can be placed in a vacuum to ensure complete saturation. Atmospheric pressure readmitted to the vacuum bell forces the resin throughout the artificial sediment plug. The tube of sample is cured at 100°C for 2 hours and allowed to cool. The glass tube is broken and a 1-in. diameter cylinder is available for thin-sectioning.

Impregnation of Moist and Dried Clay and Shale Samples

Impregnating moist clay and shale samples has generally met with less success than impregnation of coarser and dry porous materials.

In order to overcome problems of low permeability and high water content, Tourtelot (1962) modified a technique described in an earlier work by Mitchell (1956). The technique is most useful on moist samples since the impregnating medium is diffused through the natural moisture of the sample. The impregnating medium used is Carbowax 600 (Carbide and Carbon Chemical Co.), a waxlike, high molecular weight polethelene glycol compound. The wax is soluble in water in all proportions and melts at 55°C. A sample can be impregnated by soaking for three days in melted wax at about 60°C. After impregnation, the thin-section blank is prepared by sawing and grinding in kerosene. A petroleum solvent can be used to remove kerosene from the blank before mounting on the glass slide. The wax method, however, does present several disadvantages in that the wax can react chemically with pure montmorillonite, is anisotropic, and tends to crystallize in birefringent, feathery aggregates.

Dried, prehardened, silty-clayey samples can be consolidated with a mixture of CIBA Araldite liquid-coating resin F (CIBA Chemical Co., Fairlawn, New Jersey) and hardener 905 at 40°C (Müller, 1967). The viscosity of Araldite F at 20°C is about 7.3 to 13 poises. The sample in the mixture, is placed under vacuum for ½ to 1 hour, and then is placed in a drying oven at 40°C. The liquid resin hardens in about 16 days, and samples remain in the oven for an additional 48 hours at 120°C to complete the curing process. Variations to this method have been developed by Reineck (1963) and others.

Methods for obtaining surface impressions and consolidated sections of clayey sediments are described in Chapter 10.

Room-Temperature Curing Resins

Room-temperature curing resins, which harden at about 20°C without oven curing, are practical for impregnating temperature-sensitive friable samples. Müller (1967) has suggested a mixture of Araldite E with Araldite hardener 951 (ratio of 100:11) or Araldite E with hardener 943 (ratio of 100:20) with a curing period of 24 hours at room temperature.

Another cold-hardening pouring technique, developed for paleontological investigations (Merida and Boardman, 1967), can be adapted for sedimentological purposes. Loose or semiconsolidated samples are placed in an aluminum foil or silicon rubber mold. The impregnating medium used is Cobalt preaccelerated Polylite Resin

TABLE 5   Materials and suppliers

Araldite E (liquid resin form)	CIBA Chemical Co., Fairlawn, New Jersey
Araldite hardener 951	
Araldite hardener 943	
Cobalt preaccelerated Polylite 32737	Reichhold Chemical, White Plains, New York
46-700 Mek Peroxide "60"	

#32737. The resin is mixed with a liquid catalyst, Mek Peroxide 60 (0.25 to 2.0% by weight of resin or in proportion of 2 cc of resin to one drop of catalyst). The clean, dry sample in the mold is covered with the medium and is placed in a vacuum chamber for thorough penetration. The medium begins to gel in about 20 minutes at room temperature. Bubbles can be removed with tweezers by lifting the specimen from the bottom of the mold several times. The resin hardens sufficiently (hardness approximates that of calcite matrix) in about 2 hours. The impregnated sample can then be cut, polished, and mounted on a glass slide using Lakeside 70 thermoplastic (Lakeside Plastics Corp., 3325 North Shore Drive, Oshkosh, Wisconsin). Several photomicrographs of thin-sections prepared by this method are shown in Fig. 4.

Dyed Resins for Porosity and Textural Analyses

Synthetic resins can be easily dyed and, when used in this form, facilitate estimates of porosities and detection of fissures. It is essential that the dyes used be sufficiently brilliant and intense to be readily distinguishable from natural colors of minerals as seen in thin-section. The most common type of coloring material for plastics are suspended pigmenting materials that are commonly unsuitable for dyeing resins. The small suspended pigment particles tend to be filtered out in the fine pore channel system of the rock, and results in a colorless, dye-free center of an impregnated sample (Cavanaugh and Knutson, 1960). Color dyes mixed with thermosetting resins must also be heat resistant.

An oil-soluble flaming-red dye has proved satisfactory when mixed with several different types of resins (Waldo and Yuster, 1937; Lockwood, 1950). Approximately 36 mg of powdered red dye

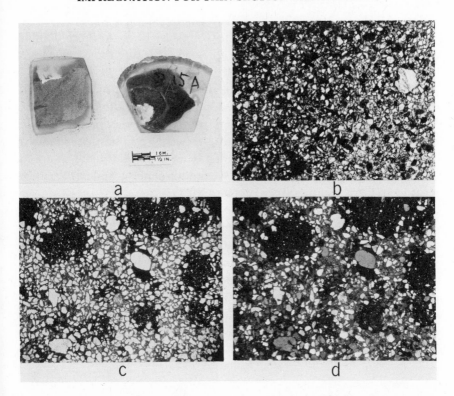

Fig. 4 Two porous fine to medium grade sandstone samples. (a) Impregnated with room-temperature cure resins. Photomicrographs of thin-sections observed under natural light (b/c) and polarized light (d).

(Flaming Red No. G-305, Ox Color Works, New York) can be mixed with 1.0 ml of monomeric styrene and, after 5 minutes of stirring, the excess dye can be filtered out on coarse filter paper. A mixture of 22 ml of liquid plastic resin and 18 ml of dyed monomeric styrene plus 0.7 ml of catalyst provide a 40-ml volume, as outlined by Lockwood (1950). Dye-filled pore patterns and microfracture systems are clearly visible in thin-sections and in color prints.

An oil-soluble blue dye can be used with thermosetting resins of the type described by Cavanaugh and Knutson (1960) and outlined in the previous section. Blue oil-soluble dye (No. 0-5-D-17, Krieger Color and Chemical Co., Hollywood, California) can be added to the resin in the following ratio: 200 ml epoxy-butyl glycidyl ether mixture to 0.8 gm soluble blue dye and 20 ml diethylenetriamine.

Red Araldite (No. DW 03, CIBA Chemical Co., Fairlawn, New Jersey) can be mixed with plastic resins such as Araldite F plus hardener Araldite 905. Approximately 4% (by weight) of the total resin-plus-hardener weight is required (Etienne, 1963).

Particularly spectacular results are obtained with a red fluorescent dye, Rhodamine B extra, when mixed with Araldite F resin and Araldite hardener 905 (resin-hardener ratio of 1:1). The amount of dye needed is 0.30% by weight of the total resin-hardener weight prepared. The dye is mixed thoroughly with the hardener at room temperature and the mixture of dye and hardener is added to plastic resin heated to a temperature of 70 to 80°C.

To facilitate porosity estimates, a method of visible reproduction of effective pore space was developed by Nuss and Whiting (1947). The method consists of impregnating a sample with thermosetting resin and leaching the hardened sample with very dilute hydrochloric acid or hydrofluoric acid. The method is particularly effective with carbonates or carbonate-cemented sandstones. Introduction of colored resins of the type described in this section produces enhancement of the plastic pore models that result from such a leaching process.

EMBEDDING AND MOUNTING OF GEOLOGICAL SPECIMENS

Elevated-Temperature Curing Resins

Specimens can be preserved and prepared for examining texture and sedimentary structure in clean, transparent mounts of artificial resin. One such method applicable for mounting both wet and very fragile samples has been described by Arthur (1949). The size of the sample is limited only by the size of the autoclave available. Before mounting, the specimen is dipped in a mixture of about 20% (by weight) of polymer (Plexiglas compression molding powder, formula 100-Y, Rhom & Haas Co., Philadelphia, Pennsylvania) and 80% monomer. The monomers used are ethyl methacrylate and methyl methacrylate, usually in a 1:1 ratio. The sample is suspended in a glass or iron mold by a thread or a sheet of Plexiglas that wedges the object in position. The plastic mixture used to encase the sample is 65% polymer and 35% monomer stirred with an electric food mixer. When the mixture begins to thicken (from 5 to 45 minutes), it is poured into the mold. The mold is placed in an autoclave and cured at 52 to 62°C under pressures of 60 to 75 psi. The time required for curing depends largely on the size of the block being prepared. In most cases, it is sufficient to leave the sample overnight in the autoclave and let it cool slowly under pressure (cooling rate about 8°C per hour).

Mounting in Room-Temperature Curing Plastic

A simple and more rapid method of mounting specimens without the necessity of applying heat or use of an autoclave and pressure equipment can be achieved with certain cold-setting epoxy resins (Gendron, 1959). Specimens are placed in natural rubber cup molds. The cups are brushed with undiluted Spraylat PM 4571, a mold release, before pouring epoxy. The mold release can be washed off after the casting is removed so that the mold can be used over again. A mixture of 100 gm Bakelite ERL 2795 Epoxy Resin and 10 gm of Hysol C1 hardener is used for samples smaller than 10 X 20 cm. The mixture is stirred for about 1 minute with a small electric food mixer in a large, waxed paper cup. A tray of sample-filled rubber molds is placed before a window air conditioner set at its coldest temperature. The epoxy is poured into the molds and set for 6 to 7 hours. The molds are then warmed to normal room temperature. The mold can be peeled off the mount, which can then be ground and polished.

Larger samples can also be prepared with this method by placing them in large Polyethelene containers with no Spraylat applied. The molds are set in a tray of water; the water line should be above that of the liquid plastic, but should not wet the sample or resin. The mixture used is 9 gm of hardener per 100 gm of epoxy resin. All mounts are ready for grinding the day after they are poured. Adhesion of the resin to most materials is excellent.

Another method of embedment with room-temperature curing resins is used at the Smithsonian Institution (J. Widener, personal communication). This method also provides a transparent, odorless mount for samples which enables them to be studied more easily and displayed. Rectangular aluminum molds of different sizes are used. These consist of a rectangular sheet bent into a U-shape that forms the base and two sides of the mold. Two separate rectangular sheets with flanged edges form the two other sides of the mold. Masking tape is used to seal the edges and corners of the open-topped box.

Aluminum frames of this type provide flexibility in the size and shape of the mold, and serve to dissipate heat resulting during polymerization. A heat-stable separator is applied to the aluminum mold as a parting film to keep the resin from sticking to the sides.

TABLE 6   Materials and suppliers

Castolite Liquid Casting Plastic	Castolite Company, Woodstock, Illinois
Castolite Liquid Hardener	Castolite Company
Polyester Parfilm (separator)	Price-Driscoll Corp., Farmingdale, New York

A low viscosity resin, Castolite Liquid Casting Plastic, is mixed thoroughly with Castolite Liquid Hardener, a methyl-ethyl-ketone peroxide, in a large, unwaxed paper cup. Hardener in the amount of 0.5 to 1.0% (by weight) is recommended. The mount is made in two steps. First, a base layer of plastic is poured. About 200 gm of resin plus 2 gm of hardener suffice to form a surface 7.5 × 7.5 × 1 cm. This base layer is allowed to gel (15 to 30 minutes), and air bubbles that form near the surface are gently drawn (teased) toward the edge of the mold. A second batch of resin (200 gm) and catalyst (2 gm) mix is prepared. A thin (about ½ cm) layer is poured on the basal-gelled surface, and the specimen is placed in position. The specimen is pushed down firmly against the basal gelled layer to eliminate the possibility of entrapping air bubbles, and is immediately covered with the remainder of the mix. Again the plastic is allowed to gel at room temperature. Bubbles are teased off before the gel has set.

The mount hardens in several hours, but it is advisable to let the embedded sample remain in the mold overnight. Polymerization, with its accompanying exothermic reaction, takes place after the resin-hardener mix has been poured. If temperatures build up to a point where the mold is uncomfortable to the touch, it is best to cool the mold by either placing it before a window air-conditioner set at a low reading (about 10°C) or in a refrigerator for a short period as the temperature reaches its peak. Excessive temperatures are to be avoided because the resultant internal stresses may cause cracking of the mount or separation of the mount from the specimen.

After the mount has hardened completely at room temperature, it is removed from the mold and milled in order to trim the six surfaces. The plastic block is wet-sanded to remove the tool marks due to milling and polished with a felt buffing wheel and jeweler's rouge.

The method is ideally suited for well-cemented and nonporous sedimentary rocks (Fig. 5a). Poor results are commonly obtained with partly consolidated sediments: entrapped air released from the sample during the cure forms bubbles which are difficult to remove from the hardening plastic (Fig. 5b). It is advisable, therefore, to impregnate porous and semiconsolidated samples prior to embedment.

ARTIFICIAL LITHIFICATION OF HIGH-PERMEABILITY SEDIMENTS UNDERWATER

Little work has been published on underwater impregnation techniques. An exception to this is an *in situ* lithification method developed by Brown and Patnode (1953). The technique involves consolidating high-permeability sediments, sands in particular, into a consolidated slab by introducing a low-viscosity liquid plastic medium via an injecting apparatus, then removing the impregnated core for further treatment and study in the laboratory. An assembled

Fig. 5 Specimens embedded in room temperature cure resin. (*a*) Polished sample of well-consolidated conglomerate. (*b*) Poorly consolidated sandy specimen showing air bubbles entrapped during cure. (Specimen should have been impregnated prior to mounting.)

plastic injection sampler is shown in Fig. 6, and a description of components and operational steps is as follows (Brown and Patnode, 1953, p. 152-153).

"Immediately prior to sampling, the piston rod assembly is clamped in position by means of the split collar (2). The volume of the cylinder is regulated by the position in which the piston rod (1) is clamped. The instrument is inverted and plastic, to which the catalyst and accelerator have just been added, is poured into the body of the cylinder (7). The end of the cylinder is closed by inserting the rubber gasket and plug (8) in the end of the cylinder. The core barrel (11) is positioned on the gasket and clamped against it by exerting tension on the bail (6) by means of the bail clamp (5). The instrument is then righted and the core barrel (11) pushed into the sand body to a depth controlled by the support flange (10). Water that is displaced from the core barrel by the sand escapes through the vent tube (4). This vent is essential to prevent disruption of the core when inserting the core barrel. When the support flange (10) is firmly seated on the bottom, the lever arm (3) is raised, releasing the piston rod clamp (2) and closing the vent tube (4). Pressure is then applied to the piston by means of the rod (1). The piston forces the plastic downward from the cylinder. The plastic passes through a small orifice in the top of the plug (8) and through a large hole in the top of the core barrel. The plastic then displaces the water in the top of the core barrel above the supporting flange and distributes evenly over the top of the sand. Continued pressure forces the plastic through the porous sand driving out most of the pore water. When all of the plastic is driven from the cylinder and the piston rests on the plug (8) the bail release (5) is lifted. This frees the core barrel from the rest of the assembly which is raised, leaving the core barrel in the sand. A rope and float previously attached to the ring (9) serves to mark the location of an underwater core and facilitate its recovery.

"The core barrels are expendable. They are fabricated from quart- or gallon-size tin cans. An opening is cut in the bottom of the can to admit the plastic and two eye bolts are inserted near the rim for attaching the bail. The cans are cut to the desired length.

"An "O" ring seal between the piston and cylinder wall is leak-proof and permits easy movement of the piston.

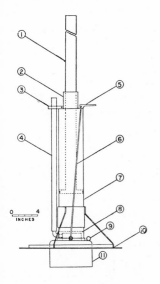

Fig. 6 Plan of plastic injection sampler showing essential elements: (1) Piston and rod. (2) Piston rod guide and clamp. (3) Vent clamp and piston rod release. (4) Vent. (5) Bail clamp. (6) Bail. (7) Cylinder. (8) Gasket. (9) Ring for attaching rope and float to facilitate recovery of core barrel. (10) Adjustable support flange. (11) Core barrel. (From Brown and Patnode, 1953)

"The instrument can be used in water to about 15-20 feet deep, the releases being operated by connecting ropes to the surface. A deep-water sampler of this type is feasible."

TABLE 7   Materials and suppliers

Selectron 5001–plastic	Paint Division, Pittsburgh Plate Glass Co., Milwaukee, Wisconsin
Lupersol DDM–catalyst	Lucidol Division, Novadel Agene Corp., Buffalo, New York
Selectron 5901–accelerator	Paint Division, Pittsburgh Plate Glass Co.

In the method outlined by Brown and Patnode, selectron 5001 resin is mixed with Lupersol DDM catalyst and Selectron 5901 cobalt accelerator. The proportions used (by trial and error) vary upon conditions of injection, and the medium will set to a stiff gel from 15 minutes to several hours. Underwater samples set very slowly because of the low temperature and presence of dissolved oxygen. Because the lower section of the sample is generally only partially impregnated and not recoverable, it is necessary to inject enough plastic so that contaminated material is pushed beyond the limits of the core barrel. Underwater samples can be recovered 1 to 2 hours after injection at temperatures of 10 to 20°C; it is better to allow the impregnated sample to cure in place for 24 hours or more. It is advisable to complete the cure by allowing samples to stand at room temperature for a few days.

Large slabs indicate very little effect of disturbance because of the injection, coring, or shrinkage on solidification of the plastic. Fig. 7 shows an impregnated slab retrieved from a submerged gravel bar.

PRESERVATION OF CORE SECTIONS

Impregnation

It is virtually impossible to examine piston and open-barrel cores without sectioning them. Slabbed faces of impregnated cores reveal sedimentary structures that would not otherwise be visible in untreated sliced or broken core section. Thin-sections can also be made from the hardened core sections. A technique developed by the Shell Development Company and described by Ginsburg et al. (1966) has been used with success by sedimentologists working with

Fig. 7 Example of a core impregnated underwater. The core slab of a submerged gravel bar deposit shows imbricated orientation of elongate pebbles. (From Brown and Patnode, 1953)

unconsolidated sediments and wishing to build up a permanent reference core library.

Extruded core sections cut in 15-cm lengths are air dried for several days, then placed in an oven at about 95°C for a few hours. Temperature is increased from about 95 to 175°C for final drying. Oven-dried cores are placed in a desicator to check moisture content (thorough drying of cores is essential). Dried cores are placed in aluminum pans and inserted in a vacuum chamber. Catalyzed resin (described at the end of this section) is introduced very slowly via a three-way stopcock after air has been removed (see Fig. 8). Caution is taken to avoid foaming. The core should be covered by about 1 to 2 cm of resin. Air is then allowed to reenter the chamber. When treating sediment cores of silt and clay, the pans are transferred to a pressure chamber (using nitrogen gas), and the pressure is increased to 160 psi. After 15 to 30 minutes, the pans are removed, and the

Fig. 8   Vacuum and pressure chambers used by Ginsburg et al., 1966. (1) Vapor trap. (2) Vacuum chamber, perforated metal-guard cylinder not shown. (3) Reservoir funnel for resin. (4) Stopcock to admit resin to vacuum chamber. (5) Three way stopcock. (6) Pressure chamber. (7) Pressure gauge. (8) Valve. (9) Exhaust valve. (10) Compressed nitrogen, 2200 psi chamber with regulator valves. (Courtesy of Shell Development Co.)

resin cured overnight. If the cores stick to the mold and are tacky, curing is not complete. The cores can be left in the sun during a day for final curing. The samples are then ready for sectioning, analysis, and display (Fig. 9).

Longer core sections (to almost 1 meter lengths) can be treated in identical fashion, but specialized vacuum and pressure chambers are required (Fig. 10). A description of these chambers, aluminum pans, saws, and other equipment used at Shell are provided by Ginsburg et al. (1966, pp. 1123-1125). Longer sections are wrapped in fiberglass cloth after extrusion, which gives added strength to the impregnated block.

Plaskon No. 0951 (Allied Chemical Co.), a low-viscosity polyester resin, is thoroughly mixed with approximately 0.03% of 6% cobalt naphthenate, a promoter. Then, 0.3% of methyl-ethyl-keotone peroxide, the catalyst (which should normally be kept under refrigeration), is added slowly and thoroughly mixed. *Caution:* Cobalt naphthenate and peroxide, if mixed, can produce a violent reaction. *These products should be stored separately in the laboratory.* The cobalt naphthenate must always be added to the resin first. *Only after these have been mixed should peroxide be added.*

The major disadvantage of this method is the necessary oven-drying which tends to distort structures and fabric of fine-grained sediments.

Embedding Thin Core Slabs

A method for encasing unconsolidated core slabs in Lucite has been developed by Esso Production Research Company and reported by Shannon and Lord (1967). The lucite mount is a clear, protective envelope that enables study of the core structure and textures from any side. Mounts, which are particularly useful for three-dimensional studies of sedimentary features, provide a permanent reference or teaching collection of selected core material to be made. The impregnation method described in the previous sections results in a completely solidifed mount. The present embedment method results only in a partial penetration of the core slab.

Cores to be mounted should be cut in half lengthwise. A cheese cutter strung with piano wire is useful for this. Half of the core is placed in a wooden trough; the sides of the core should be elevated above the sides of the box. A thin slot (½ to 1 cm) is sectioned from

Fig. 9 Aluminum pan, impregnated core section, and sliced core. (From Ginsburg et al., 1966)

205

Fig. 10 Pressure chamber for 3-ft sections of core (From Ginsburg et al., 1966).
(1) Pressure chamber of steel pipe ½-in. thick with pressure gauge. (2)
Pop-off valve 250 psi. (3) Pressure and exhaust lines. (4) Machined
flange with locking bolts. (5) Hinged door with 0-ring seal.

the cut face by using the sides of the box as a guide for the cheese
cutter (Fig. 11a). The thin slab is then transferred onto a 1-cm thick
precut Lucite base (Fig. 11b). The slab is aligned, trimmed, and a
paper label is placed at the base or top of the specimen. A coating of
Krylon plastic spray is needed to ensure adhesion of the slabs and
labels to the base. However, for fine-grained cores in which anaerobic
bacteria   may   be   present,   a   solution   of   pentachlorophenal

(C_6Cl_5-OH) and methyl methacrylate monomer (in equal parts) are applied to the sample. This coating, or that of the Krylon spray, ensures the adhesion of the core sample to the base and minimizes distortion when the liquid plastic is poured onto the core slabs. The specimens, on their base, are placed in a shallow porcelain tray with flared slides to facilitate removal of the cured mounts (Fig. 11c).

It is necessary to mount the samples in the tray before preparing and pouring the plastic. The liquid plastic consists of 65% (by weight) of Lucite-bead polymer 4F-NC-99 powder, 26% methyl methacrylate (hydroquinone 0.006%), and 9% ethyl methacrylate (hydroquinone 0.010 %). All can be purchased from Plastics Department, E. I. Du Pont de Nemours and Co., Wilmington, Delaware. Mixing should be carried out in a well-ventilated room (22°C) or under a hood. Materials can be introduced in any sequence and should be stirred thoroughly for about 5 minutes.

Care is taken to avoid pouring the liquid mix directly onto the slab surfaces. The slabs should be covered by about 2 cm of resin. The trays are placed in an autoclave and the samples cured at 65°C under 120 psi (inert gas is recommended) for at least 8 hours (overnight curing is generally convenient). The gas pressure is released slowly and samples are allowed to cool slowly (3 to 4 hours) to room temperature. The plastic mounts are then removed from the tray, sawed, and prepared for display or study.

In order to obtain serial sequence of slabs from a core section, the author has used a more rapid but somewhat less refined method than the one described above for partially impregnating half-core cylinders (Fig. 12). Liquid low-viscosity resins of the type described in this chapter are poured onto permeable sediments retained in the core liners. The liner and tape help provide a frame or mold in which to retain the resin. Excess resin is caught in the tray. Vacuum helps provide more thorough penetration of finer grained materials. Before curing, a thin board is placed on the open face of the half-core. After complete hardening, the cylinder is welded to the board. A thin (½ to 1 cm) slab can then be sliced with a saw. A second application of resin may be added on the remaining core slab, and a board is once again placed on the open face prior to gelling and curing of the resin. A second thin slab can be removed by saw. A third and fourth section may also be obtained, so that a serial group of lengthwise-cut slabs are available for examination, mounting, and storage (Fig. 12).

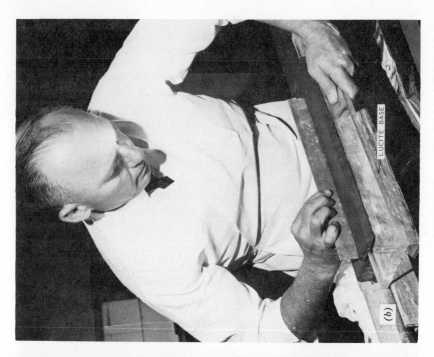

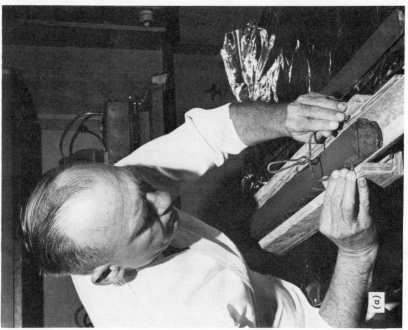

Fig. 11 Embedment and preparation of core slabs for mounting (from Shannon and Lord, 1967). (*a*) Cutting sediment core slab. (*b*) Transferring core slab to transparent lucite base. (*c*) Removing cured plastic mount from tray. (*d*) Cutting unfinished plastic mount.

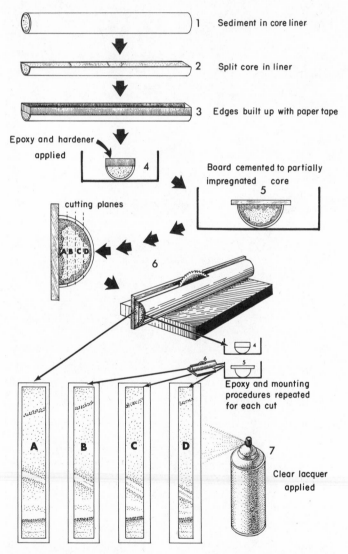

1   Sediment in core liner

2   Split core in liner

3   Edges built up with paper tape

Epoxy and hardener applied

4

Board cemented to partially impregnated core

5

cutting planes

A B C D

6

4

5

Epoxy and mounting procedures repeated for each cut

A   B   C   D

7

Clear lacquer applied

Fig. 12   Sketch showing technique for partially impregnating and mounting core slabs. Core sections shown in Fig. 14 were prepared in this fashion.

Slabbing and Finishing Core Mounts for Study and Display

Short sections of impregnated cores (Shell method) and partially impregnated cores of the type described in the previous section can be slabbed with a standard diamond saw, whereas longer core sections can be cut with a stone saw mounted on rails (Fig. 13) and rigged for automatic feed. Slabs ½ cm thick or greater can be made. If the resin has not been cured completely, the slab can be placed in the sun or a low-temperature oven to complete the hardening process. The slabs can be mounted on wood for permanent storage and handling (Fig. 14). The sawed surface can be lapped.

Fig. 13   Stone saw on rails. The saw, Type SS16, Stone Machinery Company, Manlius, New York, runs on rails as shown. The saw is pulled forward by a 100-kg block of cement as shown, and a simple boat trailer winch, at left, is used to pull the saw and weight back. (From Ginsburg et al., 1966)

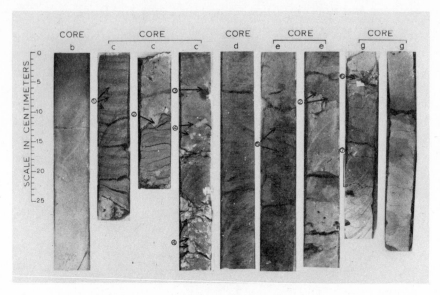

Fig. 14 Selected impregnated core slabs prepared for display (from Stanley, 1968) using method illustrated in Fig. 12. Sections of cores from the Gully submarine canyon off Nova Scotia showing mud, silt, sand, and gravel layers.

Multiple or serial sections can be prepared by cutting the core lengthwise into several thin (½ to 1 cm) sections (Fig. 12). Serial slabs are useful in studying the three-dimensional geometry of sedimentary and biogenic structures. They lend themselves well to radiographic examination (Stanley and Blanchard, 1967). A coating of clear varnish or lacquer produces a wet appearance that helps bring out subtle structures.

Core sections encased in plastic (Shannon and Lord, 1967) can be sawed with a band saw (Fig. 11d), and then finished with a cutting tool in a milling machine. Lapping on a felt lapwheel using a slurry of water and aluminum oxide produces a high polish on the mounts (Fig. 15).

If necessary, the core sections can be sampled after impregnation. Plugs of consolidated core material, for instance, can be drilled with a 1-in. hole saw and drill press to cut and punch out the plug cylinder (Ginsburg et al., 1966). Pieces of impregnated core sections can also be used to make thin-sections.

Fig. 15 Milled and polished plastic mounts prepared by Shannon and Lord, 1967. A. Clay and silt layers. B. Clay and silt lenses. C. Sand.

ACKNOWLEDGMENTS

I would like to express my appreciation to the following persons: Messrs. J. E. Merida, Department of Paleobiology, G. C. Moreland, Department of Mineral Sciences, and J. C. Widener, Office of

Exhibits (all at the Smithsonian Institution, Washington, D. C.) for kindly discussing various aspects of impregnation and embedment, demonstrating techniques, some of them as yet unpublished, and for critically reviewing the manuscript; Mr. L. B. Isham, Smithsonian Institution, for helping with the illustrations; Drs. R. J. Cavanaugh, Continental Oil Company, J. Etienne, Société Française de Recherches et d'Exploitation de Pétrole, R. N. Ginsburg, University of Miami, and J. P. Shannon, Esso Production Research Company, for providing selected photographs from their published papers to be used in this chapter; and Dr. G. Kelling, University of Wales at Swansea, Dr. G. de V. Klein, University of Illinois, Urbana, and Mrs. P. R. Brenner, Smithsonian Institution, for suggesting improvements of the text.

REFERENCES

Arthur, M. A., 1949, Method of preserving wet, fragile geologic specimens, *J. Sed. Pet.*, **19**, 131–134.

Brown, W. E., and H. W. Patnode, 1953, Plastic lithification of sands *in situ, Bull. Am. Assoc. Petrol. Geologists,* **37**, 152–162.

Cavanaugh, R. J., and C. F. Knutson, 1960, Laboratory technique for plastic saturation of porous rocks, *Bull. Am. Assoc. Petrol. Geologists,* **44**, 628–640.

Davies, P. J., and R. Till, 1968, Stained dry cellulose peels of ancient and recent impregnated carbonate sediments, *J. Sed. Pet.,* **38**, 234–237.

Etienne, J., 1963, Technique d' imprégnation de roches par des résines colorées pour l' étudé de la porosite en lame mince, *Rev. Inst. Français Pétrole et Annales Combustibles Liquides,* **18**, 611–619.

Forbes, D., 1869, On the preparation of rock sections for microscopic examination, *Monthly Microscopical J.,* **1**, 240–242.

Gendron, N. J., 1959, Mounting of geological specimens in clear, cold setting plastic, *Econ. Geol.,* **54**, 308–310.

Ginsburg, R. N., H. A. Bernard, R. A. Moody, and E. E., Daigle, 1966, The Shell method of impregnating cores of unconsolidated sediments, *J. Sed. Pet.,* **36**, 1118–1125.

Johannsen, A., 1918, Manual of petrographic methods, McGraw-Hill Book Co., pp. 599—602.

Leggette, M., 1928, The preparation of thin sections of friable rocks, *J. Geol.,* **36**, 549—557.

Lockwood, W. N., 1950, Impregnating sandstone specimens with thermosetting plastics for studies of oil-bearing formations, *Bull. Am. Assoc. Petrol. Geologists,* **34**, 2061—2067.

McGraw-Hill, Inc., 1966, Modern plastics encyclopedia, 1967, *Modern Plastics,* **44**, 1262 p.

Merida, J. E., and R. S. Boardman, 1967, The use of Paleozoic Bryozoa from well cuttings, *J. Paleon.,* **41**, 763—765.

Milner, H. B., 1962, Sedimentary petrography, methods in sedimentary petrography, 2nd ed., Vol. I, The MacMillan Co., 643 pp.

Mitchell, J. K., 1956, The fabric of natural clay and its relation to engineering properties, *Proc. 35th Ann. Highway Res. Board,* **35**, 693—713.

Moreland, G. C., 1968, Preparation of polished thin sections, *Am. Mineralogist,* **53**, 2070—2074.

Müller, German, 1967, Methods in sedimentary petrology, Hafner Publishing Co., pp. 44—49.

Nuss, W. F., and R. L. Whiting, 1947, Technique for reproducing rock pore space, *Bull. Am. Assoc. Petrol. Geologists,* **31**, 2044—2049.

Reineck, H. E., 1963, Nafshartung von ungestorten Bodenproben im Format 5 X 5 cm fur projizierbare Dickschliffe, *Senckenbergiona lethaea,* **44**, 357—362.

Ross, C. S., 1924, A method of preparing thin sections of friable rocks, *Am. J. Sci.,* **207**, 483—485.

Schlossmaker, K., 1919, Ein verfahren zur Herrichtung von Schiefrigen und lockeren Gesteinen Zum Dunnschleifen, *Zentralblatt fur Min. Geolgie Paleon.,* 190—192.

Shannon, J. P., and C. W. Lord, 1967, Preservation of unconsolidated sediment cores in plastic, *J. Sed. Pet.,* **37**, 1200—1203.

Sorby, H. C., 1882, Preparation of transparent sections of rocks and minerals, *The Northern Microscopist,* No. 17, 101—106; No. 18, 133—140.

Stanley, D. J., 1968, Comparing patterns of sedimentation in some modern and ancient submarine canyons, *Earth and Planetary Sci. Letters,* 3, 371–380.

Stanley, D. J., and L. R. Blanchard, 1967, Scanning of long unsplit cores by x-radiography, *Deep-Sea Res.,* 14, 379–380.

Tourtelot, H. A., 1962, Thin sections of clay and shale, *J. Sed. Pet.,* 52, 131–132.

Waldo, A. W., and S. T. Yuster, 1937, Method of impregnating porous materials to facilitate pore studies, *Bull. Am. Assoc. Petrol. Geologists,* 21, 259–267.

CHAPTER 10

PEELS AND IMPRESSIONS

===

GEORGE deVRIES KLEIN

University of Illinois, Urbana, Illinois

A continuing problem for the field and laboratory sedimentologist has been the need to preserve, by surface impregnation, cores and trench cuts of unconsolidated sediments and sedimentary structures. A variety of techniques has been devised, particularly since 1960, to make relief peels of these exposures. Sediment impregnation per se is covered in Chapter 9 by Stanley. This chapter reviews field and laboratory techniques for making relief peels by surface impregnation.

Relief peels of unconsolidated sediments have been made for several purposes. Peels are impregnated surface samples of sediments showing sedimentary structures and other large-scale features, and are therefore both outcrop samples and replicas which can be brought to the laboratory for study. Many measurements and observations that are commonly made in trench cuts in the field can therefore be deferred to the laboratory. In this way peel-making increases the efficiency of field work by increasing the time available to make observations that were formerly made only in the field. Relief peels permit rapid comparisons among texture, fabric, and sedimentary structures. As replicas they can be studied and restudied

217

at the convenience of the investigator. Peels can be photographed under controlled light conditions, whereas trench cuts must be photographed under natural light conditions that are favorable for only short periods of time. A peel can also be used for repeated demonstrations and discussions, thus increasing the availability to students of the field characteristics of Recent sedimentary structures.

Development of peel-making technique has come from both paleobotanists and sedimentologists. Paleobotanists pioneered the technique more than 40 years ago by making peels from etched and polished slabs of coal balls containing plant material. Peel-making was found to be an efficient way to study the cellular structure of fossil plants (review by Stewart and Taylor, 1965) and to aid plant identification. Carbonate petrologists adopted similar techniques (Buehler, 1948) to make replicas of polished and etched surfaces of limestones 20 years after Walton (1928) developed the peel method for studying fossil plants. Peels of carbonate rocks were used to study both the texture and fabric of carbonate particles.

This chapter summarizes methods for making peels of unconsolidated sediments and sedimentary rocks. Historical accounts of peel preparation are presented by Stewart and Taylor (1965), McCrone (1963), and Muller (1967).

RELIEF PEELS IN UNCONSOLIDATED SEDIMENTS

The principle of making relief peels in unconsolidated sediments depends on differential penetration by a compound that binds the sediment and cements it into an artificial sedimentary rock. Differential penetration is governed by differences in permeability which are dependent on nonuniform changes in particle size, sorting, and packing. The binding compound penetrates the sediment to a variable distance from the surface in response to these permeability differences. The resulting impregnated membrane, when removed, shows relief, particularly of stratification.

Many binding materials have been used to make peels from unconsolidated sediments, including lacquer, glue, latex, polyester resin, and epoxy. Lacquer peels were used by Voigt (1936, 1949), Van Straaten (1954, 1959, 1965), and Shell Research NV, The

Hague (1966; Dielwart, 1962), and glue by Heezen and Johnson
(1962). McKee (1966; personal communication, 1968) has used latex
to make peels since 1963. Resins have been used by Reineck (1957,
1958a, 1958b, 1960, 1961, 1962), Bouma (1964), McMullen and
Allen (1964), Maarse and Terwindt (1964), Imbrie and Buchanan
(1965), and Moiola et al. (1969), whereas epoxy peels have been
described by Burger et al. (1969) and Barr et al. (1970).

PREPARATION OF SURFACES FROM WHICH
PEELS ARE TO BE MADE

Surfaces of outcrops from which peels are to be made require a
minimum of preparation. If the peels are to be taken from a trench
wall, the surface is smoothed with a trowel so that it is nearly planar.
Some methods also require that strips of cheesecloth be placed over
the trench face and held in place with matches, nails, or L-shaped
wires. Primer coats are applied by pouring and respreading with a
paintbrush. After drying is completed, the peel can be removed from
the trench wall by placing a plywood board against it and pulling the
board and cheesecloth away from the trench wall. In other methods,
masonite boards are used as a supporting base.

If peels are to be made from core samples, the core must be split
by conventional means, such as cutting the core liner and pulling a
piano wire down the split seams. The split core should be placed in
"D" tubing and the surface must be oriented to the horizontal before
preparation of the core and application of impregnating material.
The surface is smoothened by a knife, and then cheesecloth is placed
over the surface if the method requires it. Binding compounds are
applied next. After drying the peel is removed by hand. In other
methods the binder is poured first and a masonite board is mounted
later.

When cheesecloth is used as a supporting mat for the peel, the peel
can be mounted permanently on a board by gluing the back side
onto a board. After gelling and hardening, the relief peels can be
exposed and cleaned with a dry paint brush, an air jet, or by washing
with water. In some instances, the relief peel can be sprayed with
clear Krylon for added rigidity and protection.

Lacquer Peels

Trench Samples

Lacquer peels from unconsolidated sediments were initially
made by Voight (1936, 1949). Peels were made from materials
collected in box cores, split cores, and soil-sampling trays. The
binding materials are summarized in Table 1.

TABLE 1   Materials and suppliers, lacquer peels (Voight method)

"Special Impregnation Lacquer" X 4/924	Temporal Co., Hamburg, Wandsbek, West Germany
Profieelak	Fabriek van Industriale Lackproducten "Fil,"
Cellulosverdunning V 105	Zeist, Netherlands
Note: Materials must be ordered from Europe; no United States supplier known.	

The surface to be impregnated and preserved as a peel should be
air-dried and sprayed with lacquer diluted in a thinner (dilution ratio
of four parts thinner to one part lacquer). After the thinner
evaporates, thin layers of undiluted lacquer are applied over the
surface with a paintbrush. Additional layers of lacquer are applied
until the surface is smooth. A layer of cheesecloth is then spread over
the lacquer peel before application of the last layer of lacquer. The
last layer impregnates the cheesecloth to give the peel added rigidity.
   Van Straaten (memo to author, May 1, 1958; see also 1954, 1959,
1965) refined Voight's method to account for differences in
sediment texture. Van Straaten reported that the ratios of lacquer
and thinner proposed by Voight gave excellent results when making
peels in sand. Peels from clayey materials required the application of
a diluted lacquer (four parts thinner and one part lacquer). This
diluted lacquer is poured over the clayey sediments at least three
times. The second application should occur 3 hours after the first
application, whereas the third application should occur 24 hours
after the second application.
   The peels are removed a day after applying the last lacquer coat.
The peel is then mounted permanently with glue to a board.

Shell Research NV, The Hague, has improved the Voight method for both field and laboratory use. Their binding materials are summarized in Table 2. Peels are taken from trench walls (Fig. 1), the surface being dried with a propane torch. Gauze bandages (cheesecloth is satisfactory) are hung loosely over the prepared surface and fastened with matches or nails (Fig. 2). A prime coat is applied by pouring and brushing. The lacquer is applied from the base of the bandages and is brushed upward (Fig. 3). The peel is left to dry overnight; then a coat of thick lacquer (four parts lacquer and

TABLE 2   Materials for lacquer peels (Shell Research NV method)

Cellulose Lacquer "Profielak"	Peter Schoen NV
	Zaandam, Netherlands
Thinner (cellulose Verdunning V-105)	
Krylon clear plastic spray #1303	Krylon Corp.
	Norristown, Pennsylvania

Note: Lacquer and thinner must be ordered from Europe; no United States supplier.

Fig. 1   Shell Research NV, The Hague, lacquer peel technique, Step 1. Using shovel to smoothen surface from which peel is to be made.

Fig. 2   Shell Research NV, The Hague, lacquer peel technique, Step 2. Hanging gauze bandages over surface from which peel is to be made.

one part thinner) is applied (Fig. 4). If the peel is made from clayey sediment, a second thick lacquer coat is applied. The peel is then left overnight to dry.

The peel is removed on the third day with the aid of a plywood board. The board is placed against the peel and the edges of the bandages are pulled over the top of the board. The bandage is kept tight, and the board and peel are removed carefully (Fig. 5). The complete peel is kept flat against the plywood board (Fig. 6). The peel must remain untouched at this stage. The peel can be trimmed and mounted on plywood using thick lacquer.

A day after mounting the peel on the board, excess loose particles can be removed with a jet of compressed air, but this must be done only after the peel is completely hardened.

Split-Core Samples

Similar procedures are required for making peels from split cores, as outlined by Dielwart (1962). The core is extruded (Fig. 7) into a

Fig. 3 Shell Research NV, The Hague, lacquer peel technique, Step 3. Application of lacquer with a brush, primary coat.

Fig. 4 Shell Research NV, The Hague, lacquer peel technique, Step 4. Application of second coat of lacquer. Note brushing on of lacquer from base of peel area upward.

Fig. 5 Shell Research NV, The Hague, lacquer peel technique, Step 5. Removal of lacquer peel with aid of plywood board.

Fig. 6 Shell Research NV, The Hague, lacquer peel technique. Complete lacquer peel showing alternation of sand and clay.

"D" tube and covered with another "D" tube. A stainless steel wire is drawn between the two "D" tubes (Fig. 8) and the two half-cores are separated (Fig. 9). The surface of the split core is smoothed with a knife (Fig. 10) so that the sedimentary structures are more visible. Damp clay should be lacquered 2 to 3 hours after the core is opened, but sands should be lacquered immediately. Slow drying is recommended in clay cores to prevent cracking. Wet sands can be dried adequately with filter paper in 15 to 30 minutes.

Lacquering split cores involves two steps: (a) initial priming and (b) subsequent application of thick layers of lacquer. Table 3 summarizes the ratios of lacquer to thinner for the prime coats and the number of coats needed. Lacquer and thinner can be mixed ahead of time, and should be left to stand so that air bubbles can escape.

The prime coat should always be poured onto the prepared surface (Fig. 11). A soft brush can be used to apply the second and subsequent coat. Brushes should *never* be used for the prime coat in order to avoid dislodging sediment and distorting sedimentary structures.

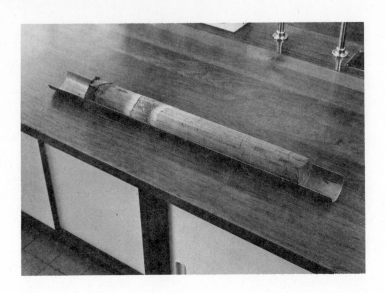

Fig. 7 Shell Research NV, The Hague, split core lacquer peel technique, Step 1. Extruded core in semicylindrical tube.

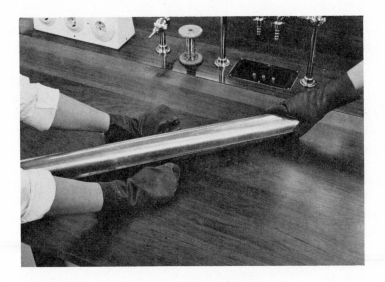

Fig. 8 Shell Research NV, The Hague, split core lacquer peel technique, Step 2. Drawing of stainless steel wire to split core.

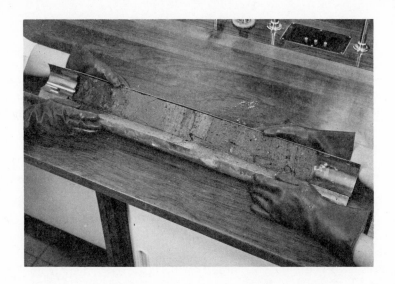

Fig. 9   Shell Research NV, The Hague, split core lacquer peel technique, Step 3. Separation of split core.

Fig. 10   Shell Research NV, The Hague, split core lacquer peel technique, Step 4. Smoothening of core surface with a knife.

TABLE 3   Ratios of lacquer to thinner: prime coat
(Dielwart, 1962, p. 2)

Sediment	Parts Lacquer by Volume	Parts Thinner by Volume	Number of Coats
Clay	1	2	2
Sand	1	1	1
Coarse or dry sand	2	1	1

After the prime coat dries, the core is covered with a strip of cheesecloth as wide as the core and 10 cm longer than the core. Thick lacquer (five parts lacquer and one part thinner) is applied by brush (Fig. 12). Sand requires only one thick coat, whereas clay requires two coats. Drying time for the last coat ranges from 1 day for sand to 3 days for clay.

Fig. 11   Shell Research NV, The Hague, split core lacquer peel technique, Step 5. Pouring of the priming coat.

Fig. 12 Shell Research NV, The Hague, split core lacquer peel technique, Step 6. Application of thick lacquer with a brush.

The peel is removed by freeing the edges with a knife (Fig. 13). It is then mounted on a plywood board with thick lacquer or a quick-drying plastic glue. Two hours after mounting on a board, the excess sediment is blown off with compressed air. The peel is sprayed with Krylon after the peel is completely dry. Cellulose-based spray should *not* be used because it will dissolve the cellulose lacquer and destroy the peel.

Glue Peels

Heezen and Johnson (1962) proposed one of the simplest methods for making peels (Table 4). A surface of a split core or box core is smoothed and leveled. A thin layer of Elmer's Glue-All (1/8-in. thick) is spread over the surface and left to stand for 10 to 15 minutes. Cotton muslin is placed gently over the glue, and another layer of glue is added and left to stand for 24 hours. The second glue layer must impregnate the muslin. Peels made in silts and clays require dilution of glue with water (Table 5).

TABLE 4   Materials and suppliers, glue peels

Elmer's Glue-All	Borden Co., New York
Krylon Crystal Clear Spray #1303	Krylon Corp., Norristown, Pennsylvania

TABLE 5   Glue-water ratios needed to impregnate different sediment types
(after Heezen and Johnson, 1962)

Sediment Type	Glue-Water Ratio	Drying Time before Applying Glue
Hard dry clays	4:1	0
Hard dry silts	6:1	0
Hard dry sands	Undiluted	0
Diatomaceous oozes and gravel	Undiluted	0
Wet clays	4:1	3—4 hours
Wet silts	6:1	3—4 hours
Wet sands	5:1	3—4 hours

Fig. 13   Shell Research NV, The Hague, split core lacquer peel technique, Step 7. Freeing peel edges from core with a knife.

Peels can be removed with a putty knife after 24 hours (Fig. 14). Excess cloth is trimmed away and the peel sprayed with clear Krylon. Three coats of Krylon are sprayed with 5 minutes of drying time between coats. Although flexible, the peel should be glued to a board for protection.

Diluton of glue with water is needed to assure impregnation of silts and clays (Table 5).

Latex Peels

McKee (1966, p. 9; personal communication, 1968) has made peels successfully using latex cement (Table 6). The following discussion is taken from a memorandum McKee circulates to visitors to his laboratory.

A smooth, even surface is cut on the face of a trench or is exposed on a box core or split core. Three or four coats of clear Krylon are

Fig. 14 Shell Research NV, The Hague, split core lacquer peel technique, Step 8. Removal of peel from core by hand.

sprayed on the surface. Five minutes drying time is required between each coat. When the surface is firm to the touch, liquid latex is brushed on the sprayed surface. A soft, wide brush is recommended. The latex should be reinforced with a large piece of cheesecloth or strips of cheesecloth. Additional latex is applied by brushing. At weak spots additional strips of cheesecloth can be added and brushed in with latex to reinforce the peel.

TABLE 6   Materials and suppliers for latex peels

Cementex #600 latex cement	Cementex Co., New York
Krylon Crystal Clear Spray #1303	Krylon Corp., Norristown, Pennsylvania

A second layer of latex is applied after the first layer is dry. On a hot, dry day, or in direct sunlight, drying requires at least ½ hour. The color of the latex changes from white to yellow-brown on drying. After applying two or three additional layers, the peel is left to dry for several hours, or as long as overnight. The peel is removed from the surface and the front side is exposed to the air to assure complete drying. A thin coat of Krylon should then be applied after the peel surface has air-dried for 2 hours. The peels can be stored flat or mounted on a board by gluing the back side. It must be emphasized that the brushes should be cleaned in plain water immediately after use or they will be completely ruined.

If the peel is made on dried mud samples, it is recommended that the peel be obtained from a box sampler or split core. The sample should be placed face-up and horizontal. Latex is diluted in water to a consistency of milk and is then brushed on with a paintbrush. Cheesecloth is added to reinforce the peel. Three to five diluted coats are applied. Each coat must be air-dried before applying the next coat. One to three coats of undiluted latex are then applied to reinforce the peel. The peel can be removed as soon as cured to a brown color. The front side will be tacky and needs to be air-dried for several days. Figures 15 and 16 are examples of latex peels.

Resin Peels

Maarse and Terwindt (1964) proposed making peels with polyester resins, rather than with lacquer. Their method is here referred to as the "Vestanox" method (Table 7). Two types of mixtures are suggested for making peels of clay, sand, and gravel (Table 8).

TABLE 7 Materials and suppliers, "Vestanox" peels

Vestanox H	Chemischer Were Huls AG, Marl, West Germany
Monostyrene	
Butanox	
Cobalt octoate	

Note: Materials must be ordered from Europe; no United States supplier known.

Peels are made from box cores, split cores, or trench cuts. Muslin is spread over the face or core and either Vestanox A or Vestanox B (Table 8) is poured over the face. After the appropriate gelling times, the peel can be removed with a knife and air-dried. After air-drying, the peel is washed to remove any excess sediment.

Fig. 15 Latex peel of foreshore beach sediments, Joulters Key, east side of Andros Island, Bahamas, BWI.

Fig. 16 Latex peel of backshore beach sediments, Island Beach State Park, New Jersey.

TABLE 8   Mixing ratios for "Vestanox" plastics
(Maarse and Terwindt, 1964)

	Parts Vestapol H	Parts Styrene	Parts Butanox	Parts Cobalt Octoate	Gelling Time	Drying Time	Sediment
Vestanox A	60	40	2	1	1–2 hours	5–6 days	Clay
Vestanox B	85	15	4	2	½–1 hour	2–3 days	Sand or gravel

Mixtures of Vestapol H and styrene can be prepared in advance in the laboratory if the field conditions are known. Butanox can be added up to 2 hours before making the peel. The cobalt octoate must be added at the last minute prior to pouring.

McMullen and Allen (1964) used polyester resins manufactured by Bakelite, Ltd., to make peels. Their method applies both to the field and the laboratory. Using Bakelite resins and catalysts (Table 9), they found that a mixture of 100 gm of resin, 4 ml of catalyst, and 2 ml of accelerator yielded satisfactory peels. The resins and catalysts can be mixed ahead of time before taken in the field. The accelerator must be added in the field shortly before application of the mixture.

TABLE 9   Materials and supplier, Bakelite Polyester Resin Peels
(McMullen and Allen, 1964)

Bakelite Polyester Resin SR 17497 or SR 19038	Bakelite Ltd., 12-18 Grosvener Gardens,
Catalyst Q 17447	London SW 1, England, U.K.
Accelerator Q 17448	

Note: Materials must be ordered from U.K.; no United States supplier known.

Peels are made by either pouring the resin mixture onto the prepared surface or by spraying the surface with a diluted solution from an insecticide spray-gun connected to a tank of compressed air. The resin solution is diluted by adding 20% by volume of monomeric styrene. The surface to be impregnated is leveled horizontally, and cheesecloth, larger than the surface, is placed over it. The resin mixture is poured over the cheesecloth and left to gel and dry. The peel is removed by hand and is placed, muslin-side down, on a flat surface for air-hardening for an additional 24 to 36 hours. The peel is then washed to remove excess sediment.

Deeper penetration is achieved by adding 20% by volume of monomeric styrene to the resin mixture. The added styrene, however, increases the gelling time.

Drying time varies with temperature and wind conditions. The most rapid gelling and drying occurs in temperatures in excess of 21°C and still-air conditions. Under cooler, breezy conditions (less than 15°C), peels can be obtained, but drying takes twice as long. Bright, sunny conditions accelerate gelling. In the laboratory, an infrared lamp also aids gelling.

Reineck (1957, 1962) and Bouma (1964) described a method for making peels from box cores and split cores (Table 10). Samples are taken into the laboratory and heated for 24 hours at temperatures of 100 to 110°C (212 to 230°F). After the sample is dried, Casting Resin F is mixed in a container with Hardener #902 by stirring at a temperature of 80 to 100°C. The resin mixture clears after 5 to 10 minutes of heating and stirring. At this stage the box core is removed from its oven and a piece of muslin is placed over the sediment surface. The resin mixture is poured over it and placed in an 120°C oven for 36 hours. The sample is cooled, and then the peel is separated with a knife. Excess sediment can be removed with washing. Similar operational instructions apply when using casting resins with Hardener #951. A curing time, at room temperature, of 24 to 36 hours, is required. If the sample is moist, good results can be obtained using Furane Epocast #530-2A and Hardener #530-2B. However, dry samples are recommended to assure excellent results. Ratios of proportions of resin and hardener are summarized in Table 11.

TABLE 10 Materials and suppliers for Casting Resin F peels

CIBA Araldite Casting Resin F	CIBA Chemical Co.,
CIBA Hardener #902	Fairlawn, New Jersey
CIBA Hardener #951	
or	
Epocast #530-2A	Furane Plastics, Inc.,
Epocast #530-2B	4516 Brazil Street,
	Los Angeles, California

TABLE 11   Mixing ratios of resin and hardener for resin peels
(after Bouma, 1964, p. 351)

Sediment Type	Araldite Resin F Hardener #902	Araldite Resin F Hardener #951	Epocast #530-2A: #530-2B
Sand	20:15	61:9	27:9
Clay	29:21	44:6	75:25

Imbrie and Buchanan (1965, pp. 150–151) described briefly a method for making peels of Recent carbonate sediments with a mixture of CIBA Araldite #502 resin and CIBA DP-118 Hardener. This mixture is poured over a horizontally oriented surface of a box core while the sample is still wet. The resin impregnates the sediment rapidly and hardens within 2 days.

TABLE 12   Materials for resin peels
(Imbrie and Buchanan, 1965)

CIBA Araldite Resin #502	CIBA Chemical Co., Fairlawn, New Jersey
CIBA DP-118 Hardener	

Before complete hardening, a masonite board, 1/8 in. thick, is cemented with resin to the impregnated surface. After the peel hardens, the board and attached sediment are separated from the box core and excess sediment is washed off with water, leaving an impregnated peel adhered to the masonite board.

A rapid field method for making resin peels was recently described by Moiola and others (1969). Trench walls from which peels were taken were smoothed with a machete and covered with cheesecloth held in place with L-shaped wires. The binding material consisted of a mixture of DuPont Elvacite 2044 dissolved in acetone (Table 13). To make the binder, 1 lb of Elvacite 2044 was mixed with sufficient acetone to make 1 gal of solution. The Elvacite solution was applied to the cheesecloth with a paintbrush. Two coats can be applied 10 minutes apart. The peel dries in 3 hours and is then removed by leaning a plywood board against the cheesecloth and pulling both the board and the cheesecloth away from the trench wall. The peels are mounted on a plywood board with Formica contact cement. The peel surface is sprayed with a clear varnish sealer or Krylon. Figs. 17 and 18 are examples of peels made from Elvacite solutions.

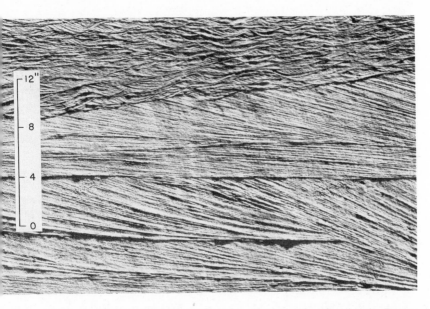

Fig. 17   Elvacite 2044 acrylic resin peel of trough cross-stratification in Recent
Point Bar. (From Moiola and others, 1969. Republished by permission
of the Geological Society of America)

TABLE 13   Materials and suppliers for Elvacite peels
(Moiola et al., 1969)

| Elvacite #2044 | DuPont De Nemours Co., Wilmington, Delaware |
| Acetone | Commonly available |

Fig. 18   Elvacite 2044 acrylic resin peel of cross-stratification in Recent Point
Bar. Height of peel is 12 in. (From Moiola and others, 1969. Republished
by permission of the Geological Society of America)

Epoxy Peels

Burger and others (1969) have used epoxy to make peels from samples collected with a box corer or from split cores (Table 14)

TABLE 14   Materials for epoxy peels
(Burger and others, 1969)

Epoxy: CIBA #6005	CIBA Chemical Co., Fairlawn, New Jersey
CIBA #6010	
Hardener: CIBA #850	
CIBA #830	

The epoxy-hardener mixture is prepared as follows. Five parts each of Epoxy #6005 and Epoxy #6010 are mixed and then four parts of Hardener #850 are added. One part Hardener #830 is immediately added. After thorough mixing, the epoxy-hardener mixture is poured over the face to be impregnated. A thin coating of epoxy-hardener mixture is put on one side of a masonite board, which is then placed face down on the poured surface. The entire core and the impregnated face and board are covered with sand and left to stand for 2 hours, by which time the impregnated face will have gelled. The sand cover is removed, and the entire box core is turned over. The corer is then removed, leaving a wedge-shaped pile of sediment on the peel. The peel and sand are immersed and washed in water so that all unconsolidated sediment is removed. After washing, a peel is left on the masonite board, and the internal stratification is emphasized by differences in depth of impregnation (Figs. 19, 20 and 21). Exposure of the tacky surface to the sun hardens the peel completely in 2 hours. Under cool and cloudy conditions, hardening takes 4 to 6 hours. The method works equally well on samples that are saturated with either fresh or saltwater, as the density of the epoxy drives out interstitial water.

Impregnation depth is a function of texture and sorting. Impregnation of clay and silt averages 1 mm, whereas silty-sand impregnation ranges from 1 to 3 mm. Fine sand is impregnated to a depth averaging 4 mm, whereas mixtures of coarse sand and granules are impregnated to depths ranging from 1 to 2.3 cm. Impregnation of cobble-sized gravel averages 4 cm.

Fig. 19 Epoxy peel of cross-stratified, fine- and medium-grained sand, Economy
Point, Nova Scotia, Canada. (Republished by permission of the Society
of Economic Paleontologists and Mineralogists and the Journal of
Sedimentary Petrology)

Gelling time is controlled by temperature and humidity. Rapid
gelling (less than 1 hour) occurs on hot, dry days (in excess of 80° F).
Up to 3 hours are required if the temperature is only 50° F. The
method is not recommended if the temperature is less than 45° F.

Barr and others (1970) have modified both the procedures and the
proportions of epoxy and hardener to take peels from trench walls.
To make such peels, three masonite boards are required, one with
dimensions of 1 ft by 4 ft and the other two with dimensions of 1 ft
by 6 in. After the trench wall is smoothed with a trowel, the base of
the large masonite board is placed in contact with the trench wall.
The two smaller masonite boards are mounted at right angles at the
ends of the large board. The side boards are needed to retain epoxy
mixture after pouring. The large masonite board is tilted outward

Fig. 20 Epoxy peel of cross-stratified, coarse-grained sand, and granule-size gravel, Big Bar, Five Islands, Nova Scotia, Canada. (Republished by permission of the Society of Economic Paleontologists and Mineralogists and the Journal of Sedimentary Petrology)

approximately 45° from the vertical trench wall (Fig. 22). Sand is then tamped behind the base of the tilted board and epoxy is poured over the lower half and allowed to flow toward the base. Once the epoxy has an equal distribution along the face, the board is tilted slowly toward the trench face. The epoxy level also rises during this tilting toward the trench wall. Sand is continually tamped against the board for support. When the epoxy level reaches 2 in. below the trench top, the board should be nearly 1 in. away from the trench wall. The fluid is allowed to stand for five to 10 minutes, permitting impregnation of the face. More epoxy can be added if needed. The peel board is then pushed flush against the trench face, and more sand is tamped against it to hold the board in place.

The epoxy will cure in 2 hours, although it will still be tacky. If time permits, a 3-hour curing period produces better results (Figs. 23

and 24). After the epoxy has gelled, the peel is removed by excavating the sand behind the peel board. The peel is pulled from the trench wall and washed with water to remove excess sand. The peel should then be stored on a flat surface for 12 hours to permit complete hardening.

The ratio of epoxy to hardener that has proved most successful in making these large peels is four parts each of Epoxy #6005, Epoxy #6010, and Hardener #850, and two parts of Hardener #830. The addition of extra Hardener #830 increases the gelling time and does not decrease the depth of penetration. Two quarts of epoxy mixture are adequate to impregnate fine- and medium-grained sand, whereas 3 to 4 quarts are required to impregnate a surface, 1 ft by 4 ft, of coarse-grained sand and granule gravel.

Fig. 21 Epoxy peel from cross-stratified, coarse-grained sand, and granule gravel, Pinnacle Flats, Five Islands, Nova Scotia, Canada.

Fig. 22 First state of epoxy-pouring in making large epoxy peels of trench wall. Operator has tilted large board 45° outward and is pouring epoxy over lower part of board. Materials shown are large board, two side boards to prevent epoxy from leaking, trowel, and epoxy in plastic tubs. Scale in decimeters and centimeters. (After Barr and others, 1970. Republished by permission of the Society of Economic Paleontologists and Mineralogists and the Journal of Sedimentary Petrology)

Fig. 23 Peel taken normal to crest of large dune with superimposed current ripples, Big Bar, Five Islands, Nova Scotia, Canada. Scale in decimeters and centimeters. (After Barr and others 1970. Reproduced by permission of the Society of Economic Paleonotologists and Mineralogists and the Journal of Sedimentary Petrology)

Fig. 24   Peel taken parallel to dune crest, at right angles to peel shown in Fig. 23.
Scale in decimeters and centimeters. (From Barr and others, 1970.
Reproduced by permission of the Society of Economic Paleontologists
and Mineralogists and the Journal of Sedimentary Petrology)

PEELS OF ANCIENT SEDIMENTARY ROCKS

Peels of etched, polished surfaces of limestones have been made by sedimentary petrologists since the technique was used by Sternberg and Belding (1942) and by paleobotanists since the technique was developed by Walton (1928) and improved by Graham (1933) and Fenton (1935). Acid etching of a limestone will bring out, in microrelief, textural and fabric features of a rock that respond differentially to the acid etching. The peel is an impression of the microrelief of the etched surface.

Since first used modifications and reviews have been published by Buehler (1948), Ives (1954), Bissell (1957), Beales (1960), Lane (1962), McCrone (1963), Stewart and Taylor (1965) and Müller (1967). Katz and Friedman (1965) and Friedman (Chapter 22) have improved on the earlier techniques by staining carbonate rocks to identify different mineral phases before making acetate peels. Differences in mineralogy are then indicated by staining of the peels.

Nitrocellulose Peels

Paleobotanists have used nitrocellulose to make peels of polished and etched surfaces of coal balls since the 1930's (Graham, 1933; Fenton, 1935). The advantages of making these peels are low cost and utility. Since preservable plant debris cannot be removed by etching, nitrocellulose peels (Table 15) were used to bring out the internal cellular structures of fossil plants to aid in identification.

TABLE 15   Materials used for nitrocellulose peels

15/20 Parlodion
Butyl acetate (commercial grade)
Tri-cresyl phosphate
Toluene or xylol

The procedure for making nitrocellulose peels is simple. A rock face is cut and ground on a glass plate with No. 600 grinding powder. After it has been washed and dried, the face is etched in 5% hydrochloric acid for 30 seconds. The sample is then washed and dried and placed in a sandbox with the etched face oriented upward and horizontal.

The peel solution is prepared by mixing 20 gm of Parlodion with 200 cc of butyl acetate. One cubic centimeter of tri-cresyl phosphate and 10 to 20 cc of toluene or xylol are added. This mixture can be stored in a widemouthed stoppered bottle and used as needed. Because tri-cresyl phosphate is *poisonous,* it must be used with caution.

The peel solution is poured over the horizontally oriented, etched rock surface and left to stand for 72 hours. It can then be removed by peeling off by hand. If the peel tears, the rest of it can be removed with a razor blade. The peel is mounted between glass and can be studied with a microscope, or a negative peel print can be made photographically.

Acetate Peels

Peels of etched surfaces of carbonate rocks are made to preserve and emphasize differences in texture, fabric, and packing of carbonate rocks. Acetate peels (Table 16) are inexpensive to make and are used by carbonate petrologists as a rapid means of determining variation in fabric and texture. Peels of carbonate rocks have proven to be popular because some of the features of carbonate rocks stand out more in peels than in thin sections. Peels and thin sections of carbonate rocks therefore supplement each other. Peels also can be stored easily and can be projected on a screen as a lantern slide. Peels can be placed in an enlarger and exposed on high contrast photographic paper to make negative prints (Figs. 25 and 26).

TABLE 16   Materials for acetate peels

Fine abrasive (No. 1000)
Glass plate for final polishing
Dilute hydrochloric acid (1%)
Acetone
Single-mat commercial acetate film (0.005 in.-thick)

McCrone's (1963) discussion is an excellent summary of acetate peel techniques and incorporates all the aspects discussed in earlier chapters. The technique requires cutting and polishing of a carbonate rock slab. After polishing, the specimen is washed and dried and the polished face is etched in dilute hydrochloric acid (1%) for 10 seconds. If the sample is known to contain dolomite, a longer etching time (1 minute) is required. The specimen is washed and dried at room temperature. *It must be emphasized that the etched surface must NOT be touched after etching* so that the microrelief on the surface is preserved.

The specimen is placed in a box filled with sand and fixed by slight rotation so that the etched surface is oriented face-up and horizontal. The etched surface is wetted with acetone and a piece of cellulose acetate is placed dull side down over the etched surface. Air bubbles are driven off by bending the acetate into a U-shape with the dull side down and then rolling the acetate onto the acetone-drenched surface. The adhesive forces at the contact between the rock and the film are such that the film adheres to the surface without outside pressure. Thus the peel should not be pressed onto the surface. The acetone dissolves film along the preferential relief produced by etching.

The peel dries within 15 minutes and can be removed by gently peeling away from one corner. Should the peel tear, the remainder can be removed with a razor blade. After removal, the peel can be mounted between two pieces of lantern-slide glass and bound.

Although the peels have been used primarily on etched limestone surfaces, the same techniques can be used with sandstones cemented with calcite to bring out textural and fabric relationships.

Katz and Friedman (1965) and Friedman (this book) have improved techniques of peel analysis of carbonate rocks by incorporating the additional step of staining rock slabs before making peels. A stained peel is the result. It has the advantage of discriminating mineralogy together with texture and fabric.

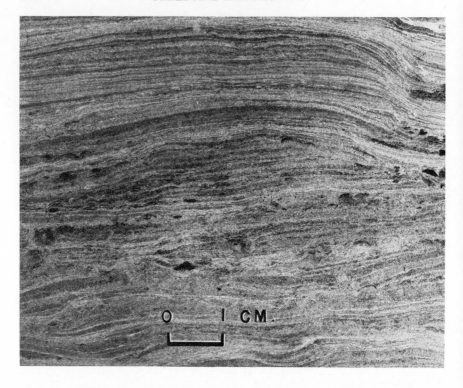

Fig. 25 Negative print of acetate peel showing laminated dolomitic and calcitic pelletal mudstone. Supratidal facies, upper Thacher Member, Austin's Glen, New York. (From LaPorte, 1967, p. 78, Fig. 6; locality 44. Republished by permission of the American Association of Petroleum Geologists)

The following procedures for making stained peels are adapted from Katz and Friedman (1965). Rock slabs are sawed and one of the cut surfaces is ground with carborundum powder. Carborundum 800 grade powder is used for the final polish. The polished surface is thoroughly washed and then etched with diluted (1 + 9) hydrochloric acid. Katz and Friedman (1965, p. 248) recommend the following etching times:

Pure dolomite: 45 seconds
Pure limestone: 30 seconds
Calcitic dolomite and dolomitic limestone: 25 seconds
Carbonate-cemented sandstone: 20 seconds

The etching is stopped by washing the polished surface with tapwater. The sample is then final washed in distilled water, and the polished surface is stained as outlined by Katz and Friedman (1965) and Friedman (this book). After staining is completed, the stained, polished surface is immersed in acetone for 2 to 3 seconds and then removed. An acetate sheet (0.005 to 0.10 in. thick) is placed immediately on the moistened, stained surface and is firmly and evenly pressed down with the fingers. This pressing should be done without sliding the film. No air bubbles will remain if the film is pressed quickly and sufficient acetone is applied. The rock slab is left to dry for 40 minutes, and the acetate film is removed carefully and placed between two glass plates. Calcite imprints the acetate film a deep red, whereas ferrous calcite imprints the film a deep blue. Dolomite remains unstained.

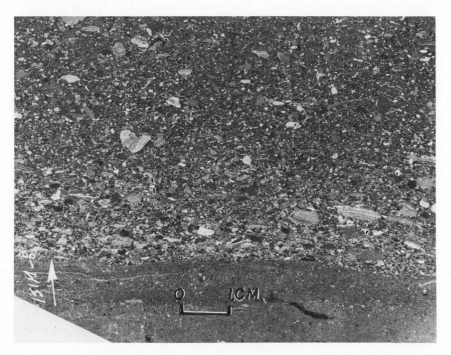

Fig. 26   Negative print of acetate peel showing contact of upper Elmwood Member and Lower Clark Reservation Member, Supratidal facies, Jamesville, New York. (After LaPorte, 1967, p. 81, Fig. 12; locality 151. Republished by permssion of the American Association of Petroleum Geologists)

Dry Cellulose Peels

Honjo (1963) followed the same general procedures as outlined by McCrone (1963), except he used a different medium for making peels. Davies and Till (1968) used this method for making stained peels (see accompanying chapter by Friedman).

Peel sheets are made of 7% solution of ethyl cellulose in trichloroethylene (Table 17). The solution is prepared by pouring 10 gm of ethyl cellulose powder in 100 mm of trichloroethylene with continuous stirring. The solution is very thick when stirring begins, but, if the solution is left to stand for 2 hours, the viscosity decreases.

TABLE 17   Materials for dry cellulose peels

Ethyl cellulose powder	Trichloroethylene	Ethyl acetate

A glass plate is bound with cellulose tape around the edges, and the ethyl cellulose solution is poured over the plate. Using a glass rod, the liquid is spread smoothly over the glass. A peel sheet, with dimensions of the glass plate, dries in 5 minutes and can be removed with a scalpel.

Rocks or impregnated sediments are cut and polished with 400, 800, and finally H-fine carborundum powder, and then etched with hydrochloric acid (1.5%) for 1 minute. The sample is then placed in a sandbox and rotated so that the polished face is horizontal. The surface is flooded with ethyl acetate, and the peel sheet is placed on the edge of the rock and lowered slowly so that a meniscus of solvent precedes the peel sheet as it progressively adheres to the polished face. The peel dries in 3 minutes and can be removed with a smooth, continuous motion. It is easier to remove if one corner of the peel sheet is lifted with a scalpel immediately after adhering the peel sheet to the rock face.

Davies and Till (1968) used these methods to make stained peels. The staining is done after the rock face is etched in acid. A series of peels can be obtained using different stains, after repolishing the etched surface, and reetching after each peel has been made. Once again, the amount and type of staining will depend on the investigation (see accompanying chapter by Friedman).

ACKNOWLEDGMENTS

I would like to take this opportunity to extend my personal appreciation to E. D. McKee, U.S. Geological Survey, for making available to me his unpublished summary of latex peel methods and granting permission to incorporate his unpublished description in this chapter. Dr. C. Kruit, Shell Research NV, The Hague, Netherlands, arranged to make available two Technical Memoranda on Shell's lacquer peel techniques. Dr. Kruit is to be thanked for his assistance. I also wish to convey my appreciation to Shell Research NV, The Hague, for their permission to incorporate the details of Shell's lacquer peel techniques into this chapter and for releasing illustrative material.

I would like to thank Dr. L. F. LaPorte, Brown University, for contributing negative peel prints to use as illustrations. Special thanks are extended to Dr. L. C. Bonham, Dr. R. E. Carver, Dr. G. M. Friedman, Dr. J. D. Howard, and Dr. D. J. Stanley for reading an earlier version of this chapter and offering constructive comments for its improvement.

REFERENCES

Barr, J. L., M. G. Dinkelman, and C. L. Sandusky, 1970, Large epoxy peels, *J. Sed. Pet.*, **40**, 445-449.

Beales, F. W., 1960, Limestone peels, *J. Alberta Soc. Petrol. Geologists*, **8**, 132–135.

Bissell, H. J., 1957, Combined preferential staining and cellulose peel techniques, *J. Sed. Pet.*, **27**, 417–420.

Bouma, A. H., 1964, Sampling and treatment of unconsolidated sediments for study of internal structures, *J. Sed. Pet.*, **34**, 349–354.

Buehler, E. J., 1948, The use of peels in carbonate petrology, *J. Sed. Pet.*, **18**, 71–73.

Burger, J. A., G. de V. Klein, and J. E. Sanders, 1969, A field technique for making epoxy relief-peels in sand sediments saturated with salt-water, *J. Sed. Pet.*, **39**, 338–341.

Davies, P. J., and R. Till, 1968, Stained dry cellulose peels of ancient and Recent impregnated carbonate sediments, *J. Sed. Pet.*, **38**, 234–237.

Dielwart, J. E. A. M., 1962, Making lacquer peels from cores, Koninkijke/Shell Exploratie en Produktie Lab. Tech. Memo T, 325.

Fenton, M. A., 1935, Nitrocellulose sections of fossils and rocks, *Am. Midland Naturalist*, **16**, 410–412.

Graham, R., 1933, Preparation of paleontological sections by the peel method, *Stain Technology*, **8**, 65–68.

Heezen, B. C., and G. L. Johnson, III, A peel technique for unconsolidated sediments, *J. Sed. Pet.*, **32**, 609–613.

Honjo, S., 1963, New serial micropeel technique, Kansas Geol. Surv. Bull. 165, Part 6.

Imbrie, J., and H. Buchanan, 1965, Sedimentary structures in modern carbonate sediments of the Bahamas, pp. 149–172, *in* G. V. Middleton, ed., 1965, Primary sedimentary structures and their hydrodynamic interpretation, S.E.P.M. Spec. Publ. 12, 265 p.

Ives, W., Jr., 1955, Evaluation of acid etching of limestone, Kansas Geol. Surv. Bull. 114, Part 1, 5–48.

Katz, A., and G. M. Friedman, 1965, The preparation of stained acetate peels for study of carbonate rocks, *J. Sed. Pet.*, **35**, 248–249.

Lane, D. W., 1962, Improved acetate peel technique, *J. Sed. Pet.*, **32**, 870.

LaPorte, L. F., Carbonate deposition near mean sea-level and resultant facies mosaic: Manlius Formation (Lower Devonian) of New York State, *Bull. Am. Assoc. Petrol. Geologists*, **51**, 73–101.

McCrone, A. W., 1963, Quick preservation of peel-prints for sedimentary petrography, *J. Sed. Pet.*, **32**, 228–230.

McKee, E. D., 1966, Structures of dunes at White Sands National Monument, New Mexico, *Sedimentology*, **7**, 3–69.

McMullen, R. M., and J. R. L. Allen, 1964, Preservation of sedimentary structures in wet unconsolidated sands using polyester resin, *Marine Geol.*, **1**, 88–97.

Maarse, H., and J. H. J. Terwindt, 1964, A new method of making lacquer peels, *Marine Geol.*, 1, 98–103.

Moiola, R. J., R. T. Clarke, and B. J. Phillips, 1969, A rapid field method for making peels of unconsolidated sands, *Bull. Geol. Soc. Amer.*, 80, 1385–1386.

Müller, German, 1967, Sedimentary petrology, Part I, Methods in Sed. Pet., translated by Hans-Ulrich Schminke, Hafner Publishing Co., 283 pp.

Reineck, H. E., 1957, Stechkasten und Deckweisz, Hilfsmittel des Meeresgeologen, *Natur und Volk*, 87, 132–134.

———, 1958a, Ueber Gefuege von orientierten Grundproben aus der Nordsee, *Senckenbergiana Lethaea*, 39, 25–41.

———, 1958b, Ueber des Harten und Schliefen von Locker-sedimenten, *Senckenbergiana Lethaea*, 39, 49–56.

———, 1960, Uber der transport des Riffsandes, *Jahrb. Forschungs. Nordeney*, 11, 21–38.

———, 1961, Versteinerte Nordsee, *Natur und Volk*, 91, 151–162.

———, 1962, Reliefsguesse ungestoerter Sandproben, *Zeits. fuer Pflanzenernaehrung, Duengung, Bodemke*, 99, 151–153.

Shell Research NV, The Hague, 1966, Making lacquer peels in the field, Koninklijke/Shell Exploratie en Produktie Lab. Unpublished Tech. Memo dated September 20, 1966, 2 pp.

Sternberg, R. M., and H. Belding, 1942, The dry peel technique, *J. Sed. Pet.*, 16, 135–136.

Stewart, W. N., and T. N. Taylor, 1965, The peel technique, pp. 224–232 *in* Bernhard Kummel and David Raup, eds., 1965, Handbook of paleontological techniques, W. H. Freeman, 852 pp.

Van Straaten, L. M. J. U., 1954, Composition and structure of Recent marine sediments in the Netherlands, *Leidse Geologische Mededelingen*, 19, 1–110.

———, 1959, Minor structures of some Recent littoral and neritic sediments, *Geologie en Mijnbauw*, 21, 197–216.

———, 1965, Coastal barrier deposits in South- and North-Holland, *Mededelingen Geol. Stichtung, ns*, No. 17, 41–75.

Voight, E., 1936, Die Lackfilmmethode, ihre Bedeutung und Anwendung in der Palaeontologie, Sedimentpetrographie und Bodenkunde: *Deutsche Geol. Ges. Zeits.*, **88**, 272—292.

———, 1949, Die Andwendung der Lackfilmmethode beider Bergung geologischer und bodenkundlicher Profile, *Mitt. Geol. Staatsinst, Hamburg*, **19**, 111—129.

Walton, J., 1928, A method for preparing fossil plants, *Nature*, **122**, 571.

CHAPTER 11

X-RAY PHOTOGRAPHY

WILLIAM K. HAMBLIN

Brigham Young University, Provo, Utah

Careful analysis of a radiograph of a rock slice commonly provides details of internal structure that can be obtained in no other way. Most studies of environments, microfacies, and paleocurrents therefore should be greatly enhanced by employing this technique in the study of "homogeneous" sediments which have yielded, up to now, little or no significant data. Standard laboratory techniques of making a statistical analysis of grain size, rounding, sphericity, and so on, have often produced disappointing results. This is due, in part, to the fact that many grain characteristics are inherited from previous sedimentary cycles. With radiography more attention can be directed to the smallest structural units, or building blocks, of the sedimentary body, which are of particular significance because their size, shape, and orientation generally reflect the physical conditions when the sediment was deposited.

In radiography, a picture is made by some form of penetrating radiation such as x-rays or gamma rays. It has been widely used in medicine, industry, and various natural sciences, but only recently has it been demonstrated as useful in sedimentary petrology. The value of this technique is that it provides an effective and inexpensive method of obtaining information on the distribution of minerals, fabric, stratification, and other structural features which may be poorly expressed or invisible in hand specimen or outcrop. Moreover, it is nondestructive so the specimens can be utilized for other types of analysis such as grain-size distribution, engineering properties, and microscopic petrography.

Since clastic sedimentary rocks develop by vertical and lateral accretion, theoretically, there should be some change, grain size, composition, fabric, cementation, and so on, between layers of grains. These variations, no matter how slight, generally produce corresponding variations in density although there may be no accompanying visible change in color or texture. If x-rays are transmitted through a slice of rock, variations between layers sufficient to cause differences in absorption of radiation can be recorded on photographic film and will outline the internal structure of the specimen. For example, in a fine-grained quartz, sandstone, zircon, garnet, clays, or other light-colored accessory minerals may be concentrated along certain cross-strata. Observed under normal conditions such a rock would probably appear to be completely structureless and homogeneous because minor concentrations of these grains could not easily be distinguished from the dominant mass of quartz grains. A radiograph of a thin slab of this rock reveals concentrations of dense, heavy minerals as light gray or white streaks on the x-ray negative in contrast to darker quartz, and delineates the size and shape of structural planes within the rock mass. The image on a radiograph depends on differential absorption of radiation. Therefore slight variations in composition, texture, cementation, and fabric are sufficient to be recorded on x-ray film. An example of the structural detail obtainable by radiography is shown in Figs. 1 and 2.

Photographs of six typical specimens of homogeneous sandstones are shown in Fig. 1. All the samples are remarkably similar in physical appearance and grain properties (i.e., composition, grain size, rounding, sphericity) and little, if any, structural detail is readily apparent in the samples. From observations on the outcrop and from standard laboratory analysis of the hand specimen, these six rocks would be considered identical for all practical purposes, and it might be interpreted that they were deposited under similar conditions. Positive prints made from radiographs of slices of the specimens in

Fig. 1 are shown in Fig. 2. Details of the internal structure of each specimen are outlined by differences in the absorption of x-rays by the various constituents of the samples. Even though the rocks appear to be homogeneous and isotropic in the hand specimen, it is clear that the individual grains in each sample are arranged systematically into definite structural units. Micro-cross-laminae are dominant in specimens B and C, whereas specimens A and D consist of horizontal laminae. Burrowing organisms have disturbed some of the stratification in specimen E, which was originally very similar to the micro-cross-laminae of specimen B or C. The internal structure of specimen F consists of larger-scale cross-stratification with concentrations of accessory minerals in a rhythmic pattern.

In essence, a radiograph is a measure of the difference in x-ray absorption by various constituents of the sample. Accuracy and sensitivity therefore fall within limits depending on the film and type of radiation used, so that internal discontinuities may not be detected at a given wavelength and intensity. Furthermore, if there are no differences in x-ray absorption between structural planes, there is no outline of the structure on the radiograph even though color contrasts between structural planes are visible on the surface of the hand specimen.

A major advantage of radiography in sedimentary petrology is that it provides a permanent visual record of the internal structure of a specimen that requires only a minimum of sample preparation. In general, the time required to prepare the specimen, expose it to x-radiation, and process the film is less than ½ hour. Another advantage is that the specimen is not destroyed, damaged, or altered in any manner, but can be preserved and used in other laboratory analyses. In effect, radiography provides visual inspection of the interior of a rock and permits study of its structural elements in three dimensions whether or not the structure is expressed on the surface.

Radiography supplements the petrographic microscope in the type of data obtainable. Generally, the field of vision in microscopic work is too small to show structural relationships, but reveals only variations in composition and texture. With radiography, however, samples over 15 in. long and 6 in. wide can be handled conveniently.

The cost of radiographic analysis is relatively low except for the initial purchase of equipment. Developing and printing expenses are

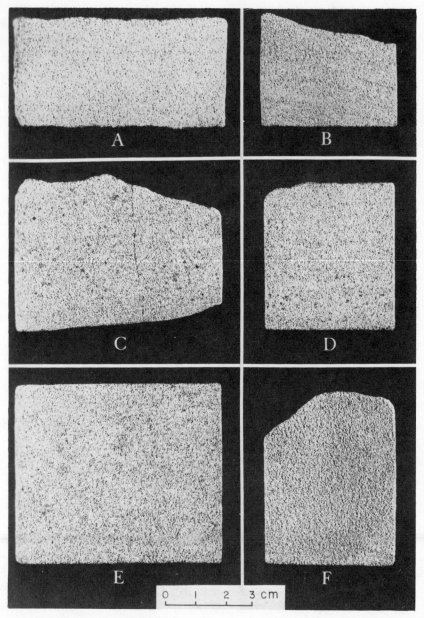

Fig. 1 Photograph of typical, seemingly homogeneous sandstones. All specimens appear to be very similar and lack definite expression of internal structures. (After Hamblin, 1962)

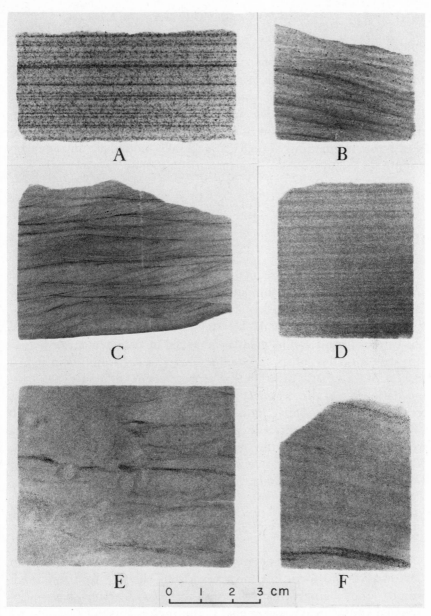

A   B

C   D

E   F

0   1   2   3 cm

Fig. 2 Positive print made from radiographs of the specimens in Fig. 1 showing the amount of structural detail obtainable from radiography. The radiographic exposure was made on Ilford nonscreen medical x-ray film with a medical x-ray unit operated at 33 kv and 30 M.A. Exposure time was 2 seconds with the film 1 meter from the tube window. Specimens are approximately 3-mm thick. (After Hamblin, 1962)

comparable to those for ordinary photographic work, and ready-pack x-ray film costs less than 75 cents per square foot.

Radiography is especially useful in the study of cores where the volume of available rock is small and weathering processes have not had a chance to etch out and accentuate structural details. In oriented cores the direction of sediment transport can be determined from cross-lamination, which is generally abundant in so-called structureless sandstones.

TECHNIQUE

Radiographs of rock are made with either a medical or industrial x-ray unit by placing a thin slice of the specimen directly on x-ray film and exposing it to adequate radiation. Positive prints from the resulting negative are made by standard photographic techniques, and certain details can be greatly enhanced by enlargement or by use of various grades of photographic paper. LogEtronic printing, a process of automatic printing that reproduces detail in both the light and dark areas of the negative, makes it possible to decrease the gross contrast of the negative and increase the detail contrast simultaneously. Good LogEtronic prints not only reproduce the finest shadow detail visible on the original radiography, but actually show shadow detail in the negative invisible to the human eye (for further details see St. John and Craig, 1957).

Specimen Preparation

A rock specimen for radiography is prepared in the form of a thin slice up to 2 cm thick. For most work rock slices, 2 to 5 mm thick produce the sharpest negatives although good results have been obtained from thicker specimens. If the rock is too thick, structural planes appear blurred unless they are perpendicular to the flat surface of the rock slice. There is also a greater chance for several structural units to be superposed, which causes their images to overlap on the radiograph. Moreover, subtle variations between structural units are unlikely to be recorded if the rock slice is too thick. Extremely thin slices are also undesirable because density contrasts may not be great enough to be recorded. The length and width of the specimen are limited only by the size of available cutting equipment. The main requirement is a minimum of variation in thickness in the specimen which would affect the intensity of radiation transmitted through the rock.

Deep saw marks should be removed, for they are commonly registered on the radiograph and may obscure structural details. Fine polishing is ordinarily not necessary although it may be desirable for some detailed work. If the specimen is polished, care should be taken to prevent particles of grinding powder from becoming imbedded in the sample, because most foreign particles produce density contrasts that will register on the radiograph. Porous samples should not be cut on a saw lubricated with oil, for the oil may impregnate the sample and affect density contrasts between the grains and voids.

Although best results are obtained from slices of rock, radiographs of cores can be made with little or no preparation. The margins of the cores, of course, are greatly overexposed because of the cylindrical shape, but the central part of the core commonly shows good detail of internal structures. Unsplit cores of unconsolidated sediment can be x-rayed through thin plastic containers or even metal liners with little loss of detail.

Exposure Factors

The quality of a radiograph is a function of several variables which must be in proper balance if a correct exposure is to be obtained. Paramount among these in radiography of rock slices are (a) target-film distance, (b) kilovoltage, (c) milliamperage, (d) exposure time, (e) type of material in the specimen, and (f) type of film.

Target-Film Distance

When x-ray output is constant, the radiation intensity reaching the specimen is governed by the distance between the tube and the specimen. The intensity of radiation varies inversely with the square of this distance. X-rays follow the laws of light and diverge when emitted from the anode so that they cover an increasingly larger area with less intensity. This example of the inverse square law is illustrated in Fig. 3 and may be expressed algebraically as follows:

$$\frac{I_1}{I_2} = \frac{d_2^2}{d_1^2}$$

I_1 and I_2 are intensities at distances d_1 and d_2, respectively.

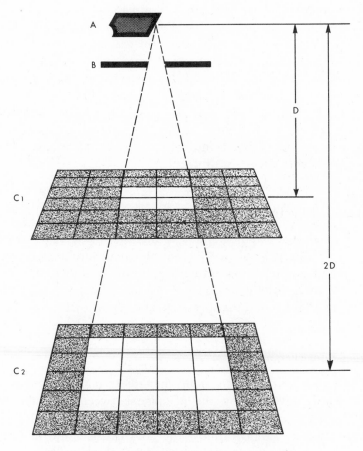

Fig. 3 Diagram illustrating the inverse square law.

In Fig. 3 it is assumed that the intensity of x-rays remains constant and that x-rays passing through the aperture cover an area of 4 sq. in. if the target-film distance (D) is 12 in. If the target-film distance is increased to 24 in. $(2D)$, the x-rays cover an area of 16 sq. in.—an area four times as great as that at C_1. It follows therefore that the radiation per square inch at distance $2D$ is only one-quarter that at D. Thus the exposure adequate at level D must be increased four times to produce a correct exposure at $2D$. Generally, this is done by increasing the time or milliamperage.

Milliamperage and Time

The x-ray output is directly proportional to both milliamperage and time. The product of milliamperage and time is constant for the same photographic effect. This relation may be expressed as

$$M_1 T_1 = M_2 T_2$$

For example, if a good radiograph is obtained at 30 ma in 2 seconds, the correct time necessary if the milliamperage is changed to 5 ma is determined as follows:

$$30 \times 2 = 5 \times T_2$$

$$T_2 = 12 \text{ seconds}$$

Medical x-ray units are designed for relatively high milliamperages and short-time exposures so that settings of 4 seconds, 30 ma, and 35 kv, at a distance of 3 ft produces satisfactory results for specimens 3 to 5 mm thick. The same exposure on an industrial unit might be 6 ma at 20 seconds with kilovoltage and distance remaining constant. Generally, most favorable results are obtained with low currents, 4 ma more or less, and long exposures, 30 seconds to 1 minute.

Kilovoltage

Kilovoltage governs the penetrating power of the x-rays and therefore the intensity of radiation passing through the specimen. It is not possible to specify a simple relation between kilovoltage and x-ray intensity because of variables such as thickness of the specimen being radiographed, the kind of material in the specimen, characteristics of the x-ray generating equipment, and film speed.

The exposure charts shown in Figs. 4 to 7 show the relation between thickness, kilovoltage, and exposure (milliamperage times time) for sandstone, shale, calcareous shale, and limestone. These graphs are adequate for determining the general exposures, but cannot be used for different x-ray machines without a suitable correction factor. Considerable time and film may be saved, however, by using the exposures recommended here as a starting point. The

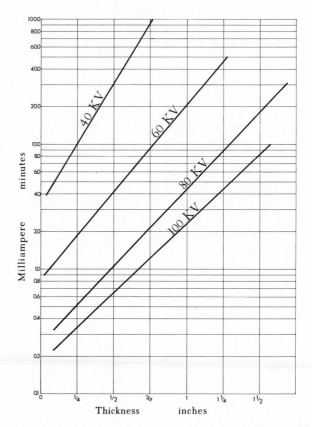

Fig. 4 Exposure chart for sandstone based on Kodak Industrial x-ray film type AA. Focus-film distance, 40-in. Development, 5 minutes in Kodak Liquid x-ray Developer and Replenisher at 68°F. (After Hamblin, 1967)

graphs were made from a series of specimens ranging from ¼ to ½ in. in thickness. Radiographs were made using several different exposure times at each kilovoltage and the proper exposure for each thickness determined.

Each exposure chart applies to a set of specific conditions: (a) x-ray equipment used, (b) focus-film distance, (c) film type, (d) film-processing conditions. Different machines operating at the same kilovoltage and milliamperage may not only produce different intensities but also different qualities of radiation, therefore it may be difficult to determine the correction factor necessary to make an exposure chart prepared for one x-ray machine applicable to another.

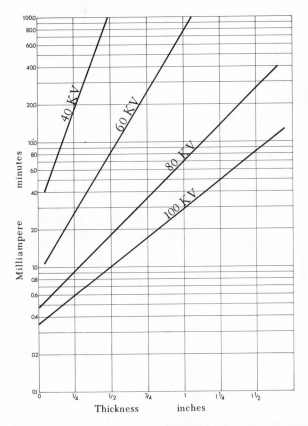

Fig. 5 Exposure chart of shale based on Kodak Industrial x-ray Film Type AA. Focus-film distance, 40-in. Development, 5 minutes in Kodak Liquid x-ray Developer and Replenisher at 68°F. (After Hamblin, 1967)

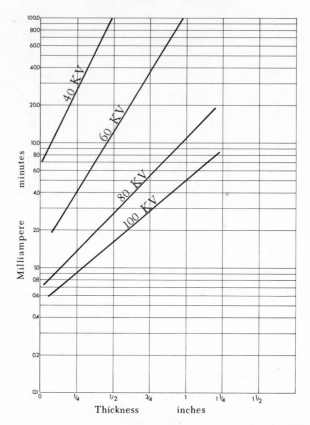

Fig. 6 Exposure chart for calcareous shale based on Kodak Industrial x-ray Film Type AA. Focus-film distance, 40-in. Development, 5 minutes in Kodak Liquid x-ray Developer and Replenisher at 68°F. (After Hamblin, 1967)

Certain amount of experimentation must be used in order to modify these charts for a particular x-ray unit.

A change in focus-film distance may be accomplished by using the inverse square law described above.

Different film types can be corrected by adjusting for the film speed. Table 1 gives the relative film speeds for Kodak industrial x-ray films. Effective film-speed is modified by changes in film processing. If the processing differs from that used in making the exposure chart, a correction factor is necessary.

Film

A variety of x-ray films is available that varies in contrast, speed, and developed grain size. As in photography the choice of film depends on the nature of the work and limitations of the equipment. X-ray film is generally coated on both sides with photographic emulsion, which is an advantage in viewing the negatives, but is a disadvantage if the radiograph is to be printed or enlarged.

Both medical and industrial x-ray film are available in ready-packed individual envelopes or as uncovered film sheets that must be loaded into film-holders. Medical x-ray film used in hospitals is satisfactory for most routine work, although the large grain size makes it

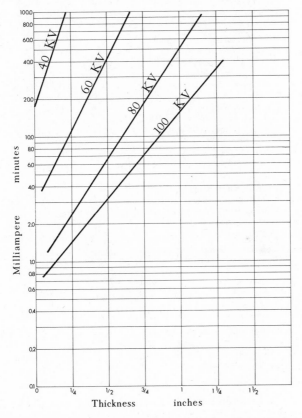

Fig. 7 Exposure chart for limestone based on Kodak Industrial x-ray Film Type AA. Focus-film Distance, 40-in. Development, 5 minutes in Kodak Liquid x-ray Developer and Replenisher at 68°F. (After Hamblin, 1967)

TABLE 1 Exposure and development of Kodak industrial x-ray film[a]

Film Type	Film Characteristics	Contrast	Relative Speed (5 minute) Development	Development for Maximum Speed	Speed Increase with 8 minute Development
Kodak industrial x-ray film Type R (Single-coated)	Ultra fine grain and high contrast. The single-coated film has Type R emulsion on one side of the base only. Single-coated Type R is particularly useful if radiographs are viewed under high magnification.	High	7	8 minutes	35%
Kodak industrial x-ray film Type R	Ultra fine grain and high contrast. Recommended when very fine detail is required. For use direct or with lead foil screens.	High	10	8 minutes	35%
Kodak industrial x-ray film Type M	Extra fine grain and high contrast. Used to obtain high quality. Recommended when critical radiography is required. For use direct or with lead screens.	High	27	8 minutes	40%
Kodak industrial x-ray film Type T	Extra fine grain and high contrast. Speed approximately halfway between Type M and Type AA. Particularly recommended for use in multiple film radiography and in single film techniques in which a film of both extra-fine grain and higher speed than Type M is indicated or a film of lower graininess than Type AA is required without too great a sacrifice in speed.	High	60	8 minutes	30%
Kodak industrial x-ray film Type AA	Fine grain and high contrast. Its grain is not quite as fine as that of Type M, but its higher speed makes it more widely usable. For use direct or with lead foil screens.	High	100	8 minutes	15%
Kodak industrial x-ray film Type F	Highest available speed and high contrast when used with fluorescent screens. Lower contrast when used direct or with lead foil screens. Records large range of thicknesses in relatively small density range.	Medium	200		
Kodak industrial x-ray film Type KK	Highest speed available when high voltage x-rays or gamma rays are used. For use direct or with lead foil screens.	Medium	630		
Kodak no-screen medical x-ray film	Speed and contrast similar to Type KK. Can be used direct or with lead foil screens.	Medium	630		

[a]Modified from *Radiography in Modern Industry Supplement No. 3*, Eastman Kodak Co., Rochester, New York.

undesirable for delicate work requiring enlargements. Industrial x-ray film provides a higher contrast, which is desirable when fine detail is required. A list of Kodak industrial x-ray films showing general characteristics, relative contrast, and film speed is given in Table 1.

Contrast

Radiographic contrast depends on contrast in absorption of radiation by materials within the specimen, film contrast, and wavelength of the x-rays.

Unless there are inhomogeneities within the specimen, there will be very low contrast in the negative or print. Greatest contrasts are obtained from rocks containing quartz and calcite, quartz and clay minerals, or quartz and ferromagnesian minerals.

In a given specimen, low contrast is produced by short wavelength (hard) x-rays and high contrast by longer wavelength (soft) x-rays. This principle is clearly illustrated in Fig. 8, which shows three radiographs of sandstone samples ranging from 1 mm to 2 cm in thickness. Radiograph C was exposed at a high voltage and radiographs A and B at a lower voltage. It is apparent that more subtle details can be detected in thin specimens exposed at low voltage (A) than with thicker specimens where higher voltage is necessary for penetration.

High voltage produces short wavelength x-rays that penetrate the specimen too readily. This decreases the contrast by increasing penetration and scatter. Therefore it is necessary to use the lowest voltage that will penetrate the specimen and properly expose the film. Greater contrast can be produced by using an x-ray machine with a beryllium window which allows transmission of only the longer or soft wavelengths.

In general, greatest detail in radiography of rock specimens is obtained from high contrast film with low speeds and minimum grain. Slow-speed films require greater exposure times, but are superior to films adaptable to shorter exposures that produce fuzzy, grainy radiographs.

Processing X-Ray Film

Procedures for developing x-ray film are essentially the same as those for photographic film, except that most x-ray film has

Fig. 8 Radiographs of sandstone stepped-wedge having a thickness range from 0.3 to 0.8 in. The exposure was 4 kv in A, 80 kv in B and 100 kv in C. Greatest detail is shown in the thin specimen exposed at lowest voltage. Detail and contrast in the thickest specimens (top of Fig. C) are obscured because of hard radiation and distortion resulting from thickness of specimens. (After Hamblin, 1967)

emulsion on both sides and tank developing is advised. Proper chemicals, recommended time, and temperature are supplied by the film manufacturer. For a concise treatment of this subject the reader is referred to *Radiography in Modern Industry*, by Eastman Kodak Co., 1957, pp. 53–86.

Contact prints or enlargements can be made from x-ray films and reproductions of radiographs as positives, or negatives, can be made using Kodak autopositive materials. Radiographs may be projected on a screen with an overhead projector or part of the negative may be mounted between glass and shown as a lantern slide. Standard 2 X 2 in. slides can be copied from radiographs by photographing them on a light table with either Kodachrome film for color slides or a slow-speed copy film such as Panatomic-X. If Panatomic-X is used, it can be processed either as a normal or reverse negative.

DEVELOPING SUITABLE TECHNIQUES

Many of the variables described above can be held constant in a given laboratory so that much of the complexity of making radiographs of rock specimens is eliminated for routine work.

For each general rock type the best exposure (MAS) is determined for a particular film at a given target-film distance. Once this product is determined it should not be varied. If the film developing process is standardized, the only remaining variable is voltage. The exposure graphs (Figs. 4 to 7) show that the voltage increases linearly as thickness increases. If the thickness of specimens is constant, voltage need be varied only as the density of the rock varies. High voltage produces short x-rays that penetrate the specimen readily and decrease the contrast of the radiograph. Therefore it is necessary to use the lowest voltage that will allow x-rays to penetrate the specimen and properly expose the film.

Thus for routine work on a specific rock type the focus film distance, MAS, thickness of specimens, developing procedures, and type of film should be determined by a certain amount of experimentation. Voltage can then be selected by using the exposure charts as a guide.

X-RAY EQUIPMENT

A great variety of x-ray equipment is available, of which most is suitable for radiography of sedimentary rocks. Medical and dental x-ray units are satisfactory for most purposes and used equipment can be usually obtained economically. Medical equipment, however, is designed specifically for human radiography, which requires limited range of penetration and short exposure times. For sedimentary petrology a greater range of penetration and longer exposure times are a distinct advantage. Industrial units rated at 0 to 150 kv designed for continuous operation up to 6 ma are well suited for most sedimentalogical purposes. A portable industrial unit is recommended for those purchasing new equipment. As a general rule an x-ray unit with a small focal spot with low voltage and a beryllium window is recommended, especially for studies of subtle density differences.

SPECIAL TECHNIQUES

Stereoradiography

Stereoscopic radiography has been used for many years in medicine and various fields of nondestructive testing. The technique is based on the same principles as stereoscopic photography and involves radiographs made from two positions of the x-ray tube, separated by a normal interpupilary distance. When the radiographs are oriented in their proper relative positions and viewed under a stereoscope, the two images are fused into a stereoscopic model in which the internal structure of the specimen appears in three dimensions. In most respects the stereoradiography technique is analogous to stereoscopic aerial photography. The major difference is that the stereo-aerial photography represents surface relief, whereas stereoradiography employs some form of penetrating radiation which produces an image of the internal structure.

Stereoscopic radiographs are made by moving either the x-ray tube or the specimen (Fig. 9). Stereo shift is usually accomplished best by moving the specimen. Collimation marks are made on the table or stand on which the specimen rests. The x-ray tube is thus maintained

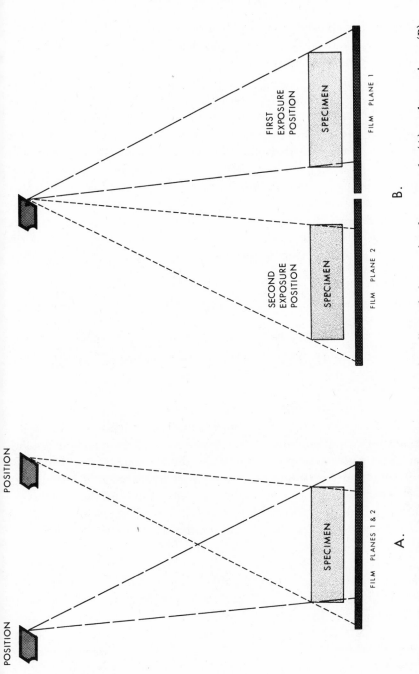

Fig. 9 Diagram showing methods of making stereoscopic radiographs by moving the x-ray tube (A) or the specimen (B).

269

in a vertical position at a constant focus-film distance, and the specimen is moved, between exposures, a distance representing the average interpupilary distance.

Stereoscopic relief is much more distinct if the radiograph shows well-defined structural features. If well-defined structures do not exist in the rock, it may be necessary to provide reference surfaces with which isolated structural features can be related. A wide-spaced wire mesh placed on both the top and bottom of the specimen will generally suffice. These not only add references for the stereo relief but help in securing satisfactory register of the two films.

The exact position of a structure within the specimen may be determined by measuring parallax from the two exposures. The geometry of this method is shown in Fig. 10. The distance of the structure above the film is given by the equation

$$d = \frac{bt}{a + b}$$

where d = distance of the structure above the film,
  b = change in position of structure,
  a = stereo shift,
  t = focus-film distance.

If care is taken in measuring target-film distance, stereoshift, and tilt, accurate photogrammetric measurements can be made using parallax bars, stereocomparagraphs, and other photogrammetric instruments. Negatives may be mounted in glass plates and used in a Kelsh plotter or other high-order photogrammetric instrument if high precision photogrammetric work is desired. Three-dimensional drawings and contour maps may be made of the internal structure if quantitative information is desired.

Stereoscopic radiography is capable of greatly supplementing standard techniques of analyzing texture and structure of sedimentary rocks such as serial sectioning, thin-section orientation, and point counting. In effect, it provides a three-dimensional model of the internal structure and enables one to look into the rock as if it were transparent and to view inhomogeneities in three dimensions. Although only a few publications have reported the use of stereoscopic radiography, the potential of this technique is very exciting. Rutledge (1966) utilized time-lapse stereoradiography in sandstones to study the nature of fluid flow in sandstones and the relation of fluid flow to the internal structures of the rocks. He injected opaque radiographic fluids under various pressures into small specimens of sandstones and made a series of time-lapse stereoradiographs with a medical x-ray unit that permitted very short exposures. Examples of flow patterns recognized are shown in Figs. 11 and 12. Howard (1967) used a similar system of time-lapse radiography to observe burrowing organisms disrupting internal structures of sediment in aquariums (Fig. 13).

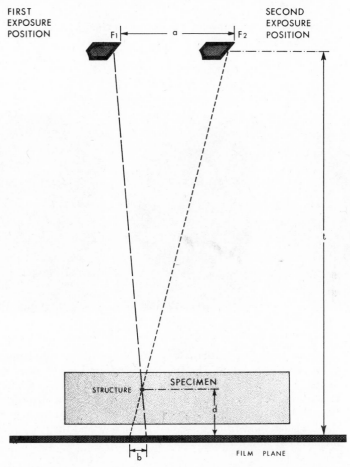

Fig. 10 Diagram showing the geometry of parallax measurements.

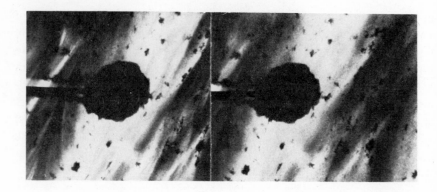

Fig. 11   Stereo radiographs showing the migration fluid in a sandstone. (After Rutledge, 1966)

Fig. 12   Types of fluid flow patterns in sandstone as revealed by radiography. (After Rutledge, 1966) The specimens are sandstone blocks approximately 1-in. thick. Fluids opaque to x-rays have been injected into the sand through a needle at the top of the sample. The pattern of fluid flow is shown as the dark area in the specimen around the needle. Note the irregular fluid front in several specimens and the control of fluid migration by internal structures.

272

Stereoscopic radiography holds great promise for studies of diagenesis, for selective cementation can be viewed within a rock specimen (Fig. 14).

MICRORADIOGRAPHY

In microradiography x-rays are directed through an extremely thin specimen onto fine-grained photographic film. The image recorded on the film is enlarged through a microscope or by ordinary optical projection. This method generally employs soft x-rays generated in the range 5 to 50 kv. The window of the x-ray tube must be made of suitable low-absorbing material such as beryllium. Extremely good contact between the specimen and film is necessary if maximum enlargement that the film is capable of is to be achieved. A simple vacuum exposure holder can be constructed from aluminum plate or other rigid material with a milled recess 1/8 to 1/4" deep (Fig. 15). The specimen and film are placed in the recess and covered with a thin, flexible x-ray-transparent cellulosic sheet. Plastic materials containing chlorine should not be used because of their high absorption of soft x-rays. Evacuation of the recess with a vacuum pump or water aspirator causes the atmospheric pressure on the cover to press the specimen into close contact with the film. Target-film distance may range from 3 to 12 in. since there will be minimum blurring resulting from the thickness of the specimen.

The type of film or plate used in microradiography depends on the desired enlargement. Table 2 lists photographic materials suggested for this work together with relative speed and degree of magnification permissible. Special precautions are necessary only with Spectroscopic Plates Type 548-O, which should be processed as soon as possible after exposure to avoid fading. A stop bath (3.5% acetic acid) between developer and fixer is especially important in processing films of extremely high resolution.

Exposure techniques are difficult to specify in microradiography because of wide variation in equipment used, speed of film, and thickness and composition of specimen. Exposure time may range from a few minutes to several hours. A significant amount of experimentation is therefore necessary before optimum results are obtained.

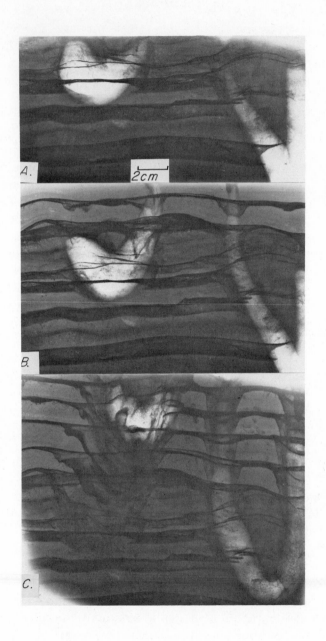

Fig. 13 Time lapse x-ray radiograph of two burrowing holothurians (*Thyone*). As shown in these three radiograph prints the holothurians have moved first vertically downward to bury themselves in the sediment and then vertically upward to keep pace with sedimentation as new sediment was added to the aquarium. The light-colored areas are the bodies of the holothurians. An apparent three-dimensional view was achieved by slightly tilting the aquarium out of the vertical plane and shows the way in which the animals have cut across the planes of lamination.

(A) Time 12 hours after the holothurians were placed on the sediment surface in the aquarium.

(B) Time 18½ hours. The holothurians have moved upward in response to addition of 2.5-cm sediment which was added following the exposure at time 12 hours. The holothurians have reestablished contact with the sediment-water interface through the new layer of sediment.

(C) Time 113 hours. As shown by the radiographic print additional increments of sediment have been added. Other radiograph exposures not shown here record the response as each additional sediment layer was added. By moving up, the animals have produced a cup-in-cup pattern of biogenic structures. The two individual holothurians have moved up in different ways. The one on the left moved its whole body upward, whereas the one on the right maintained contact with the sediment surface by continually extending (stretching?) its body. (After Howard, 1967)

TABLE 2   Photographic materials for microradiography
(after Kodak, Radiography in Modern Industry, 1957)

Material	Relative Speed	Approximate Maximum Optical Enlargement (diameters)
Kodak Industrial x-ray film Type M	100	12
Kodak Spectroscopic Plate Type V-O	40	35
Kodak fine-grain positive film	7	75
Kodak Spectroscopic Plate Type 548-O	0.2	400

A new approach to the study of fabric of fine-grained material can be made utilizing the principles described in the previous section on "stereoradiography." This technique provides a more accurate three-dimensional picture of microconstituents within the rock. An example of a microradiograph of a sandstone is shown in Fig. 16. This radiograph is part of a stereo-pair in which stereorelief appeared to be more than 1 in. Enlargement is approximately 52 diameters.

Fig. 14   Stereoradiograph of a specimen of the Potsdam Sandstone showing nature of selective cementation by calcium carbonate (black areas on radiograph). No surface expression of the cement concentrations was apparent on the surface of the sawed specimen. Compare the nature of distribution of cement with the fluid-flow patterns shown in Fig. 12.

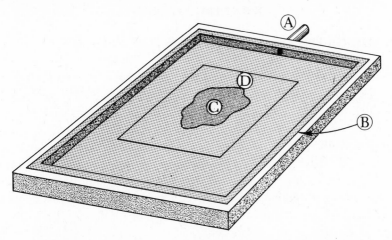

Fig. 15   Diagram of a vacuum exposure holder for microradiography. (A) Vacuum pump connection. (B) Opaque cover (nonopaque plastics may be used if exposures are made in total darkness). (C) Specimen. (D) Photographic emulsion on glass plate.

Fig. 16   Microradiograph of a sandstone specimen (enlargement approximately 52x).

X-RADIOGRAPHY WITH IODINE-125 POINT SOURCE

Weiss (1966) has demonstrated that excellent radiographs of rock samples can be made with a small apparatus charged with iodine-125, a radioisotope that emits x-ray and gamma-ray photons with energies in the narrow range of 27 to 35 kev. Because the I-125 x-ray source contained only a few millicuries, exposures were made at a working distance of 12 to 25 cm. As a consequence, radial scale distortion of the image is large, but not very important in qualitative studies. Exposures suggested by Weiss for I-125 radiograph at a source-sample distance of 12.5 cm are listed in Table 3.

TABLE 3   Exposures suggested for I-125 radiography
(after Weiss, 1966)

Composition	Sample Thickness, in mm	Curie-minutes
Low Z anions (SiO_4 & CO_3)	5 ± 1	0.3
Intermediate Z cations (Ca, Mg)	5 ± 1	0.4
Fe-bearing rock with SiO_2	2 ± 0.5	0.1
Pb-Zn-bearing rock with SiO_2	3 − 5	0.1 − 0.3

The small size, modest cost, and simplicity of control and safety measures combine to make the I-125 source an attractive instrument for petrography. Except for length of exposure, the working time required to make an I-125 x-radiograph is hardly more than that required with electrical equipment. Moreover, a portable I-125 source and container could be easily constructed for field radiography.

X-RAY SCANNING TECHNIQUES

Continuous x-ray scanning techniques have been developed for evaluations of inhomogeneities in solid rocket fuel and other materials and have recently been introduced in studies of sedimentary rocks. This method involves somewhat more sophisticated equipment than radiography, but holds great promise for nondestructive testing concerned with density variations, which reflect porosity and fabric and directional sedimentary structures. Variations of this technique have been described by Bunker and Bradley (1961); Davison et al. (1963); Foster, et al. (1965); and Evans, et al. (1965). Gamma-radiation was used in addition to x-rays. Essentially, all employ the same principles as illustrated in Fig. 17.

The components of the x-ray scanning system consist of a commercial x-ray unit (59 to 150 kvcp) used as a radiation source with a collimator array, which restricts the primary beam to an area as small as 0.78 in. in diameter. A rock core is held by a mechanical system which can move the specimen as the x-ray beam is passed through. A detector measures the amount of radiation passing through the specimen which is recorded on a strip chart plotter. Two types of scanning are possible. The core can be rotated in the x-ray beam, and density variations in a horizontal section are determined or a longitudinal scan can be made which measures variations along the length of the core. The basic measurement made with this sytem is variation in the intensity of x-rays transmitted through the rock as a result of variation in density of the specimen. Density variations, in turn, are influenced by porosity, composition and fabric of the rock sample and by the size, shape, and contents of the pore space (air, or gaseous hydrocarbons, oil or water, etc.). If a sandstone is cemented with material such as calcite or iron both of which have high density contrast compared to the clastic particles, density variations can be measured which reflect directly the pore space within the specimen.

Variations in composition can produce differences in absorption of radiation if there is a significant contrast in the electron density of the materials. Quartz, feldspar, kaolinite, and collapsed montmorillonite have a density between 2.60 and 2.65 and for all practical purposes can be considered as equal. Potassium and sodium feldspars and chlorite have a density between 2.50 and 2.55, and calcite and dolomite have densities ranging from 2.71 to 2.82. Hydrated clays have lower densities than those cited above and amphiboles as well as many ferromagnesium minerals commonly have densities in excess of 3.0. In a quartz rich sandstone, calcite, dolomite, hydrated clays, amphiboles, pyroxenes, and ferromagnesium minerals would likely be registered as an anomaly in an x-ray scan.

An example of tests on a variety of sandstone types (Hamblin, in press) is shown in Fig. 18. These specimens were analyzed with equipment at the Oak Ridge National Laboratories by S. D. Snyder. The specimen consisted of fine- to coarse-grained sandstones. Slight suggestions of cross-strata were apparent in some of the specimens, but the rocks did not contain clear evidence of internal structure or stratification. The x-ray scans represent a 360° rotation of the core within the x-ray beam. Strong indications of high and low densities are indicated at positions A and B.

The greatest utilization of x-ray and gamma ray scanning techniques have been in the studies of porosity variations. However, utilization of these techniques in studies of fabric and other inhomogeneities holds great promise.

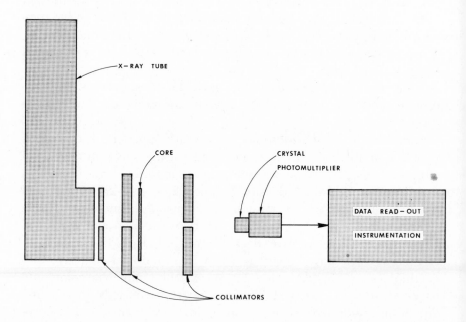

Fig. 17 Diagram showing arrangement of equipment for x-ray scanning. (After Foster et al., 1965)

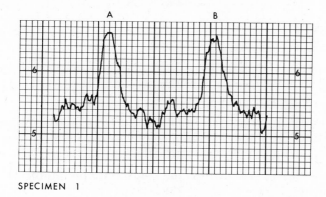

SPECIMEN 1

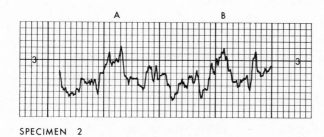

SPECIMEN 2

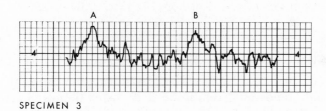

SPECIMEN 3

Fig. 18 Examples of scanning cores of sandstone with a small x-ray beam. Specimen 1 is a coarse sandstone and specimens 2 and 3 are fine-grained sandstone. All three specimens show directional anomalies indicated at points A and B. These anomalies can be correlated with the orientation of obscure cross-strata in specimens 1 and 2 and the trend of ripple marks in specimen 3.

281

REFERENCES

Ayer, N., and A. R. Richards, 1967, Stereoradiography of sediment cores within large-diameter plastic core tubes (Abs.), *Geol. Soc. Amer. Spec. P. 115*, p. 462.

Bunker, C. M., and W. A. Bradley, 1961, Measurement of bulk density of drill core by gamma-ray absorption, *Geol. Surv. Res.*, U.S. Geol. Surv. P.P. 424-B, B-310-313.

Bouma, A. H., 1964, Notes on x-ray interpretation of marine sediments, *Marine Geol.*, 2, 278–309.

Calvert, S. E., and J. J. Veevers, 1962, Minor structures of unconsolidated marine sediments revealed by x-radiography, *Sedimentology*, 1, 287–295.

Clark, G. L., ed., 1963, Encylopedia of x-rays and gamma rays, Reinhold Publishing Co., 1149 pp.

Davidson, J. M., J. W. Biggar, and D. R. Nielsen, 1963, Gamma-radiation attenuation for measuring bulk density and transient water flow in porous material, *J. Geophysical Res.*, 68, 4777–4785.

Eastman Kodak Co., 1957, Radiography in modern industry, Eastman Kodak Co. X-Ray Division, 136 pp.

Evans, H. B., 1965, GRAPE. A device for continuous determination of material density and porosity, *Proc. 6th Ann. Logging Symposium, Soc. Prof. Well-Log Analysts*, 2, B1–B25.

Foster, B. E., S. D. Snyder, and R. W. McClung, 1965, A continuous x-ray scanning technique for determining minute inhomogeneities in solids, *Materials Evaluation*, 23, 191–195.

Fraser, G. S., and A. T. James, 1969, Radiographic exposure guides for mud, sandstone, limestone and shale, Illinois State Geol. Surv. Circ. 443, 19 pp.

Haase, M. C., 1967, X-radiography of unopened soil cores, U.S. Army Engineer Waterways Experiment Station, Misc. Paper No. 3-918, 24 pp.

Hamblin, W. K., 1962, X-ray radiography in the study of structures in homogeneous sediments, *J. Sed. Pet.*, 32, 201–210.

———, 1963, Radiography of rock structures, pp. 940–942, *in* Encylopedia of x-rays and gamma rays, G. L. Clark, ed., Reinhold Publishing Co., 1149 pp.

———, and C. A. Salotti, 1964, Stereoscopic radiography in the study of ore textures, *Am. Mineralogist*, **49**, 17—29.

———, 1964, Internal structures in homogeneous sandstones, *Bull. Kansas Geol. Surv.*, **175**, 1—37.

———, 1967, Exposure charts for radiography of common rock types, Brigham Young Univ. Geol. Studies, **14**, 245—258.

———, X-ray scanning techniques for studying inhomogeneities in sandstones, *Bull. Kansas Geol. Survey* (in press).

Harms, J. C., and P. W. Choquette, 1965, Geologic evaluation of a gamma-ray porosity device, Proc. 6th Ann. Logging Symp., Soc. Prof. Well-Log Analysts.

Howard, J. D., and V. J. Henry, Jr., 1967, Use of radiography in the study of bioturbate textures (preprint), 7th Inter. Sedimentological Congr., Reading, U. K.

Klengebiel, A., A. Rechiniac, and M. Vigneaux, 1967, Etude radiographique de la structure des sediments meubles, *Marine Geol.*, **5**, 71—76.

Krinitzsky, E. L., 1967, X-radiography for engineering-geological research in clays and clay shales (Abs.), Geol. Soc. America Spec. Paper 115, p. 124.

McNeil, G. T., 1965, Some quantitative fundamentals of x-ray stereo photogrammetry, Tech. Papers, Am. Soc. Photogrammetry Semi-Annual Convention, pp. 195—227.

Moore, M. C., 1967, A preliminary comparison of aerial photogrammetric determinations with information obtainable from stereoradiographs (in press).

Patchen, D. G., 1967, Developing suitable x-ray radiographic techniques for thin rock slabs, *J. Sed. Pet.*, **37**, 225—227.

Richards, A. F., 1968, Nondestructive testing methods applied to sedimentological studies (Abs.), 23rd Inter. Geol. Congr., Prague.

Rutledge, R. J., 1966, A study of fluid migration in porous media by stereoscopic radiographic techniques, Brigham Young Univ. Geol. Studies, **13**, 89—104.

Reoult, M., and R. Ribg, 1963, Examen radiographique de quelques minerais de fer de l'Ordovicien normand. Importance des Rayons x en sedimentologie, *Bull. Soc. Geol. France*, **7**, 59—61.

Schmidt, R. A. M., 1948, Radiographic methods in paleontology, *J. Am. Sci.*, **246**, 615—627.

Sopp, O., 1964a, X-radiography and soil mechanics: localization of shear planes in soil samples, *Nature,* **202,** 832.

_____, 1964b, X-radiography and geotechnique: a method of investigation: Norwegian Geolichnical Institute, Internal Report F. 259, 9 pp. (unpublished report)

Sorauf, J. E., 1965, Flow rolls of upper Devonian rocks of south-central New York State, *J. Sed. Pet.,* **35,** 553—563.

St. John, E. G., and B. S. Craig, 1957, LogEtronography, *J. Am. Roentgenology, Radium Therapy and Nuclear Medicine,* **78,** 124—133.

Stanley, D. J., and L. R. Blanchard, 1967, Scanning of long unsplit cores by x-radiography, *Deep-Sea Res.,* **14,** 379—380.

Weiss, M. P., 1966, X-radiography of rocks with I-125 source, *Bull. Am. Assoc. Petrol. Geologists,* **50,** 1507—1510.

Zangrel, R., 1965, Radiographic techniques, pp. 305—320, *in* B. Kummel and D. Raup, eds., Handbook of Paleontological Techniques, W. H. Freeman and Co., 852 pp.

CHAPTER 12

MEASUREMENT OF GRAIN ORIENTATION

LAWRENCE C. BONHAM

JOHN H. SPOTTS

Chevron Oil Field Research Company, La Habra, California

This chapter summarizes techniques of oriented sample collection and grain orientation measurement in the field and in the laboratory.

Directional properties of sedimentary rocks have been studied extensively over the past two decades. Several new concepts and trends in geological thinking and technology have contributed to this increased interest in directional properties and the concomitant increased volume of literature. Some of these are the turbidite concept and the numerous directional features associated with turbidity-current sediments, paleocurrents and basin analysis, increased interest and need for quantitative data, application of statistical analysis to geological data, and new instrumentation capabilities for field and laboratory analysis of geological materials.

A successful investigation of directional properties in general, or of grain orientation specifically, requires an unusually well-coordinated effort of several operations including (a) recording of pertinent field data, (b) accurate field sample collection, (c) sample preparation, (d) orientation measurements, and (e) data handling and interpretation. The often-repeated comment that "the analysis is no better than the sample on which it is made" is particularly appropriate for grain orientation. Accurate, time-consuming laboratory measurements can be quickly invalidated by sloppy collection and sample handling which result in inability to reproduce field orientation in the laboratory.

This summary emphasizes the grain orientation measurement techniques commonly used and essentially available to all geologists. Nevertheless, we anticipate that some of the rapid analytical tools mentioned may soon be in more general use. Several of these show sufficient promise that further experimentation and development could make them technologically valid and economically feasible.

FIELD MEASUREMENTS OF GRAIN ORIENTATION

It is usually most convenient to make field orientation measurements on large particles (pebbles to boulders), fragile (charcoal), sparsely distributed (fossils), or any other particles difficult to collect and transport to the laboratory. For general paleogeographic and stratigraphic studies, it is sometimes adequate to estimate or roughly measure the orientation of elongate pebbles, fossils, or other particles (such as charcoal) in a bed by either "eyeballing" the direction or taking a compass bearing on the estimated mean direction. Detailed sedimentological studies require more accurate statistical analysis of particle orientation, and several methods for field measurement have been devised.

As discussed by White (1952), it is often convenient to ignore pebbles with apparent elongation ratios of less than some arbitrary ratio. This policy discriminates against more equant pebbles that are presumably less well oriented and more difficult to measure. White chose a threshold of 1.5:1 for apparent length to apparent breadth for Keweenawan conglomerate, but the threshold should be adjusted to the sediment under study.

At the time of pebble orientation measurement, it may also be worthwhile to record the length of the axes, the gross particle size, lithology, and type of shape (tabular, pyramidal, rod, or ellipsoid).

Three-Dimensional Pebble Orientation

Kalterherberg (1956) described a three-dimensional method of plotting the orientation of pebbles axes in loosely indurated sediments, where pebbles can be removed from the outcrop. With this method it is possible to determine the orientation of the longest, shortest, and/or intermediate axes as desired.

1. Carefully remove a pebble from the outcrop; disturb the matrix as little as possible.
2. Mark the longest, shortest, and intermediate axes as desired.
3. Replace the pebble in its original position in the outcrop.
4. Determine the axial directions with a Brunton compass.
5. Plot the axial orientations and the bedding plane on a stereographic net.
6. Contour the stereographic plot.

Krumbein's Pebble Orientation Method

Krumbein (1939) described a method for three-dimensional plotting of the longest axis alone, or both the longest and shortest axes of pebbles in loosely consolidated sediments. The method, modified from an earlier technique used by Wadell (1936) utilizes a frame (5 X 6-in. photoprinting frame) with thin brass rods and a spirit level attached (see Fig. 1).

1. Select or prepare a planar, vertical exposure, and record the azimuth of the plane.
2. Holding the collecting frame upright, level, and parallel to the exposure, with the cross-rods in front of pebble, mark an "L" on the pebble in the upper right quadrant.
3. Repeat the marking on all pebbles in the sample (100 or more).
4. Remove the marked pebbles from the exposure.
5. In the laboratory, tape a piece of polar graph paper to a level table top, and clamp the collecting frame at the edge of the graph parallel to the azimuth of the field exposure surface.
6. Mark the longest and shortest axes of each pebble.
7. Mount a pebble on modeling clay at the center of the polar plot.

8. Sight through the collecting frame, and orient the pebble so that the reference "L" coincides with the guide bars of the frame.

9. Measure the dips of the long and short axes with a protractor or the clinometer of a Brunton compass.

10. Sight down on the pebble, and determine the azimuths of the long and short axes.

11. Plot the axes on a stereographic net.

12. Count and contour the stereographic plot.

Two-Dimensional Pebble Orientation

Several methods have been used to obtain two-dimensional data in sediments ranging from unconsolidated to highly indurated, including beach boulder beds, glacial till, and metamorphosed conglomerates. Depending on the problem and the exposures available, the two-dimensional method can be used for determining grain orientation in the bedding planes and imbrication.

One method for loosely consolidated pebbly beds involves the preparation of a plotting board of plywood or masonite approxi-

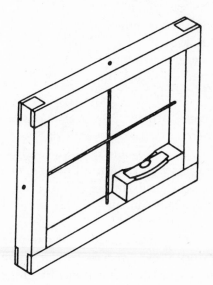

Fig. 1 Frame for labeling orientation of pebbles to be collected. (After Krumbein, 1939)

mately 12 in. square. A Brunton compass is attached parallel to one margin and polar graph paper is attached to the board.

1. As the observer moves about the outcrop, hold the board horizontally and with fixed azimuthal orientation.
2. Remove the pebbles carefully from the outcrop and move to the center of the graph paper.
3. Plot the longest axis directly on the paper.
4. Record the strike and dip of the outcrop on the plot.

The obvious danger in this method is possible rotation of the pebble as it is moved to the center of the graph; thus this operation should be done very carefully or repeated several times.

Another technique, which is applicable in both unconsolidated sediments and indurated rocks, involves marking pebbles directly with a line approximating the direction of the longest axis. The direction is either measured with a compass and tabulated in field notes or plotted directly on polar graph paper.

Photographic Measurements

If suitable outcrops are available, photographic methods may be useful, or even preferable. Photographs provide a permanent graphic record, and time-consuming orientation measurements can be done in the office or laboratory, thus economizing field time. Also, applicability of the photographic method is independent of the induration of the sediment.

1. Select an appropriate surface, one parallel to the bedding for two-dimensional orientation or perpendicular to the bedding for imbrication
2. Indicate the dip and strike on the outcrop with chalk, marking pens, or paper strips (also indicate dip-strike data and station number, if possible; otherwise, record these data in field notes).
3. Photograph an outcrop with the camera axis perpendicular to the exposure surface.
4. Measure the orientations of particles on photographic prints or projected color transparencies.

COLLECTING ORIENTED SAMPLES

Careful collecting and labeling of oriented field samples is one of the most critical procedures in obtaining reliable grain orientation measurements. It is advisable to establish a routine of labeling samples and recording field notes, so that field orientations can be unequivocally reconstructed in the laboratory. Simple rotational and inversional errors creep into such operations very easily and can produce questionable or uninterpretable, and thus useless, results.

If the rocks are poorly indurated, it may be necessary to cement the sample before removing it from the outcrop. This can be done by spraying with latex or clear plastic, or by impregnating with shellac, thinned Duco cement, film-base dissolved in acetone, or one of the epoxy cements. Allow the impregnating material to set up completely. If the sediment is very loose, a tin can with the ends removed or a short plastic tube can be forced into the sand, which is then cemented as above. Care should be taken that there is minimum disturbance of the grains as the sample-holder is inserted.

For indurated rocks and cemented sediments, a suggested procedure is as follows.

1. Break the selected sample from the outcrop; then replace it carefully in its original position.

2. Place a dip-strike symbol on the sample, using tape, felt tip pens, or colored pencils. Place the symbol on a bedding plane if one is well exposed; if not, place it on any convenient surface, such as a joint.

3. Measure the dip and strike of the bedding, and of any other surfaces on which dip-strike symbols are placed. It is advisable to mark the bedding with a line around the sample, because bedding is often less obvious on hand samples in the laboratory than it was in the field.

4. Mark other fabric elements on the sample (these may be less obvious on the small sample in the laboratory).

5. Mark the sample number and the *top* of the sample and remove it from the outcrop.

6. Record other orientation data as follows:
    (a) Sample number
    (b) Location, formation, age, and so on

(c) Nature of surface marked with dip-strike symbol (bedding, joint, weathered surface)

(d) Dip and strike of bedding (note if overturned)

(e) Dip and strike of surface marked, if not bedding plane

(f) Orientation of other fabric elements (grooves, casts, cross-bedding, etc.)

(g) Sketch or photograph of outcrop or sample to show position in beds, adjacent features, and so on.

LABORATORY RECONSTRUCTION OF FIELD ORIENTATION

For various reasons, it may be desirable to reconstruct the field orientation of the sample in the laboratory. This is particularly desirable if the field orientation marks are on surfaces other than bedding planes, and oriented samples parallel to bedding are needed. Several devices have been designed for this purpose, but the following is a simple, reliable laboratory technique that we have found useful.

1. Tape a sheet of polar graph paper to a horizontal table top, and label N, S, E, and W.

2. Draw a heavy line across the plot in the strike direction of the orientation marks, as measured in the field.

3. Cut a flat surface on the rock parallel to bedding if visible or marked.

4. Mount the sample in the center of the graph paper with modeling clay.

5. Rotate the sample until the strike line is horizontal and parallel to the line marked on the polar plot.

6. Set the field dip-angle of the orientation mark on the clinometer of a Brunton compass. Hold the compass against the sample, and rotate the rock around the strike line until the dip line is the same as it was in the field. The sample is now oriented exactly in its original field position.

7. If the bedding was not marked or is not visible, but was measured in the field, use the field values. Ink a line around the sample to indicate bedding. Saw the sample parallel to the marked

plane, and return the lower part of the sample to the mount, in the same orientation.

8. Place a strike-dip mark on the bedding plane surface.

Oriented thin-sections, polished slabs, or peels can be prepared from the oriented sample, and in some cases observations can be made directly on the sawed and smoothed surface.

MICROSCOPIC METHODS

The precise orientation of particles in sediments is a three-dimensional problem that is increasingly difficult with decreasing grain sizes. Three-dimensional orientation of pebbles is relatively simple, but similar manipulations and measurements on sand and silt are not feasible. A statistical approach in measuring several hundred grain orientations in two or more planes has been followed by several workers (Dapples and Rominger, 1945; Spotts, 1964). This method usually involves determining the apparent long axis orientations in the plane parallel to bedding and (usually) two other planes mutually perpendicular to the bedding plane. From these the true three-dimensional orientation can be derived.

Microscopic grain orientation measurements are made on petrographic thin-sections, peels, and polished slabs. The types of samples are listed in decreasing order of precision, which is related primarily to the observer's ability to identify discrete grains and recognize grain boundaries. We have found that petrographic thin-sections are probably worth the extra effort in sample preparation as compared to polished surfaces and peels.

Preparation of Oriented Thin-Sections

Oriented thin-sections require much greater care than ordinary petrographic sections. It is advisable for the geologist who collected the samples to prepare the thin-sections or to supervise closely the process. The two greatest sources of error are angular errors in placing the rock on the slide and inversion or rotation of the sample with respect to the proper field orientation. For these reasons it is necessary to establish a routine of sample labeling, cutting and

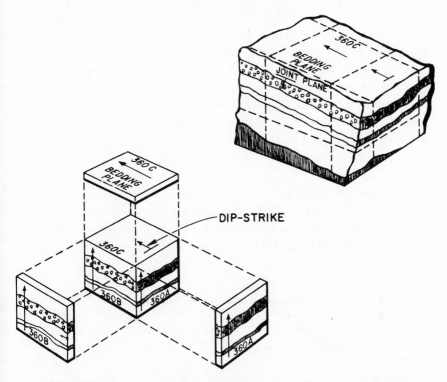

Fig. 2 Preparation and labeling of oriented thin-sections from oriented outcrop sample. Cube is sawed from oriented hand sample as indicated.

mounting, and to follow it faithfully. The following is a example for thin-section preparation.

1. Mark an area about 1 in. square on the surface selected for an oriented sample. Sides of the square should be parallel to dip and strike (see Fig. 2).

2. Within the square, mark the sample number, the orientation letter C (for bedding-plane surfaces), and an arrow in the dip direction (or in the north direction if there is no dip).

3. Prepare a sketch of the entire sample, showing the marked area, all labels, and any fabric features such as joints and veins.

4. Saw a 1-in. cube from the sample so that the marked area is one face of the cube.

5. Ink the sample number, the orientation mark B, and an arrow pointing up on one of the two faces of the cube cut parallel to the dip and perpendicular to the bedding.

6. Ink the sample number, the orientation mark A, and an arrow pointing up on one of the two faces of the cube cut parallel to the strike and perpendicular to the bedding.

7. Prepare a sketch of the cube and all its markings in relation to the original sample.

8. Cut a slab from the cube parallel to the bedding plane (see Fig. 3).

9. Grind and polish to a plane surface the side of the slab opposite C (i.e., the bottom of the bed; see Fig. 3).

10. With a diamond pencil, scratch on a petrographic microscope slide the sample number, the orientation mark C, and an arrow.

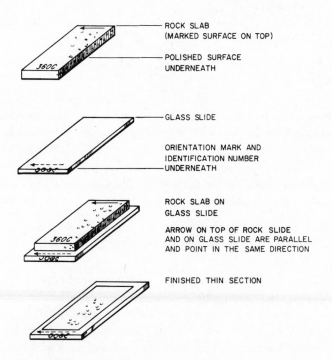

Fig. 3 Recommended procedure for labeling and orientation of oriented thin-section for grain measurement.

11. Cement the slide to the polished rock surface. Make sure that the arrow on the top of the rock slab and the arrow on the glass slide point in the same direction.

12. Cut the rock slab from the slide, except for a thin rock slice.

13. Finish the thin-section by normal procedure.

14. If orientation measurements in sections perpendicular to the bedding are to be made, make additional sections from the cube in the same manner, but parallel to the A and B faces.

15. Add to the sketch the position, orientation, and markings of all sections prepared.

ORIENTATION MEASUREMENTS

Dapples and Rominger (1945) discussed the three principal methods of measuring elongation directions of grain projections.

1. Long dimension elongation. The elongation direction is that of the longest line which can be drawn on a grain projection (Fig. 4). This method is simple in application, but very susceptible to irregularities of grain projection shape.

2. Least projection elongation. The elongation direction is assumed to be that of two parallel lines with minimum separation which can be drawn tangent to the grain projection (Fig. 4). This assumption may be theoretically valid because the least projection would probably offer the least resistance to the depositing medium.

3. Center-of-area elongation. The elongation direction is that of the longest straight line that can be drawn through the center of area of the grain projection. Center of area is equivalent to the two-dimensional center of mass; therefore the method requires planimetering of the grain projection. This method may be the most valid assumption because it approximates the center of mass, but the planimeter requirements preclude its use on other than a limited scale.

Dapples and Rominger indicated a preference for method 2 (least projection elongation), because it represents a compromise between theoretical accuracy and operational facility; most workers have subsequently used that technique.

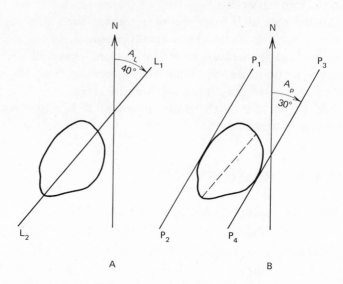

Fig. 4 Comparison of two ways of measuring grain elongation direction. (A) Long dimension $(L_1 L_2)$ makes angle of $40°$ from north. (B) Least-projection tangents $(P_1 P_2, P_3 P_4)$ on same grain make $30°$ angle from north. (After Dapples and Rominger, 1945)

Grain elongation measurement with a petrographic microscope can be done with either conventional cross-hairs or a radial reticule with 10 or $20°$ rays. With a radial recticule a grain is brought to the center of the field, and the elongation direction is read directly, according to the ray to which it is parallel. No rotation of the stage is required. With conventional cross-hairs, the centered grain is rotated until the elongation direction is parallel to one of the cross-hairs. The rotation method is probably more accurate.

Hand (1961) describes a rapid method that uses a wide-field binocular microscope and an oriented tablet of the sample. Grain length and breadth are measured with an ocular micrometer, and the direction of the major axis is determined by turning the ocular to bring a cross-hair into alignment with the axis. The position of the aligned ocular is read from a calibrated collar attached to its rim.

Measurement of orientation by a projection of grain images was described by Curray (1956). The image of the field of the thin-section is projected up onto the underside of the glass top of a microscope light-table. The microscope is operated by remote control cables attached to the fine focus, the mechanical stage, and the analyzer prism.

Grains may be selected for measurement in several ways. Selection by point-counter method, as described by Chayes (1949), gives distribution related to volume or weight percent. The point-counter method will bias the distribution in favor of the larger grains, which is probably desirable for grain orientation. A population or number distribution is obtained by measuring all the grains in randomly selected small fields of the section. The measurement of all grains along a series of linear scans or traverses leads to erroneous results by giving more weight to grains oriented normal to the scanning direction (Curray, 1956).

IMBRICATION ANALYSIS

Imbrication studies are merely a special case of three-dimensional grain orientation. Spotts (1964) described imbrication studies in which orientation sections were cut parallel to bedding. After the preferred orientation direction was determined, two sections were cut perpendicular to bedding, one parallel to the orientation direction and one normal to that direction. Microscopic imbrication studies are feasible only when the bedding can be defined accurately, because the trace of bedding on the vertical sections is used as a reference direction. Because sand-grain imbrication angles are usually less than 10°, it is desirable to define the bedding traces with an accuracy of ±2° at least. Results of a three-dimensional study showing imbrication are illustrated in Fig. 5.

FLYING SPOT SCANNER

Several electronic instruments for rapid measurement of grain or particle counts, grain size, and grain shape and fabric have been

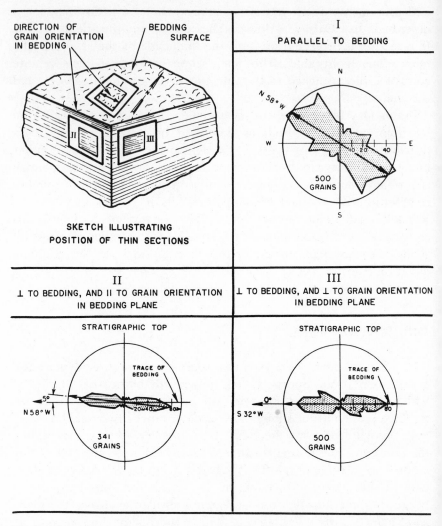

DIRECTION OF
GRAIN ORIENTATION
IN BEDDING

BEDDING
SURFACE

I
PARALLEL TO BEDDING

SKETCH ILLUSTRATING
POSITION OF THIN SECTIONS

500
GRAINS

II
⊥ TO BEDDING, AND ‖ TO GRAIN ORIENTATION
IN BEDDING PLANE

III
⊥ TO BEDDING, AND ⊥ TO GRAIN ORIENTATION
IN BEDDING PLANE

STRATIGRAPHIC TOP

TRACE OF
BEDDING

341
GRAINS

STRATIGRAPHIC TOP

TRACE OF
BEDDING

500
GRAINS

Fig. 5 Three-dimensional study of grain orientation, indicating preferred orientation of N 58° W in bedding plane (I) and imbrication of 5° in section perpendicular to bedding and parallel to mean orientation direction in bedding plane (II). (After Spotts, 1964)

described in the literature. Zimmerle and Bonham (1962) adapted a flying spot scanner type of electronic device for the specific purpose of rapidly measuring dimensional grain orientation. The principle of the flying spot scanner is illustrated in Fig. 6. A tiny beam of light scans a photographic transparency prepared from thin-sections or polished surfaces. A photoelectric cell measures the number of times a grain is encountered by the scanning beam. Evenly spaced scans parallel to elongate axes thus record fewer encounters than scans perpendicular to the long axes. Scans taken in several different directions, for example, at 10° intervals, can produce a statistical plot of the numbers of grain encounters in each scan direction. The reciprocal of this plot is equivalent to a frequency plot of long axes orientation. A block diagram of Zimmerle and Bonham's device is shown in Fig. 7; further details on instrumentation are provided in the 1962 paper.

Transparencies used for scanning can be either positive or negative, but must have high contrast between grains and all other areas of the rock. Rocks containing transparent cement or matrix, or altered or poorly transparent grains are sometimes difficult to prepare, because high contrast on film may not be easily achieved.

High-contrast transparencies of most sandstone types have been obtained by using fluorescent dyes and ultraviolet illumination for photography (Fig. 8). The sample is sawed, polished moderately, and then coated with fluorescent paint. A quick second polishing

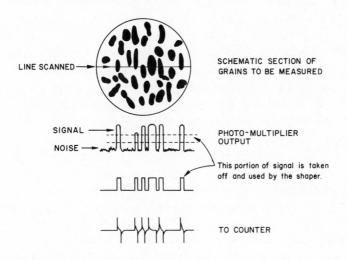

Fig. 6   Principle of flying spot scanner: signals generated by one high-frequency sweep. (After Zimmerle and Bonham, 1962)

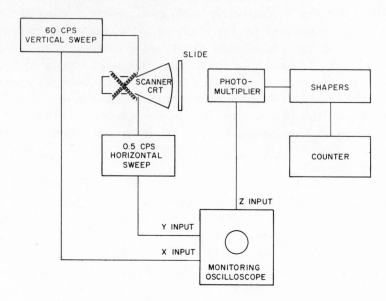

Fig. 7   Block diagram, flying spot scanner for grain orientation measurements. (After Zimmerle and Bonham, 1962)

removes the paint from grain surfaces and leaves it in pores and matrix which have slightly lower relief. Cemented sandstones are etched slightly to produce relief—hydrochloric acid etch for carbonate cements or hydrofluorosilicic acid for silica cement. The painted areas fluoresce intensely under ultraviolet illumination, and high-contrast films can be produced.

The transparency is then placed in a film-holder, and the number of grain-matrix contacts is read on the counter. The film is rotated through small angle (10°) increments, and new counts are made at each position until the holder has been rotated through 180° or 360°. Each count requires about 2 seconds. Any type of conventional grain orientation diagram can be prepared from the data.

Zimmerle and Bonham's experimental results show good correlation between scanner and manual methods. Comparative results from the flying spot scanner and conventional microscopic methods are shown in Fig. 9. The diagrams illustrate the coincidence of not only general trends but also secondary maxima of grain orientation derived from the two methods. The authors conclude that electronic scanning devices are suitable for rapid grain orientation measurements and can provide accurate, objective diagrams in a very short time.

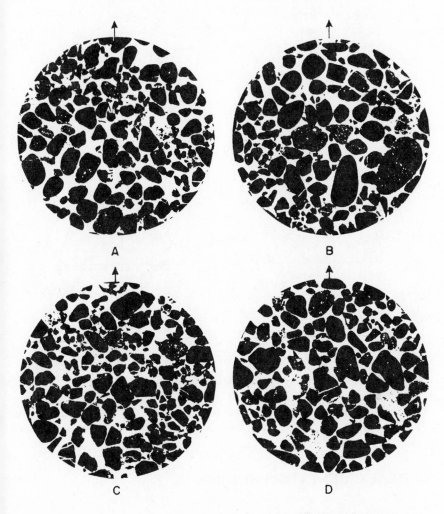

Fig. 8 Ultraviolet-light photographs of four areas of polished Viking Sandstone. Film transparencies are scanned by flying spot scanner for preferred grain orientation direction. (After Zimmerle and Bonham, 1962)

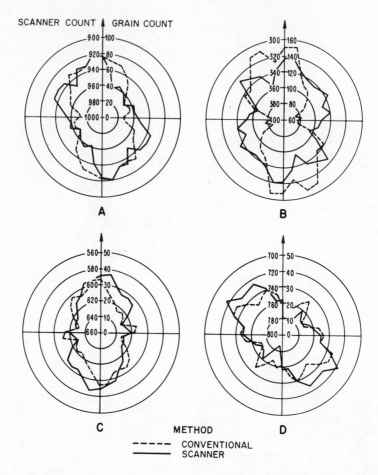

Fig. 9 Comparison of results from scanner and conventional microscopic methods of grain orientation measurement. (After Zimmerle and Bonham, 1962)

OPTICAL ORIENTATION METHODS

The photometer method for determining the direction of quartz-grain orientation is based on the premise that there is a consistent relationship between grain morphology and optic axis (crystallographic) orientation. However, the several workers who investigated grain orientation photometrically do not unanimously agree on the validity of this premise.

Basically, all photometric methods are similar; light is transmitted through a thin-section and crossed polarizers, and as the section is rotated, variations in light intensity are recorded. For single quartz crystals or aggregates with unique preferred optical orientation, no light is transmitted at the extinction position when principal optic vibration directions are parallel with the polarization directions. Samples with some degree of preferred optical orientation will give minimum transmission when the preferred direction is parallel to the planes of polarization. If the relationship between the crystallographic and morphological axes is known, the morphological grain orientation direction can be determined. Martinez (1958) and Pierson (1959) describe photomultiplier photometric devices for obtaining optic orientations in quartz.

Results from optical crystallographic and photometric studies of quartz and quartz aggregates have led to a diversity of conclusions regarding the relationship between c-axis orientation and either the longest axis or the maximum projection area of elongate grains. An early study of St. Peter and Jordan Sandstones made by Wayland (1939) indicated that the direction of elongation coincides with the direction of the optic axis of quartz. Wayland concluded that quartz grains are longer and harder in the direction of the optic axis. An experimental fracturing study by Ingerson and Ramisch (1942) showed that (a) there was a decided tendency for some samples to fracture parallel to the unit-rhombohedron, but no sample showed pronounced fracturing parallel to the c-axis; (b) igneous and metamorphic quartz grains show a tendency to be elongate parallel to prism-and unit-rhombohedral faces; and (c) abrasion tests on oriented prisms show that quartz is harder than normal on prism faces.

Rowland (1940, 1946) compared optic-axis orientation and three-dimensional grain shape fabrics, and concluded that clastic quartz grains tend to be elongate parallel to the unit-rhombohedron and, to a lesser extent, to the prism, which is similar to Ingerson and Ramisch's experimental results. Rowland's frequency distribution curves for the angle between the optic axis and the longest dimension show three peaks: (a) between 35 and 55°, (b) between 60 and 70°, and (c) between 75 and 90°. These results indicate at least a complex relationship between the optic axis and grain morphology.

Bonham (1957) measured the optic and elongation axes of 200 quartz grains in a Modelo Sandstone sample, and found that the long axes were oriented preferentially parallel to the rhombohedral direction, that is, 38° from the optic axis (c-axis). Ingerson and Ramisch (1942) reported similar observations. Bonham (1957) suggested that quartz grains elongate parallel to rhombohedral directions would be deposited with their crystallographic c-axes oriented in isolated maxima within 38° of the bedding plane. Optical-axis measurement on Pico rocks, in the same study, showed a majority of the maxima to be within 35° of the bedding plane, which was interpreted by Bonham as indicative of depositional fabric.

Martinez (1958) reports on photometric quartz-grain orientation studies of a sandstone in the Bell Canyon Formation in Culberson County, Texas. Although the agreement among photometric measurements, ripple mark trends, and fusulinid orientation was not striking on the maps, statistical data indicate fairly good agreement between c-axis orientation and grain elongation. Martinez concludes that in similar sands in which there are no oriented elements other than quartz grains, the photometer method described might furnish evidence for grain orientation.

Nevertheless, it appears that we still do not have unequivocal proof that quartz grains tend to be elongate parallel to the c-axis. More work will be required on photometric measurement of sand aggregates or, preferably, on statistical studies of the optical orientation of individual elongate grains.

Graphic Display

Results of grain orientation measurements may be displayed in several ways. The choice of an appropriate display is dictated by the objective of the display, the type of measurement made, and the data reduction techniques.

Detailed results showing the axial direction of each measured grain in a sample are given in Fig. 10. The a-and c-axes are plotted stereographically. Stereographic plots of this type of data on one or more samples can also be contoured, as in Fig. 11, which represents a

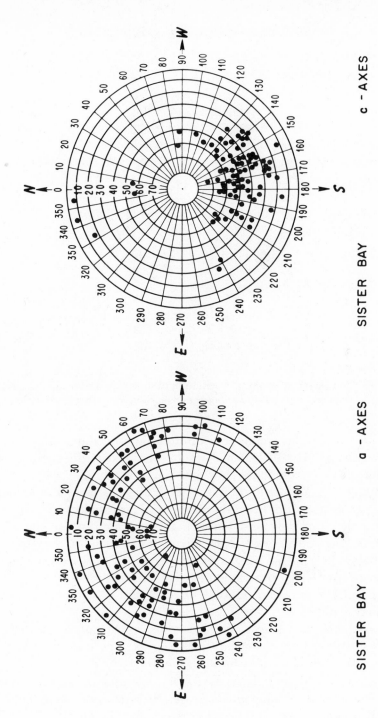

a - AXES   SISTER BAY   **c - AXES**

SISTER BAY

Fig. 10   Stereographic plot of distribution of *a*-axes and *c*-axes of beach pebbles. (After Krumbein, 1939)

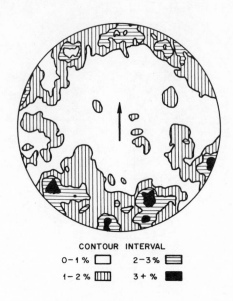

CONTOUR INTERVAL

0-1 % ☐   2-3% ☰

1-2 % ⅢⅢ   3 + % ▆

Fig. 11   Petrofabric diagram of orientation of long dimensions of sand grains by universal stage. Two hundred grains measured. (After Rusnak, 1957)

further step in data reduction. Contouring procedures are described by Fairbairn (1949). Detailed results obtained in different sections within a simple sample are shown in Fig. 5.

Details of grain orientation for samples from various levels in a bed and composites for the entire bed were illustrated by Hand (1961) with polar stick diagrams (Fig. 12). This method can also be used for representing mean orientation directions determined by statistical analysis: the relative lengths of the sticks, or vectors, indicate the degree of preferred orientation. In Fig. 13, data from other types of directional features are plotted for comparison with the results of grain orientation.

Directional data from grain measurements can also be plotted with cartesian coordinates, as shown in Fig. 14. The most common and effective method of illustrating mean grain orientations on maps is with lines or doubly barbed arrows, where only the direction is known, or with single-barbed arrows, where both direction and sense of current movement have been interpreted. (Fig. 15.)

OTHER METHODS

The principal problem in grain orientation is data acquisition, that is, obtaining a sufficiently large number of reasonably accurate orientation measurements. Most of the direct observational methods have proven to be too tedious to permit the accumulation of large amounts of data; the various rapid, indirect methods have uncertainties related to reliability or difficulties in sample handling. Sedimentologists should seek new methods of obtaining dimensional data, and several possibilities merit watching and further development. Advances in several types of instrumentation, particularly electrooptical devices, may provide apparatus applicable to grain orientation measurements.

Two recent developments may have potential applications: (a) microscopes with computerized analytical television attachments and (b) laser technology. The Quantimet, which is manufactured in England by Rank-Cintel, is an example of the television-microscope in which the field of view can be displayed on a television screen and also scanned and analyzed by a scanning spot to obtain several measurements (grain count, areas, grain size, and form factor). A

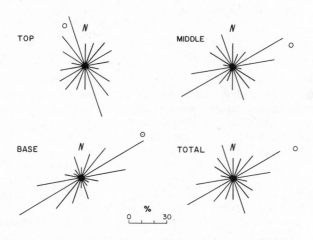

Fig. 12 Vector diagrams of grain orientation azimuth frequencies for three samples from one bed (each represented by 150 grains) and a composite for the whole bed. (After Hand, 1961)

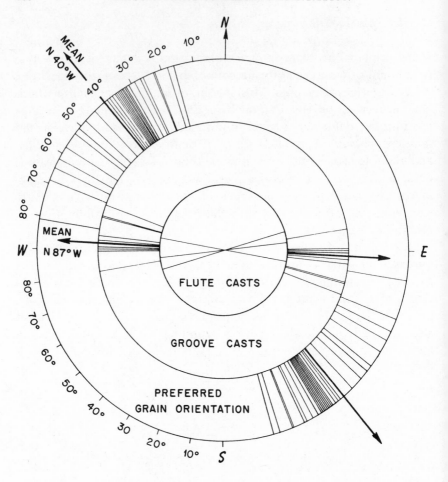

Fig. 13   Comparison of several directional properties of single outcrops. Grain orientation measurements represent 500 grains for each sample. Each groove and flute cast line represents average direction on single beds in cliff exposure. (After Spotts, 1963)

discriminator enables image and contrast manipulation. Such a device may have some potential in orientation measurements.

Grain orientation is given by Pincus and Dobrin (1966) as one of several geological examples of laser processing. For grain orientation, laser light is passed through a transparency of a photomicrograph.

The diffraction pattern produces provides information on the statistical distribution of directional and spatial frequencies associated with all elements of the photograph. Preliminary results appear promising, and certainly all promising new methods of this type should be actively pursued.

Some other indirect techniques that may have applications for grain orientation are (a) magnetic properties, assuming a relationship between magnetic domains and grain shape, (b) dielectric measurements, and (c) various physical measurements on flooded or impregnated samples, assuming a relationship between pore shape and preferred grain orientation.

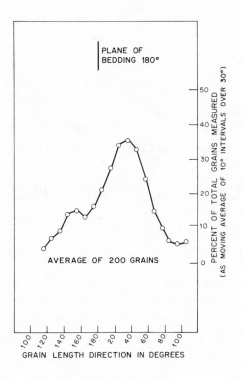

Fig. 14 Grain-orientation measurements on core; section was oriented perpendicular to bedding, parallel to plane of maximum permeability. (After Griffiths, 1950)

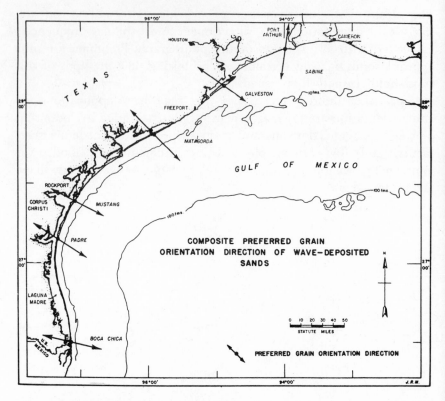

Fig. 15 Composite preferred grain orientation directions of wave-deposited sands. (From Curray, 1956)

ACKNOWLEDGEMENTS

The writers acknowledge the assistance of numerous colleagues who have contributed in many ways to the study of grain orientation over the past several years. Dr. Z. V. Jizba and Dr. W. Zimmerle have been particularly helpful through comments, discussions, and mutual investigations.

REFERENCES

Bonham, L. C., 1957, Structural petrology of the Pico Anticline, Los Angeles County, California, *J. Sed. Pet.*, 32, 251–264.

Chayes, F., 1949, A simple point counter for thin-section analysis, *Am. Mineralogist,* **34,** 1–11.

Curray, J. R., 1956, Dimensional grain orientation studies of Recent coastal sands, *Bull. Am. Assoc. Petrol. Geologists,* **40,** 2440–2456.

Dapples, E. C., and J. F., Rominger, 1945, Orientation analysis of fine-grained clastic sediments: a report of progress, *J. Geol.,* **53,** 246–261.

Fairbairn, H. W., 1949, Structural petrology of deformed rocks, Addison Wesley, 344 pp.

Griffiths, J. C., 1950, Directional permeability and dimensional orientation in Bradford Sand, *Producer's Monthly,* 14, 26–32.

Hand, B. M., 1961, Grain orientation in turbidites, *Compass,* 38,133–144.

Ingerson, E., and J. L. Ramisch, 1942, Origin of shapes of quartz sand grains, *Am. Mineralogist,* 27, 595–606.

Kalterherberg, J., 1956, Uber Anlagerrungsgefuege in grobklastischen Sedimenten, *Neues Jahrb. fur Geologie u. Palaontologie, Abh.,* **104,** 30–57.

Krumbein, W. C., 1939, Preferred orientation of pebbles in sedimentary deposits, *J. Geol.,* 47, 673–706.

Martinez, J. D., 1958, Photometer method for studying quartz grain orientation, *Bull. Am. Assoc. Petrol. Geologists,* 42, 588–608.

Pierson, A. L., 1959, A photomultiplier photometer for studying quartz grain orientation, *J. Sed. Pet.,* 29, 98–103.

Pincus, H. J., and M. B. Dobrin, 1966, Geological applications of optical data processing, *J. Geophysical Res.,* 71, 4861–4869.

Rowland, R. A., 1940, Petrofabric determination of quartz grain orientation in sediments, *Bull. Abs., Geol. Soc. Am.,* **51,** 1941–1942.

Rusnak, G. A., 1957, The orientation of sand grains under conditions of "undirectional" fluid flow, *J. Geol.,* 65, 384–409.

Spotts, J. H., 1964, Grain orientation and imbrication in Miocene turbidity current sandstones, California, *J. Sed. Pet.,* **34,** 229–253.

Wadell, H., 1936, Volume, shape, and shape position of rock fragments in openwork gravel, Geografiska Annaler, 74–92.

Wayland, R. G., 1939, Optical orientation in elongate clastic quartz, *Am. J. Sci.,* **237**, 99–109.

White, W. S., 1952, Imbrication and initial dip in a Keweenawan conglomerate bed, *J. Sed. Pet.,* **22**, 189–199.

Zimmerle, W., and L. C. Bonham, 1962, Rapid methods for dimensional grain orientation measurements, *J. Sed. Pet.,* **32**, 751–763.

CHAPTER 13

MATHEMATICAL ANALYSIS
OF GRAIN ORIENTATION

ZDENEK V. JIZBA

Chevron Oil Field Research Company, La Habra, California

MOTION OF GRAINS

Clastic particles in sedimentary rocks are known to be preferentially aligned in certain directions. This preferential alignment can be best explained by studying the phenomena of tumbling in free fall and of hydrodynamic drag.

The phenomenon of tumbling can be clarified by an illustrative analogy. Let us consider a closed book as a rectangular prism with three principal axes. The thickness of the book is its shortest axis (called the z-axis); the height of the book is its longest axis (called the y-axis); the width of the book is its intermediate length axis (the x-axis).

If we tape the book shut and throw it into the air with a twisting motion, it will rotate around its center of gravity and we can study the tumbling motions of the book around its three principal axes. If we throw the book so that it rotates around its shortest axis, or z-axis (Fig. 1), the book will continue to rotate in a stable manner. If

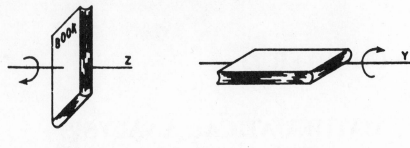

Fig. 1   Fig. 2

we throw the book so that it rotates around its longest axis, or y-axis (Fig. 2), we find that this motion is also stable and rotation continues. However, if we attempt to throw the book in such a way that it rotates around its intermediate length, x-axis as in Fig. 3, we find this difficult, if not impossible. The book does not rotate steadily about the x-axis, but tumbles irregularly.

The explanation hinges on the concept of angular momentum. When we throw the book into the air, we impart to it an angular momentum, which is equal to the product of the moment of inertia about the axis of rotation and the angular velocity. The moment of inertia of a rectangular prism of dimensions, a, b, and c, and mass m is given by the following relationship (Selby, 1964 p. 586):

$$I_a = m \cdot \frac{b^2 + c^2}{12} \tag{1}$$

where I_a is the inertia of the prism when it rotates around an axis parallel to dimension a.

The moment of inertia of an ellipsoid with dimensions a, b, and c is given by

$$I_a = m \cdot \frac{b^2 + c^2}{20} \tag{2}$$

Except for a different number in the denominator, Equations (1) and (2) are the same. This is important, since it means that had we tossed an ellipsoid instead of a book, we would have had essentially the same result, (but ellipsoids are not nearly as available as books). It is important to note that Equations (1) and (2) are valid for arbitrary choice of axes *a, b,* or *c,* so that *a* can represent either the long dimension, the short dimension, or the intermediate dimension. Also it is important to note that these equations imply right angles between *a, b,* and *c.*

Equations (1) and (2) which give a clue to the tumbling stability of irregular objects must be evaluated for three cases: (a) when rotation axis is the short dimension; (b) when rotation axis is the long dimension; (c) when rotation axis is the intermediate dimension.

Case 1. When *a* is the short dimension, the moment of inertia is maximum. If the axis were to be tilted, then the orthogonals to *a* would become smaller, and I_a would decrease. Therefore, to maintain angular momentum, the axis of rotation cannot change.

Case 2. When *a* is the longest dimension, the moment of inertia has the smallest value. If *a* were to be tilted in either the *b* or *c* direction, say *a′*, then either the *b-* or *c*-axis would get longer, and I_a would increase. Angular momentum will remain constant only with an invariant axis of rotation.

Case 3. When *a* is the intermediate dimension, let us assume that *b* is the short dimension and *c* is the long dimension. If we now tilt the axis of rotation from *a* to *a′* in the *ab* plane (Fig. 4), the moment of inertia will be

$$I_a{}' = \frac{m}{K}\left(c^2 + b'^2\right)$$

Since *b′ > b,* the moment of inertia will *increase.* (*Note: K* is a constant whose value would depend on shape; that is $K = 12$ for a prism and $K = 20$ for an ellipsoid.) Similarly, when we tilt the axis of

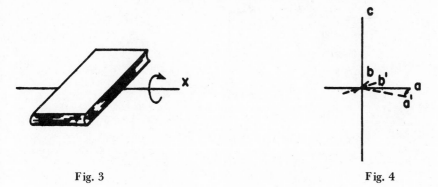

Fig. 3   Fig. 4

rotation from a to a'' in the ac plane (Fig. 5), the moment of inertia will be

$$I_a' = \frac{m}{K}(c'^2 + b^2)$$

Since $c' < c$, in this case the moment of inertia will *decrease*.

Since tilt in the plane of short axis increases the moment of inertia, and tilt in the plane of long axis decreases the moment of inertia, it is reasonable to expect that there is a tilt in a direction that has components both in the b and c directions in which the moment of inertia will remain constant.

In summary, a fairly elementary (and not very rigorous) argument suggests that a rock that has an ellipsoidal (or prismoidal) shape generally tumbles when in free motion. The only exception occurs when the rock rotates either along its longest dimension or along its shortest dimension. These are special cases corresponding to smallest and largest moment of inertia, respectively. In these two special cases, the dimension along which rotation occurs is steady and unaffected by translational motion. (i.e., the rock behaves as a gyroscope).

The tumbling phenomenon described in the preceding paragraphs is valid for particles in free fall only; a moving particle that is constrained by a viscous or turbulent fluid exhibits radically different behavior. If the viscosity or turbulence is sufficiently great, drag and lift play a greater role than angular momentum in determining grain orientation.

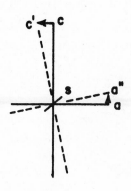

Fig. 5

Experiments on drag for bodies of various shapes in wind tunnels suggest that long bodies show less drag when their long axes are parallel to the motion of the fluid. Furthermore, the orientation of a grain does not depend on its kinetic energy, but rather on its relative velocity with respect to the moving fluid. This fact has an important corollary: a grain's orientation tendency due to free fall stops when its motion (linear and angular) stops, but the orientation tendency due to drag forces continues after the grain has reached a final destination. These forces continue as long as fluid moves around the grain and the grain is not constrained by contact with other grains.

The drag of a particle is given by the following equation (Freiberger, 1960 p. 264):

$$D = \frac{1}{2} C_D \rho V^2 S_1 \qquad (3)$$

where D is the drag, C_D is the drag coefficient (related to shape), ρ is fluid density, V is the fluid velocity, and S_1 is the area of the grain as projected on a plane perpendicular to the fluid flow. For different orientations of a grain, both C_D and S_1 will vary. However, for long grains such as ellipsoids, drag is likely to have its minimum value when a grain's long axis is aligned with the direction of flow. Thus equal-sized and equal-shaped particles are less likely to move the more closely they are aligned with stream flow.

Drag, however, is not the only force acting on a grain. The force acting on the grain and in a direction perpendicular to fluid flow is called *lift*. The amount of lift is given by (Freiberger, 1960 p. 541):

$$L = \frac{1}{2} C_L V^2 S_{LL} \qquad (4)$$

where L is the lift, C_L is the lift coefficient of the particle (related to shape), and S_{LL} is the plan area of the grain as projected horizontally. The most stable particle orientations would be those that have a negative (downward) lift. Generally, this position corresponds to a particle with its long axis tilted slightly downward in an upstream direction.

The critical factor, then, in determining how a grain may be deposited in a sediment is the relative significance that these three phenomena (angular momentum, drag, and lift) may have under varying conditions. We could think of this in two ways: we could consider equal-sized grains in different fluid media such as air and water; or we could consider deposition in water (or air) for two different grain sizes.

Consider equal size small grains (sand). In air a grain has a trajectory that is essentially a free fall. In water viscosity is too great to permit small grains to move independently of motion in surrounding fluid. Any angular momentum that a grain may acquire in water tends to be dissipated by viscous drag. This effect is a function of grain mass. Free-fall tumbling is a greater factor in the depositional orientation of windborne grains than waterborne grains. Wind deposition, however, is not this simple, since elastic rebound at the depositional surface may have quite a large randomizing effect on grain orientation.

The effect of grain size is more important in waterborne sediments. A large pebble can acquire considerable angular momentum when rolled along the bottom by a current. This rolling motion may orient the long axis of a pebble at right angles to the current. After settling, a large pebble is not so likely to be turned parallel to the current as a small sand grain.

In summary: in a depositional sedimentary environment drag due to viscosity, turbulence and inertial forces must be taken into account when studying grain orientation. The complexity is such that it would probably not be practical to develop a theory of clastic grain deposition that would explain orientation of all sizes and shapes of grain in sediments. It is possible, however, to give heuristic arguments that will clarify the type of phenomena that might have a bearing on grain-matrix analysis.

Many studies of particle motion in air and in water during sediment transport have been made. These phenomena have a significant influence on large-scale sedimentary structures. However, the ultimate orientation of a grain does not depend so much on its trajectory in the fluid medium as on its final motion just prior to ultimate placement in the sedimentary matrix.

To study the final orientation of grains in a sediment, let us assume that grain shape can be approximated by ellipsoids with orthogonal axes $a < b < c$. Just prior to final deposition, individual grains will have a certain amount of kinetic energy. Part of this is due to a grain's translational motion in the fluid medium and part to angular momentum. Here again we can discuss the three principal motions about the three axes of the ellipsoid.

Some grains will rotate around their shortest axes in a stable manner. As these grains slow down and come to rest, either the long axis or the intermediate axis will be parallel to stream flow. Thus the preferred orientation in the plane of deposition would be expected to be in the direction of motion. However, just as an egg cannot be made to stand on its long axis, there is a certain probability that such grains will settle in an unpredictable direction. If there were no drag, this would happen, but drag has a chance to reorient the long dimension parallel to the motion of current.

Other grains rotate around their intermediate axes. As we have seen, this motion is unstable, so the orientation of the grains at any one moment is quite unpredictable. We should not expect these grains to settle in any preferred orientation. If they settle in a stable position, drag may not have a great effect, but the direction of elongation of the grain may be in the plane of deposition.

Finally, some grains will be rotating around their major axes. This rotation is stable and contains rest positions that are inherently stable. It is reasonable to assume that a percentage of the grains with this motion will, in fact, come to rest with their major axes in the plane of deposition and at right angles to the final grain motion. If this reasoning is correct, then one would expect several corollaries that might be subject to experimental verification.

First, the phenomena described here apply to individual grains in motion. Any sediment that is formed as a result of mass movement (e.g., slumps on lee slopes of sand dunes or turbidites) would not be expected to satisfy these conditions. In those sediments in which grain motion prior to deposition might lead to preferred orientation, it is reasonable to expect a dependence of degree of orientation on size. The larger the grain, the more likely that it will have a preferred orientation at right angles to the current direction (greater influence of inertial forces). The smaller the grain, the more likely it will have a preferred orientation parallel to the current direction (greater influence of drag forces). For very fine grains the depositional roughness of the settling surface is likely to have a completely randomizing effect on their final attitudes.

Observations reported in the literature tend to substantiate some of these conclusions. See, for example, Rusnak (1957a,b) or Spotts (1964).

POSTDEPOSITIONAL FABRIC CHANGES

As a grain settles onto a depositional surface, it loses its kinetic energy, part of which may be imparted to its immediate neighbors. In sand deposited under aeolian conditions, the kinetic energy of a grain may be considerable, whereas sand particles deposited under water may have relatively small kinetic energies. The kinetic energy refers to that of the sand particle, and not that of the fluid medium in which the relationship may be completely reversed.

Few will deny that a grain of sand on a beach is moved thousands of times by the surf and that tremendous kinetic energy is expended in this motion. But the grain of sand that we actually study in thin-section was put there by only one wave: the last one to move the grain! Even that last placement, however, is not the one we observe, for as successive layers of sand are repeatedly superimposed (and removed), the underlying grains (those that have been permanently placed) are likely to be tilted, compacted, and realigned by pressure of the overlying sediment.

As the sediment becomes more deeply buried, further fabric transformation may occur as a result of compaction, grain-contact solution, and recrystallization.

SINGLE-GRAIN ORIENTATION

Although clastic grains come in many physical shapes, in practice it is convenient to idealize a shape into a form amenable for analysis. From the preceding discussion, it would seem appropriate to deal with orthogonal ellipsoids as generalized grain shapes. This abstraction is useful as an intermediate step between the physical grains that make up a sediment and numbers we wish to derive that provide us with a measure of orientation.

Clearly, a mental picture of an ellipsoid is helpful in understanding spatial relationships, but it does not supply numerical values. In other words, vector analysis is required. A vector is a pair (or a triplet) of numbers that expresses the coordinates of a point in space. The most common way of representating a vector is by drawing an arrow that connects the origin of coordinates with the point represented by the pair (or triplet) of coordinates.

Vectorial representation of a single ellipsoid is a relatively simple abstraction for grain orientation studies. In three dimensions, we would approximate a grain by an ellipsoid. We would then determine the longest axis of this ellipsoid, and call this direction the orientation of the grain. The fact that the other two axes of the ellipsoid may not be of the same length can be neglected if we accept the rationale for grain orientation discussed at the beginning of this chapter. How to properly define the length of the orientation vector, however, is an open question. On the one hand, it would seem logical to make the orientation vector length proportional to the size of the grain (a pebble should have more significance than a small grain). On the other hand, we might consider making the vector length proportional to the grain elongation. (A needle-shaped grain should be more significant than a nearly spherical grain.) It may be that different orientation problems require different treatment.

In a two-dimensional section of a sediment, the orientation problem is somewhat simplified. The idealized shapes of grains are no longer three-dimensional ellipsoids, but rather simple ellipses and vectorial representation of grain orientation is more straightforward. The major axis of the ellipse best fitting the grain shape becomes its orientation vector, but how to determine the length of this vector is still unanswered. Should it be proportional to grain area or its length-width ratio?

The techniques that automatically scan grain contacts resolve this problem by default, but this is not necessarily an acceptable answer. Automatic scanning techniques for studying fabric and grain orientation rely on entirely different principles (Zimmerle and Bonham, 1962). Nevertheless, it is still useful to think of grains as idealized ellipsoids.

A scanning device is usually adjusted to discriminate between grains and matrix. As long as the scanning beam is passing over a grain, it produces one type of response, but when it enters nongrain region (matrix, pore, or clay) it produces a different type of response. Determination of grain orientation is based on the count of the number of times the scan has crossed the boundary between grain and matrix (Fig. 6). First, we must remember that the scanning lines are parallel and equally spaced. A long grain that is aligned so that its long axis is at right angles to the scanning lines, as grain a (Fig. 6), will be cut by the scanning lines more times than a grain whose long axis is parallel to the scanning beam. Thus the grain boundaries of a (Fig. 6) have been crossed ten times, and the grain boundaries of b have been crossed only four times. Note that it is the number of crossings of grain boundaries that is important, *not* the total length of the grain interval. It is true that there are five traces across grain a and only two traces across grain b, but the traces across a are shorter than those across grain b. The total length of the traces indicates size, not orientation (although size also affects the number of intercepts).

The relationship between vector notation of grain orientation and the scanning technique of counting grain boundaries can be best illustrated by reference to Fig. 7. Here an elliptical grain is rotated with respect to the scanning beam. The number of grain-boundary intercepts changes with the grain orientation. When the grain is so placed that its long axis lies at right angles to the scanning lines, it has the maximum number of intercepts. In this position the apparent size of the grain, as seen by the scanning beam, is equal to its long diameter. As the grain is rotated, the number of intercepts decreases and its apparent size diminishes. For any particular scan, a vector representing apparent grain size is oriented at right angles to the scanning direction. The magnitude of this vector is proportional to the number of scan intercepts. Note that when the grain is oriented at an oblique angle, the diameter r of the ellipsoid at right angles to the scanning lines is not necessarily the same as its apparent height.

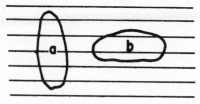

Fig. 6

Fig. 7

PREFERRED GRAIN ORIENTATION

In the previous section we considered how a single-grain orientation can be described as a vector. When there is a need to compare the relative orientation of two or more grains, a new problem arises: an ellipse rotated 180° around one of its principal axes appears unchanged.

If we were to represent grain orientation by vectors, this 180° rotation would produce a two-way ambiguity (which way should the arrow in the vector point?). For a single grain this is not a serious problem, but for more than one grain the problem is unsolvable and some additional approach is required.

The manner in which the ambiguity of orientation is resolved depends to a large extent on the model and the hypothesis that one is interested in investigating. For example, if the model assumes that a single direction of orientation is expected, then the approach may be different from that taken if more than one preferred orientation is expected.

We now consider vectorial techniques for measuring preferred two-dimensional orientation in one direction; in two perpendicular directions; and in two or more nonperpendicular directions. We then consider the use of grain boundary contact counts. Finally, methods of study of preferred orientation in three dimensions are briefly mentioned.

UNIMODAL GRAIN ORIENTATION

The unimodal model is based on the assumption that grains have a tendency to become aligned in a specific direction. The strength of this tendency determines the percentage of grains that will actually be aligned in that direction. In statistical terms the model assumes the existence of a relationship between an angular direction and the probability that a randomly selected grain actually possesses that particular orientation. The function describing this relationship is called the *probability distribution.*

Three specific probability distributions have been described as convenient models for unimodal orientation: the wrapped normal distribution, the circular normal distribution, and the Rayleigh distribution (Curray, 1956). The general shape of these three functions is sufficiently similar so that for most practical applications they can be treated as one.

The investigator studying grain orientation is usually interested in answering three questions.

1. Is a preferred orientation of grains statistically significant?

2. If there is a preferred orientation, what is its azimuth?

3. How strong is the tendency for preferred orientation?

In practice, an angle of preferred orientation (2) is estimated first; from this estimate must be decided whether the preference is large enough to exclude mere random scatter (1). If a preferential orientation is shown, the intensity of orientation can be determined (3).

Before discussing the actual equations used to compute the directions of preferred orientation, it is useful to relate them to well-known and conventional statistical techniques. In most applications, some sort of assumption is made about the type of distribution with which one is dealing. However, even if such an assumption is not made or cannot be made, descriptive measurements can be obtained.

The most frequently used descriptive measurements are the mean and the standard deviation. These two quantities summarize and, in many instances describe fully the probability distribution. In the evaluation of circular distributions, we are interested in finding a measure analogous to the mean and another measure analogous to standard deviation. These measurements are descriptive, and can only incidentally be called estimators of specific parameters of specific frequency distributions.

The analog of a mean (or average) in circular distribution is the *resultant vector*, sometimes also known as the *mean vector*. In the shorthand of vectorial notation, the mean vector, $\mathbf{R}$, is related to the individual vectors, $\mathbf{r}_i$, by the following equation (**ave** indicates vector average).

$$\mathbf{R} = \text{ave} \,(\mathbf{r}) = \frac{\Sigma \mathbf{r}_i}{\Sigma \,|\mathbf{r}_i|} \tag{5}$$

Note that if all vectors entering this computation are of unit length, then the denominator reduces to the value N (number of vectors). Unfortunately, there are two things wrong with this equation. First, it does not state explicitly how one should proceed with the computation, and second, it does not provide for the ambiguity of single-grain orientation. If we resolve a vector into its orthogonal components, we can write

$$r_i = a_i \sin \alpha_i \, \mathbf{i} + a_i \cos \alpha_i \, \mathbf{j} \qquad (6)$$

where r_i is the radius vector of length a_i oriented in the direction α_i. Note, however, that if we rotate the ellipse 180° around its center, the angle α_i will become $\alpha_i + \pi$, which changes this equation to

$$r_i' = a_i \sin \alpha_i \, \mathbf{i} + a_i \cos \alpha_i \, \mathbf{j} \qquad (7)$$

Now both r_i and r_i' give the orientation of the same ellipse in a plane. Equation (5) can be written as follows:

$$\mathbf{R}_\alpha = \text{ave}(\mathbf{r}) = \frac{\Sigma \pm a_i \sin \alpha_i}{a} \, \mathbf{i} + \frac{\Sigma \pm a_i \cos \alpha_i}{a} \, \mathbf{j} \qquad (8)$$

The first term of Equation (8) represents a vector in the y-direction and the second term represents a vector in the x-direction where

$$a = \Sigma \, |a_i| \qquad (9)$$

The ambiguity of signs, however, makes this approach unfeasible. If a grain i is oriented at an angle α_i, this orientation is equivalent to one at an angle $\alpha_i + k$, and the sign arbitrarily depends on which of these choices is taken. We can convert a half-circle periodicity to a full-circle periodicity by simply multiplying all angles by two. Thus, if a grain is oriented at an angle α_i, we record it as being oriented at an angle $2\alpha_i$. Then the alternate orientation at $(\alpha_i + \pi)$ becomes $2(\alpha_i + \pi)$, or $(2\alpha_i + 2\pi) = 2\alpha_i$, and instead of computing $\mathbf{R}_\alpha$, we compute $\mathbf{R}_{2\alpha}$ as follows:

$$\mathbf{R}_{2\alpha} = \frac{\Sigma \, |a_i| \sin 2\alpha_i}{a} \, \mathbf{i} + \frac{\Sigma \, |a_i| \cos 2\alpha_i}{a} \, \mathbf{j} \qquad (10)$$

The angle of preferred orientation α can be readily computed from this result.

BIMODAL GRAIN ORIENTATION WITH MODES AT RIGHT ANGLES

In the discussions dealing with the physics of grain orientation, arguments were given postulating not one but two distinct preferential alignments of elongated grains. One of these alignments was parallel to the fluid flow and the other at right angles to it. If we accept this as the basic hypothesis, then it is possible to evaluate the resulting preferred orientation in much the same way as unimodal distribution.

The only difference is that the periodicity of orientation of an arbitrary grain is not $180°$ but $90°$. Therefore, if a grain has an orientation angle α_i, we consider this angle equivalent to $(\alpha_i + \pi/2)$, or $(\alpha_i + \pi)$, or to $(\alpha_i + 3\pi/2)$. If we simply multiply each angle by 4, we get $4\alpha_i = 4(\alpha_i + \pi/2) = 4(\alpha_i + \pi) = 4(\alpha_i + 3\pi/2)$. Consequently,

$$R_{4\alpha} = \frac{\Sigma \mid a_i \mid \sin 4\alpha_i}{a} \, i + \frac{\Sigma \mid a_i \mid \cos 4\alpha_i}{a} \, j \qquad (11)$$

The value of α, computed from $R_{4\alpha}$, should have approximately the same direction ($\pm 90°$) as α computed from $R_{2\alpha}$. The absolute magnitude of the vector depends on the presence or absence of the secondary orientation peak.

MULTIMODAL GRAIN ORIENTATION

Although it is possible to develop methods for the analysis of complex orientation patterns, their utility is somewhat doubtful. The basic procedure for evaluating multiple modes of preferred orientation involves subtraction. This is begun by estimating the most prominent direction of preferred orientation. Next, using a frequency distribution, such as a circular normal, the component frequencies of this most prominent orientation are subtracted from the observations. The residuals are then analyzed for other components of preferred orientation. At each step care must be taken that the residuals and the frequencies just removed are sufficiently nonrandom.

Much of the complexity and difficulty of analytical techniques is avoided by using graphical methods for displaying orientation such as the rose diagram. This is simply a circular histogram, where the observed frequency within angular class boundaries is plotted twice, once at the class mark for the angle α_i, and again, as its mirror image with respect to the origin, at the class mark for the angle $(\alpha_i + \pi)$. Thus a rose diagram is symmetrical about its center of gravity. Orientation characteristics of directional data produce striking patterns when plotted on a rose diagram. This method often indicates subtle orientation patterns that may be overlooked with statistical techniques.

GRAIN-BOUNDARY CONTACT METHODS

This method of measuring preferred orientation differs in a fundamental way from vectorial techniques. Orientation of individual grains is not measured and moment statistics based on summation of data can not be computed. Instead, grain-boundary intercepts are counted for all grains for different azimuths. Thus in the grain-boundary approach the data consist of a table relating number of grain contacts as a function of azimuth. The grain-boundary contact method requires that the area over which counts are made be the same for all azimuths. Discrimination of grains on basis of size is possible. A size threshold can be established so that if the intercept length in a grain (or matrix) is too small (or too large), the count can be rejected. Interestingly enough, in this method it is possible, without further sophistication, to count certain portions of a grain that satisfy a threshold and to reject other portions of the same grain that do not.

The mathematical technique for determining preferred orientation by the grain-boundary approach is essentially the same as that used in the vectorial model. Each grain-matrix (and matrix-grain) contact becomes a vector of unit length directed at right angles to the scan. Except for the interpretation on unit vectors, the computations are the same as in the single-grain orientation approach.

GRAIN ORIENTATION IN THREE DIMENSIONS

The vectorial techniques for studying grain orientation can be extended into three dimensions. Data from these three-dimensional studies are usually represented on stereographic projections and the results evaluated by conventional fabric studies. These methods were discussed in Chapter 2.

SIMPLIFIED CHARTS FOR GRAIN ORIENTATION EVALUATION

To simplify the process of measuring grain orientation, it is convenient to develop standard forms and methods of recording the data. Figures 8 and 9 are examples of such forms. Figure 8 (top) is a representation of a circular histogram. The angular observations are recorded in much the same way as on an ordinary histogram. Ten degree-class boundaries are marked (with 5° subintervals for more refined work). The concentric half-circles provide facility to record separately grain specimens with different weighting characteristics. The class boundaries are labeled from 0 to 180° to emphasize that the choice of baseline is immaterial in this particular method of analysis.

The left center of Fig. 8 contains a convenient table for computing preferred orientation. The frequencies obtained in the circular histogram (from left to right) are entered in column C. The columns labeled A and B show the appropriate values of the direction cosines for 2α. The observed frequencies in column C are multiplied by the respective values in columns A and B, and the products are entered in columns AC and BC together with the correct signs. When all the products have been entered, the three columns (C, AC, and BC) are each summed, and the totals are entered in the appropriate boxes at the bottom. The sums of columns AC and BC are then divided by N, the sum of column C, and the results are recorded at Y and X, respectively.

The three numbers, N, Y, and X, are then used in Fig. 9 to evaluate the preferred orientation as follows. Note that N is an integer, and Y and X are positive or negative numbers between 0 and 1. Treating X and Y as coordinates, plot the point X, Y on the circular diagram of

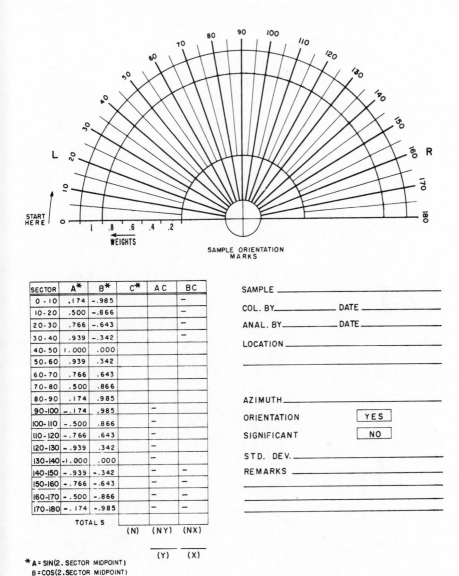

SECTOR	A*	B*	C*	A C	B C
0 - 10	.174	-.985			−
10 - 20	.500	-.866			−
20 - 30	.766	-.643			−
30 - 40	.939	-.342			−
40 - 50	1.000	.000			
50 - 60	.939	.342			
60 - 70	.766	.643			
70 - 80	.500	.866			
80 - 90	.174	.985			
90 - 100	-.174	.985		−	
100 - 110	-.500	.866		−	
110 - 120	-.766	.643		−	
120 - 130	-.939	.342		−	
130 - 140	-1.000	.000		−	
140 - 150	-.939	-.342		−	−
150 - 160	-.766	-.643		−	−
160 - 170	-.500	-.866		−	−
170 - 180	-.174	-.985		−	−
TOTALS					
			(N)	(NY)	(NX)
				(Y)	(X)

* A = SIN(2 . SECTOR MIDPOINT)
B = COS(2 . SECTOR MIDPOINT)
C = NO POINTS

SAMPLE _____

COL. BY_____ DATE _____

ANAL. BY_____ DATE _____

LOCATION _____

_____

AZIMUTH _____

ORIENTATION   [YES]

SIGNIFICANT   [NO]

STD. DEV. _____

REMARKS _____

_____

_____

Fig. 8 Preferred orientation of two dimensional data. (Z. V. Jizba)

Fig. 9. Note that care must be taken for appropriate sign control. The line connecting point X, Y with the origin is the resultant vector $\mathbf{R}_{2\alpha}$.

If the length of this vector is such that it cuts through the concentric circle indicating the value of N, then the orientation is significant at the 5% level. (This diagram is based on an assumed Rayleigh distribution.) If the arrow fails to cut that circle, then its length can be used to estimate the additional sample size required to

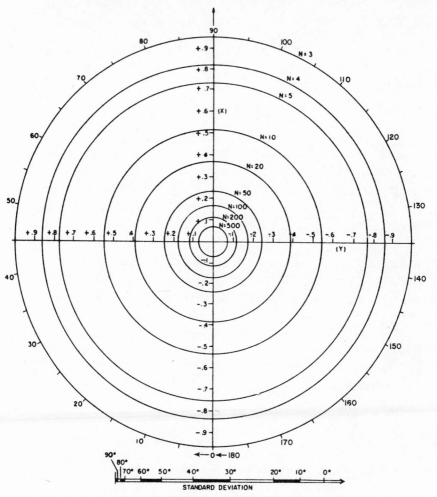

Fig. 9   Rayleigh test on 5% significance level.

test the hypothesis that preferred orientation exists, but is very weak. The azimuth of this orientation can be obtained by extending the plotted vector to the outer circle $(N-3)$ and reading off the angle. The strength of orientation, a measure equivalent to the standard deviation in linear statistics, can also be obtained; at the bottom of Fig. 9 is a scale that gives the standard deviation in degrees as a function of the length of the resultant vector.

MATRIX ANALYSIS

A discussion of preferred orientation of grains in a sediment would be incomplete without a mention of the "nearest-neighbor" concept. The measurement of preferred orientation produces an index of the tendency of an "average" grain to be aligned in a specific direction. However, the physics of grain deposition also suggests that the final orientation of a grain may depend not only on inertial, drag, and lift forces but also on the shape of the immediate neighborhood where that grain is being lodged. If the grain is large compared to its surroundings, this effect may be small. If, however, the grain is roughly the size of depositional surface irregularities, its placement and orientation with respect to neighboring grains may be of interest.

A fairly useful matrix analysis can be made with information obtained through the grain intercept method. The technique consists of computing an autocorrelation function of the grain-matrix sequence for each of the various azimuths at which a grain contact count is made. The computed autocorrelation functions then can be contoured into a two-dimensional grain-matrix autocorrelation pattern.

The numerical technique operates as follows. A trace is made as for counting grain-matrix contacts, but a count is made at regular intervals along the trace. The presence of either grain or matrix is recorded digitally. Let us represent "grain" by the digit 1 and "matrix" by the digit 0. The sequence of grain and matrix occurrence would then be converted into a sequence of zeros and ones:

$$\ldots 000110001111111101000000001111000 \ldots, \text{ etc.}$$

If X_i denotes the ith occurrence and Y the jth occurrence, where $j = i + k$, and N as the total number of points measured, the autocorrelation function A_k is given by the following equation:

$$A_K = \frac{N\Sigma x_i y_j - \Sigma x_i \Sigma y_j}{\sqrt{\left[N\Sigma x_i^2 - (\Sigma x_i)^2\right]\left[N\Sigma y_j^2 - (\Sigma y_j)^2\right]}} \tag{12}$$

If N is sufficiently large, the end effect $(N - K)$ can be neglected and $\Sigma x_i = \Sigma x_i^2 = \Sigma y_j = \Sigma y_i^2$ (since X and Y can have only two values, 1 and 0, and $1^2 = 1$, $0^2 = 0$). Furthermore, the sum can be interpreted as the number of grain-grain overlaps, since $1 \times 0 = 0$, $0 \times 1 = 0$, and $0 \times 0 = 0$. If N_{GO} is the number of 1's in the sequence and N_{GK} is the number of overlaps of grain-grain, when the sequence has been shifted K-spaces with itself, then the autocorrelation function is given by

$$A_K = \frac{NN_{GK} - N_{GO}^2}{N_{GO}(N - N_{GO})} \tag{13}$$

A possible application of this type of matrix analysis might be in the study of massive sands in which bedding planes are poorly defined. In such cases the autocorrelation pattern may give clues on the attitude of depositional surface.

REFERENCES

Curray, J. R., 1956, The analysis of two dimensional orientation data, *J. Geol.*, **64**, 117–131.

Freiberger, W. F., Editor-in-Chief, 1960, The international dictionary of applied mathematics, D. Van Nostrand Co., 1173 pp.

Rusnak, G. A., 1957a, The orientation of sand grains under condition of "undirectional" fluid flow 1. Theory and experiment, *J. Geol.*, **65**, 384–409.

——, 1957b, A fabric and petrologic study of the Pleasantview sandstone, *J. Sed. Pet.*, **27**, 41–55.

Selby, S. M., Standard mathematical tables 14th Ed., Chemical Rubber Co., 632 pp.

Spotts, J. H., 1964, Grain orientation and imbrication in Miocene turbidity current sandstones, California, *J. Sed. Pet.*, **34**, 229–253.

Zimmerle, W., and L. C. Bonham, Rapid methods for dimensional grain orientation measurements, *J. Sed. Pet.*, **32**, 751–763.

CHAPTER 14

MEASUREMENT OF POROSITY AND PERMEABILITY

BRUCE F. CURTIS

University of Colorado, Boulder, Colorado

PORE EVALUATION

A body of sediment or rock often is of interest chiefly because it may hold and transmit fluids in its pores. Examination of the rock matrix is of limited value for appraising pore properties; instead the openings themselves must be the principal objects of study. Qualitative representations of the shapes, sizes, and the continuity of pores are very helpful for understanding the evolution of sedimentary materials. The observations for such porosity descriptions usually are made with the microscope, and they result in classifications such as intergranular, vuggy, intercrystalline, pinpoint, interoolitic, oomoldic, or fractured. These descriptions, in turn, may reflect episodes in the geologic history of the rock and are an essential part of any complete examination of a sedimentary material. Notwithstanding their usefulness, these descriptive terms do not serve to appraise either the quantities of fluids contained within bodies of rock material nor the ways in which these fluids may move.

The need to evaluate these other characteristics has generated a field of quantitative study of rock pore systems and their dependent physical properties which has been called *petrophysics* (Archie, 1950).

The numerical values most commonly measured in petrophysical studies are porosity (the fractional pore space) and permeability (the fluid conducting property) of a rock. Capillary pressure curves are also of interest as indicators of pore sizes. The porosity and permeability, of course, are mass properties which ideally should be measured on sedimentary materials *in situ,* for removal to the laboratory may seriously change the pore geometry. Partly in response to this condition, methods have been developed for at least semiquantitative evaluation of porosity and permeability from subsurface data measured in boreholes. Porosity is calculated from electrical, radiometric, or acoustic logs or from gravimetry, whereas permeability measurement in the field generally requires pumping or pressure-recovery tests in wells. Geophysical logging of deep wells is common practice and the records are quite generally available. Pumping or pressure test records are more often considered confidential, but many of them may be obtained from public agencies concerned with water or petroleum resources as well as from private companies.

Laboratory determination of porosity and permeability has become more or less a standard process because it is more convenient and less expensive than field measurement and because data from small samples often reveal patterns of interest. With adequate care very precise values may be determined. However, it should be remembered that one of the most prominent characteristics of sedimentary rocks, and especially of carbonate rocks, is variability in properties from point to point, horizontally as well as vertically. There is no purpose, then, in seeking great precision of laboratory measurement when it is almost certain that the pore properties of any sample will differ somewhat from those of the rock volumes that surrounded that sample in place. Determinations accurate to the nearest 1% of porosity and to the nearest 10% of the measured value of permeability should be regarded as adequate for representing geological materials.

POROSITY

Total Porosity

The common measure of free fluid content of a sediment or rock is the *total porosity* which is defined as the ratio of pore volume to bulk volume of a specimen:

$$\phi = \frac{V_P}{V_B}$$

where ϕ is porosity and V_P and V_B are the pore and bulk volumes, respectively. The porosity has a value of less than one and it generally enters as a decimal in calculations, but quite commonly it is multiplied by 100 and stated as pore percentage of the bulk volume. In engineering work the ratio of pore volume to its associated solid rock matrix is employed frequently as *voids ratio*. If the voids ratio is v and decimal porosity is ϕ, the relation between the two expressions of comparative pore volume is

$$\phi = \frac{v}{(1 + v)}$$

A porosity or voids ratio is a dimensionless number which is expressed without units if (as commonly is true) the volumes entering the ratio are measured in consistent units. Sometimes, however, geologists deal with hybrid porosity relations for which the volumetric units are stated, as in the case of gallons of pore per cubic foot of rock (occasionally used in hydrology) or barrels of pore per acre foot of rock (used in petroleum reserve work).

Effective Porosity

A distinction commonly made among porosity values is that of *effective porosity* (symbol ϕ_c), the fractional volume of connected or "available" pores as contrasted with *total porosity* (symbol ϕ) which includes in its ratio not only the connected but also the isolated pore volume. Effective porosity is seldom more than 3% porosity less than the total value that is denoted when the unqualified word porosity is used.

Ground water hydrologists employ the term *effective porosity* in a quite different fashion. In their studies effective porosity is synonymous with *specific yield* (symbol S) which is a ratio of drained liquid to bulk volume calculated as

$$S = \frac{W_d}{V_B}$$

In this equation W_d is the volume of water or other wetting liquid which may be drained under specified hydraulic conditions from the fully saturated porous medium, and V_B is bulk volume of the medium. Another volume of liquid W_r is retained in the porous medium by capillary and adhesional forces, not only in the isolated pores but in the interconnected ones too. The volume of W_r is therefore considerably larger than that of isolated pores. Consequently, if consistent volumetric units are used,

and

$$S = \phi - \frac{V_{pi} + W_{r2}}{V_B}$$

$$\phi_c = \phi - \frac{V_{pi}}{V_B}$$

where V_{pi} is the volume of isolated pores, W_{r2} is the liquid volume retained in connected pores under drainage, and $V_{pi} + W_{r2} = W_r$. The symbol ϕ means the total porosity and S and ϕ_c denote effective porosity as the term is used in hydrologic and in petrophysical work, respectively. The value represented by ϕ_c is used in this chapter as effective porosity, for it is a property of the rock alone, whereas the value of S depends partly on physical properties of the rock grain surfaces and the contained fluid, the temperature, and the specific drainage conditions.

Measurement of Porosity

Because the bulk volume of a specimen equals the volume of solid material plus the volume of pore, the value of porosity may be determined by measuring any two of those three volumes. However, some laboratory techniques which depend on filling or emptying of pores ascertain either V_{pc} or $(V_{pi} + V_S)$ where V_{pc} and V_{pi} are the volumes of connected and of isolated pores, respectively, and V_S is the volume of solids. In these instances only the effective porosity may be determined by relating one of these measured values to the bulk volume. Since $V_{pc} + V_{pi} + V_S = V_B$ (where V_B is bulk volume), three of these individual quantities must be found in order to calculate the total porosity.

Laboratory Determination of Bulk Volume

The most direct method of bulk volume measurement is to cut a specimen into a regular geometric shape. Dimensions are calipered and the volume determined geometrically.

Several fluid displacement techniques are also used. The most prominent ones among them utilize some form of mercury pycnometer. All such instruments employ a carefully designed vessel whose fluid capacity may be measured when completely filled with mercury and when it contains mercury plus the completely submerged specimen. The difference between the first and second measurements, of course, represents the bulk volume of the sample. One of the most useful of these pycnometers is attached directly to a mercury metering pump that measures the liquid volumes placed into the pycnometer chamber. This method is especially recommended for rocks with fairly large pores, because the nonwetting mercury does not significantly penetrate the specimen.

If the pores in a carefully dried rock specimen have been saturated with a suitable wetting liquid such as tetrachloroethane, water, or kerosene, and the surface of the sample has been blotted carefully to remove excess fluid, the piece may be immersed in a volumetrically graduated vessel partly filled with the saturating liquid. The increase in volume reading on the scale of the container provides a direct measure of the volume of the specimen.

A procedure may be arranged easily so that a test specimen is submerged in a filled container that overflows into a tared weighing vessel. The liquid displaced by the specimen may then be weighed and its volume (which equals the specimen volume) may be determined from its known density. When mercury, which is a most convenient liquid, is used for this purpose, the sample is held submerged with prongs of known volume. Another technique utilizing weights alone is to saturate the sample with a wetting liquid of low volatility and known or easily determined density (e.g., water). The liquid-saturated sample is first weighed rapidly in air and then suspended and weighed submerged in the same liquid. The difference between the two weights divided by the density of the liquid used is equal to the bulk volume of the sample.

In any of the fluid displacement methods it is possible to coat the specimen with a thin wax, rubber, or plastic cover before it is submerged and to correct the measurement later for the volume of the cover material. The presence of very large pores often requires this precaution. In addition, fluid displacement techniques require care to prevent air bubbles from clinging to the submerged specimens and to be sure that the liquids employed do not react to change the volume of a dry specimen. The method adopted must be determined after considering the nature of the samples and the sizes of their pores.

Laboratory Determination of Solid Volume

The weight of a dry rock specimen divided by the density of its solid material is equal to its matrix volume. Therefore, if the average density of the solids is known with fair accuracy, an approximate value of their volume may be determined easily by weighing. Some density values ascribed to common sedimentary materials are quartz, 2.65; calcite, 2.71; dolomite, 2.86; and illite, 2.76. Frequently, the average solid density of one specimen in a suite is measured directly, and the resulting value is then applied to other samples of similar lithology. To find the density, the weight of a crushed sample is determined and liquid displacement or gas displacement in a Boyle's law chamber (described below) may be used conveniently for measuring solid volume. Similarly, the solid volumes of individual samples that are friable may be measured directly by displacement after the pieces are crushed finely enough to eliminate any aggregates enclosing pore space. Alternatively, the volume of crushed matrix may be determined by weighing it first in air and then suspended and submerged in a liquid. The difference between the two weights divided by the density of the liquid is equal to the volume of the solids. Measurement of solid volume permits calculation of total porosity.

Laboratory Determination of Combined Isolated Pore and Solid Volume

A chamber whose volume may be varied as corresponding pressures are measured permits determination by Boyle's law of combined isolated pore and solid volume. A pycnometer attached to a volumetric mercury pump is most often used for the purpose. One compression from atmospheric pressure P_a to a gauged absolute pressure P_2 is made with only air in the chamber at the outset. The volume of mercury pumped into the chamber to effect the pressure change is V_{Hg1}. With the porous sample placed in the emptied closed chamber, a volume of mercury V_{Hg2} is then injected to compress the air once again from P_a to absolute pressure P_2. Assuming the temperature remains constant, a volume $(V_{pi} + V_S)$ may be calculated for the specimen by the formula

$$(V_{pi} + V_S) = \frac{P_2(V_{Hg1} - V_{Hg2})}{P_2 - P_a}$$

The units of pressure and volume may be any convenient ones, provided, of course, that they are used consistently in the various terms of the equation.

Two connected chambers of known volumes may be used for measurement of combined isolated pore and solid volume by the Boyle's law method, if an accurate gauge and a pressure source are attached to one vessel (Fig. 1). Once the two chamber volumes V_1 and V_2 have been determined, the sample is placed in the chamber of volume V_1 and the cell is closed, with valve A open and both chambers at atmospheric pressure P_a. The valve A is then closed and valve B opened to admit compressed gas and raise the pressure in chamber V_1 to pressure P_1 absolute. Then the valve B is closed and valve A is opened. The pressure in the two chambers is thus allowed

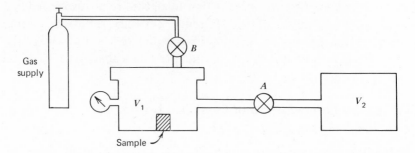

Fig. 1   Diagram of one form of Boyle's law apparatus for determining combined isolated pore and solid volume.

to equalize at a lower value P_2 absolute. Using any serviceable units consistently, the combined isolated pore and solid volume of the sample is easily obtained by the relation

$$(V_{pi} + V_S) = V_1 - V_2 \frac{P_2 - P_a}{P_1 - P_2}$$

Many modifications of this apparatus have been built and have given satisfactory results (American Petroleum Institute, 1960).

Laboratory Measurement of Connected Pore Volume

Procedures similar to those just described are used for measuring the volume V_{pc} of connected pores in a sample. The measured volume of liquid which may be injected into a porous specimen at high pressure (of the order of 50 atmospheres) provides a fair measurement of the connected pore volume. A metering displacement pump filled with mercury lends itself conveniently to this process, but this fluid is expensive and a mercury-impregnated sample is not useful for later tests. Other nonreacting liquids in suitably designed and calibrated pressure vessels can be used to carry out the same basic process. It is recognized that an error is involved in the trapping of an indeterminate amount of air in the sample pores, but the method has proved to be a good way to get an approximate effective porosity value rapidly.

A weighed sample that has been extracted carefully and dried may be impregnated with liquid more successfully if it is placed first under vacuum and then, while still at low pressure, is covered with a low viscosity and low vapor-pressure liquid. Pressure is then applied to the liquid, and after thorough saturation the sample is removed and weighed. The difference between the dry and saturated sample weights divided by the density of the liquid is the connected pore volume. Alternatively, the fluid in a saturated rock sample may be retorted, condensed, and measured in a closed system to determine the effective volume of pore. Water is useful for this purpose if it does not react with the rock solids.

A mercury pump and an attached pycnometer equipped with a vacuum gauge is another device frequently used for measuring connected pore volumes. The piston of the mercury pump is withdrawn after the sample in the chamber has been surrounded by mercury and the chamber sealed. The air in the pores of the specimen is the only gas present in the pycnometer and it expands as mercury is withdrawn. If the original air pressure in the sample pores was atmospheric (P_a) and the withdrawal of a volume V_{Hgw} of mercury results in a lower absolute pressure P_2, the volume of interconnected pores may be calculated by

$$V_{pc} = \frac{P_2 V_{Hgw}}{P_a - P_2}$$

As in the other porosity equations any units may be employed consistently.

One of the most widely used pore-volume devices is an apparatus modified after that described by Washburn and Bunting (1922). In this instrument (Fig. 2) the sample is surrounded with mercury at essentially atmospheric pressure. The system is then closed and the mercury level is lowered to allow the pore air to expand at low pressure. The air rises into a measuring burette and by subsequent equalizing of the mercury levels in the instrument the entrapped air is brought to atmospheric pressure again. The observed air volume is then approximately equal to the connected pore volume. Depending on the accuracy required, the direct reading may be corrected for air adsorbed on the surface of the sample and for a low pressure volume of air remaining in the pores of the sample when it is again surrounded by mercury after the expansion step. However, if the

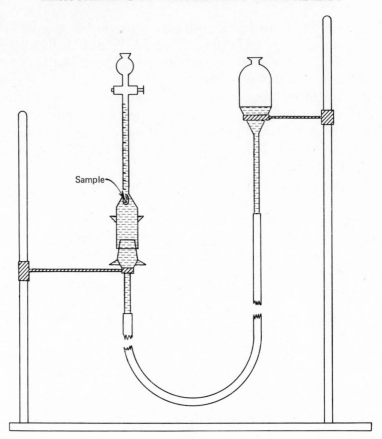

Fig. 2   Modified form of Washburn and Bunting apparatus as used for geological
materials.

total volume of the burette tube is of the order of twenty times the
bulk volume of the samples used, the residual air correction is
acceptably small for ordinary geological materials. In practice,
therefore, a well-designed apparatus of this kind may be read
directly, provided that several readings taken on each sample are
averaged and that an adsorption correction (determined by using a
nonporous blank specimen) is applied.

Laboratory Measurement of Total Porosity

If the pore distribution within a sample is rather uniform,
fractional pore volume may be calculated by observing and meas-

uring the actual pore area as a fraction of the total area of a sliced section of the rock. Usually this requires projection of an enlarged image and measurement with a planimeter. Alternatively, a large number of linear scans of a section may be made under the microscope using a mechanical stage. The lengths of pore and matrix encountered on each line are recorded, and the summations give average relative pore and solid volumes for the sample. The recording of observations on randomly scattered points may yield similar results (see "Point Counting"). An interesting variation on this method is attributed by Scheidegger (1960) to Chalkley whose technique involves throwing a pin many times at a photomicrograph of a sample. The resulting hit and miss statistical data reveal the total porosity.

Undoubtedly the most rapid way of measuring the porosity of cores from wells is the gamma-ray attenuation device (Evans, 1965; Harms and Choquette, 1965) which yields a continuous record of porosity. The reduction in the intensity of a gamma-ray beam, caused by passage through the automatically calipered core, is measured. Provided the densities of the matrix material and the pore fluid are known, or fairly closely estimated, this measurement may be converted continuously to porosity values by a computer. About 150 ft of core per day may be analyzed by this instrument with good accuracy.

Laboratory Measurement of Effective Porosity

A simple method of determining an approximate porosity value depends on measuring the electrical resistivity of a sample whose pores are filled with a brine of known salinity. The simple formula $R_o/R_w = F$ derives a quantity called the formation resistivity factor F from the measured resistivity R_o and the resistivity of the brine R_w. In turn, the value F has been shown, empirically, to be related to the decimal effective porosity ϕ_c approximately as set out in the Archie formula $F\phi_c^m = 1$. The value of m is selected within a range from 1.3 for unconsolidated materials to 2.2 for tightly cemented rocks. The even more commonly used Humble formula $(F\phi_c^{2.15} = 0.62)$ relates the porosity to the formation resistivity factor in an equation which has proved satisfactory for a variety of rock types. If the resistivity values are in consistent units, the remaining quantities in these equations are dimensionless and not expressed in units.

Selection of Laboratory Procedure

Plainly the diverse techniques for measuring the porosity percentage or the bulk, solid, connected pore, or isolated pore plus solid volumes may be combined in many different ways. The methods adopted normally will be those that permit handling the specific material with suitable accuracy and with as few measurements as possible. The mercury displacement pump and pycnometer used as a bulk volume device and as a Boyle's law apparatus is perhaps the most compact and useful single instrument, although it is rather expensive. The accurate laboratory determination of porosity in unconsolidated materials requires considerable precaution. A cementing substance (such as plastic) injected into the material to be sampled while it is still in place develops sufficient consolidation to allow removal and handling. Of course, the volume of cementing material must be corrected for in the subsequent laboratory measurements.

Field Measurement of Porosity

A field measurement of the specific yield of an unconfined aquifer may be derived from a well-pumping test and suitable drawdown observations (Davis and DeWiest, 1966). The specific retention or retained liquid volume per unit bulk volume must also be measured, or estimated from average data, in order to relate specific yield to total porosity.

Geophysical well logs provide the only practical means of *in situ* porosity measurement involving confined aquifers. The techniques are given detailed treatment in books on well log analysis (e.g., Pirson, 1963). Three classes of logging instruments have been used extensively and a fourth is now appearing. The resistivity of a fluid-filled rock mass may be determined from logged electrical measurements suitably corrected for effects introduced by the borehole. Furthermore, the resistivity of the fluid in the pores may be ascertained by direct measurement on a fluid sample or calculated from log data. These values allow calculation of the formation resistivity factor and hence of the effective porosity.

A second group of logging instruments record the effects arising from subjecting rock masses to radiation. From experimental stuides it is known that total porosity is proportional to logged quantities if allowance is made for extraneous factors. Once the lithology is known, the porosity may be determined with fair confidence.

Another type of log automatically times the passage of acoustic waves through rocks in a borehole wall and supplies a record that is the basis of a porosity calculation using a time average formula (Wyllie, 1956). Provided the lithology and nature of the contained fluids are known and data on the degree of consolidation are available, a rather accurate value of total porosity may be found (Sarmiento, 1961). Recently borehole gravimeter instruments have been developed and have shown great promise for measuring porosity in the field (McCulloh, 1967).

All the field methods of measuring porosity have uncertainties resulting from the physical surroundings in which the measurements are made, but at best they give results accurate within a very few units of percent porosity. They provide the notable advantage of a continuous porosity record for the entire thickness of a bed and are especially valuable for displaying the considerable variations in porosity within many beds that appear superficially to be quite uniform. Any reduction in porosity resulting from load stress is included in the field-measured value. However precise the individual laboratory data may be, in many instances they represent only semiquantitatively the porosity of a whole bed of sedimentary rock. Thus in practice the field measurements are often as dependable as laboratory results or more so.

PERMEABILITY

The concept of permeability, or the fluid transmitting capacity of a porous rock, is founded on the observations of Henri Darcy more than a century ago. In general form, Darcy's equation states that $v = K(-dh/dL)$. The symbols represent a rate of transmission v of water through the porous material, a loss of head $(-dh)$ per unit length (dL) in the direction of flow, and a coefficient of permeability K (also called hydraulic conductivity). The value v is not a true rate of fluid flow within the winding pore channels, but rather an apparent rate of flow resulting from dividing the rate of water discharge by the cross-sectional area normal to the flow path through the medium. In porous material this so-called Darcy velocity will always be less than the true velocity within the pores, the two being related as $v_t = v/\phi_c$, where v_t and ϕ_c refer to true velocity and decimal effective porosity, respectively.

The permeability coefficient K defined in the Darcy formula has the dimensions of velocity and it applies specifically to water at a given temperature, resulting in units such as cm/sec (i.e., $cm^3/(sec)(cm^2)$ at 20°C or gal/(day)(ft^2) at 60° F. Even though the physical conditions specified for measuring such permeability coefficients may not prevail in the geological setting of the rock, the values are standardized, readily compared, and widely used in dealing with ground water.

For many purposes it is desirable to arrive at a permeability value dependent solely on the pore geometry and solid-surface characteristics of the porous medium and not contingent on temperature or the fluid that is transmitted. To achieve this, a value $k = K\mu/\rho g$ is defined in which μ is the viscosity and ρ is the density of the transmitted fluid, g is the gravitational acceleration, and K is the coefficient of permeability. In this equation the value k has the dimensions of area and is known as the intrinsic permeability, absolute permeability, or the specific permeability. Rationally it may be looked on as a sort of average pore area that determines the rate of fluid flow in the rock.

The permeability defined in this way is considered to be a property of the rock itself and it is implied that the moving fluid does not act in any fashion to change the volume or surface characteristics of the rock solids. Phenomena of fluid-rock interaction sometimes create problems in selecting the liquids for laboratory determinations of permeability. If water is used, it is best to employ an actual sample of liquid from the rock formation, or a replica, in order to obtain a permeability value consistent with subsurface conditions. In clayey samples, distilled water freed of air and salts has been found to give permeability values that are lower by as much as two orders of magnitude than those obtained using saline water. This is thought to reflect an increase in structured water on clay particle surfaces when ion concentration is low. Because the trouble of obtaining a compatible water is seldom warranted, it is common to employ a nonreacting fluid for laboratory tests. Permeabilities derived from gas flow are undoubtedly the most consistent and widely measured values.

In careful usage the unqualified term *permeability* denotes the specific permeability—a practice which is followed in this chapter. However, caution must be exercised by a geologist dealing with ground water literature where the word permeability is often employed loosely to designate the coefficient of permeability K.

On the basis of the stated relations, a permeability in square centimeter units may be defined as

$$k_{cm^2} = -\frac{Q \mu \Delta L}{A \rho g \Delta h}$$

When the discharge Q is in cubic centimeters per second the viscosity μ is in poises, the length of the sample in the flow direction ΔL is in centimeters, the cross-sectional area of the sample A is in square centimeters, and the density ρ, gravitational acceleration g, and head-change Δh in the flow direction are measured in cgs units, the resulting value of k is in square centimeters. The negative sign in the equation reflects the fact that head h decreases in the direction of increase in length L. In many fluid flow calculations this vector convention is important, but for laboratory and field determinations of permeability the negative sign becomes unnecessary because the direction of lowering of head is obvious. In the equations following, the minus has been removed.

The square centimeter unit of permeability is very large and inconvenient; consequently, the darcy has been defined as a common unit. It is obtained from the above equation by employing centipoises for the units of viscosity, and by converting the product $(\rho g \Delta h)$, which has the dimensions of pressure to atmospheres. For consolidated rocks even the darcy is rather a large unit and the more convenient millidarcy (0.001 darcy) is more commonly used (American Petroleum Institute, 1956). A simple equivalency serves to convert a permeability expressed in square centimeters to one in units of darcys:

$$k_{cm^2} = 9.869 \times 10^{-9} \, k_{darcys}$$

If a gas is employed as the fluid for measuring permeability, the values obtained are usually higher than when a liquid is used. The effect has been ascribed largely to a slippage resulting from molecular activity of the gas, and it is particularly noticeable in testing materials with small pores. Klinkenberg's (1941) study of this phenomenon led to the simple means of reconciling gas and liquid permeabilities noted in the following discussion.

Laboratory Measurement of Permeability

When laboratory permeability measurement is made, the specimen must be fitted firmly into a sample holder. Consolidated material is cut into a geometric shape and mounted with open ends through which the test fluid is made to flow. Permeabilities may be estimated by making capillary pressure measurements on irregularly shaped specimens, but there are some uncertainties as discussed under that heading. Poorly consolidated, superficial sediments or soils may be accommodated in liquid permeameters, but these materials are almost always disturbed during transfer to the instrument, and the results, although useful for comparative purposes, are questionably representative of the natural conditions. In fact, even in consolidated rocks, permeability is dependent partly on the imposed confining pressure. Studies by Fatt and Davis (1952) showed that permeabilities at conditions corresponding to a depth of 15,000 ft may be, on the average, only about 60% of the permeabilities measured on the same materials at surface conditions. Caution is always appropriate when projecting laboratory measurements to the natural environment.

In liquid permeameters, water or brine is most commonly used as the test fluid, but other nonreacting liquids of low viscosity and moderately high boiling point, such as kerosene, are also satisfactory. In any such device there must be provision for measuring head difference across the sample. This is easily accomplished in equipment of the variety indicated in Fig. 3 which represents a constant head permeameter. With such an instrument it is possible to ascertain the volume of fluid discharged during measured time under a head differential Δh. The equation presented for specific permeability may be applied directly, for all the variables are known from direct measurement and from tabulated properties of the liquid.

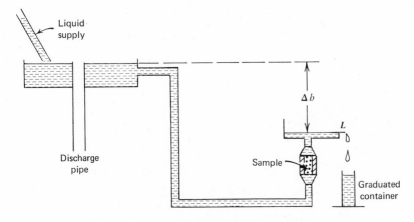

Fig. 3 A constant-head permeameter.

A falling head permeameter (Fig. 4) in which the rate of inflow may be observed by the descent of liquid level in a tube is appropriate for materials of lower permeability. The formula applicable to this apparatus is

$$k_{cm^2} = 2.3 \ \frac{d^2 \, \mu \, L}{D^2 \, \rho g \, \Delta t} \ \log \frac{h_1}{h_2}$$

in which the quantities not previously defined are d and D, the internal diameters of the measuring tube and sample, respectively, L the length of the specimen along the flow path, and Δt the number of seconds between the observations of fluid heights h_1 and h_2 above the free surface, all expressed in cgs units. If the ratio of D to d is large, the apparatus is quite sensitive.

Observations on materials of low permeability commonly employ instruments that impose a driving pressure on a fluid in order to produce an easily measurable flow rate. In a liquid permeameter of such design, Q in the Darcy relation is determined by a flow meter or by the timed rise of liquid in a calibrated receiving vessel. Fluid temperature must be measured to provide tabulated density and viscosity values. Outlet pressure is commonly atmospheric, and the input pressure imposed by a pump or by compressed gas is measured

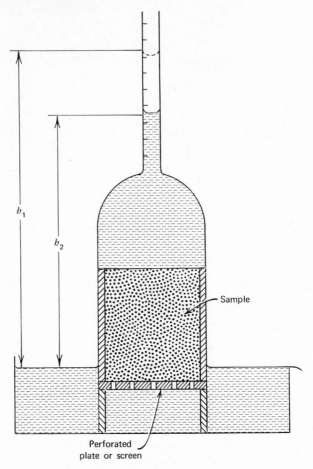

Fig. 4 A falling-head permeameter.

to provide a pressure differential value (ΔP). When a liquid is used, care must be taken to include in the calculation not only the gauged pressure differential but also the head pressure differential resulting from difference in elevation of the input and outlet ends of the sample. Then

$$k_{cm^2} = \frac{Q\mu L}{A(\Delta P + \Delta z\rho g)}$$

where the symbols not previously defined are the pressure difference ΔP across the specimen as indicated by gauges at the input and outlet ends of the specimen and Δz, the elevation of the input end minus the elevation of the output end of the specimen. All units are cgs system.

Although liquid permeameters are used when thought necessary, it is far more common to measure permeabilities using gas as the moving fluid. Compressed air or nitrogen supplied from a pump or cylinder with a pressure regulator works quite satisfactorily. The pressure at each end of the specimen must be known. In many instruments only an input gauge pressure and temperature need be measured directly because the outlet pressure is atmospheric. The flow rate must also be measured at the input or outlet with an orifice meter, a liquid displacement meter, or with calibrated, bored tubes containing floats. The orientation of the sample in a gas permeameter does not affect the calculations because the head differential of gas within the specimen is a negligible quantity. Assuming the gas temperature remains constant through the specimen, only two modifications of the liquid flow formula are necessary for calculating permeability. The first can be incorporated in the equation to take account of the fact that the gas expands as the pressure is reduced across the specimen. Applying Boyle's law the equation becomes

$$k_{cm^2} = \frac{2P_i Q_i \mu L}{A\left(P_i^2 - P_o^2\right)}$$

where cgs units are used with P_i and P_o referring to absolute input and outlet pressures and Q_i to the flow rate measured at the input end of the sample. If the flow is measured at the outlet, the subscripts of P and Q in the numerator change accordingly.

The second modification is the correction for slippage devised by Klinkenberg (1941). A plot of the permeabilities k_i measured on the instrument is made versus the reciprocals of P_m, the corresponding mean pressures $(1/P_m = 2/P_i + P_o)$. These data plot in a straight line (Fig. 5) and if projected to $1/P_m = 0$ a quantity k is obtained. This quantity is the absolute permeability and corresponds to a reading at theoretically infinite pressure. The average experience with various

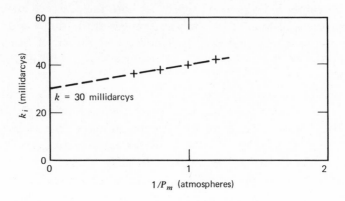

Fig. 5 Determination of absolute permeability (k) from measured permeabilities to gas.

oil reservoir sandstones charted by the American Petroleum Institute (1956) allows an estimate of the magnitude of error that might be imposed by neglect of this correction. At pressures normally used in laboratory instruments, the error may be less than 10% for permeabilities of several hundred millidarcys, but for permeabilities of approximately a few millidarcys the error is 60% or more, and at a few tenths to a few hundredths of a millidarcy the uncorrected gas permeabilities will be several times those measured with liquids. Routine analyses of geological samples usually report uncorrected permeability to gas. As long as the information is used simply for general comparison of materials, this is not bothersome. When the values are to be used for calculations, it is well to estimate the error.

An unusual method, developed by Teodorovich, which permits visual estimating of permeability in carbonate rocks was reviewed by Aschenbrenner and Chilingar (1960). Observations of the pores displayed in thin-sections are related to some empirical numbers that characterize the shapes, sizes, degrees of interconnection, and abundance of pores. The product of these numerical values, which have been developed from experience with many samples, gives the permeability in millidarcys. The reviewers' studies indicated that permeabilities thus determined may be expected, on the average, to be within 10% of the instrumentally measured value.

Directional Permeability

Permeability is treated by the linear-flow Darcy formula as if it were uniform regardless of direction. Actually it is a directional property of most natural materials, varying with heterogeneity and with anisotropy or "fabric" in the rock. The so-called vertical permeability, measured normal to the bedding, is commonly considerably lower than any value parallel to the bedding surfaces, and even within the bedding planes there may be substantial directional variation. Scheidegger (1960) especially has studied this property, developed a tensor theory of directional permeability, and examined its consequences.

To determine an average permeability value within bedding planes (often called the horizontal permeability) a radial flow instrument is sometimes employed in the laboratory. A cylindrical sample is cut with its axis normal to the bedding and an interior is hole drilled along the axis. Leaving one end of the hole open, the ends of the sample are sealed and the fluid is made to flow either into or out of the central well. The resulting value of average permeability in planes perpendicular to the axis is especially valuable for characterizing the carbonate rocks, many of which display notable directional variations in permeability. As applied to such a flow pattern, the steady state Darcy formula becomes

$$k_{cm^2} = \frac{2.3 \, \mu Q \log R/r}{2 \pi h \, (P_i - P_o)}$$

where h is the axial length of the cylindrical sample, R its external radius, r the radius of the axial hole, P_i and P_o are the input and outlet pressures, respectively, and the other values are as previously defined. All are in cgs units. In the event a gas is used in the measurements the corresponding formula is

$$k_{cm^2} = \frac{2.3 \, Q_b P_b \, \mu \log R/r}{\pi h \, (P_i^2 - P_o^2)}$$

where the new symbols Q_b and P_b are corresponding flow rate and pressure values measured at any selected pressure base—usually at the outlet or input pressure, depending on the location of the flow meter.

If there is need to examine the orientation of maximum permeability in an anisotropic rock, the permeabilities must be measured in three directions. This is usually done by shaping the sample into a cube and providing for closing off four faces at a time while the measurement is made between the remaining two opposing faces. Mast and Potter (1963) recently investigated the relation between rock fabric, maximum permeability, and the direction of maximum imbibition of a wetting liquid against air contained in the rock pores. Theoretical evidence that the direction of maximum imbibition rate and that of maximum permeability should agree received some experimental confirmation, suggesting a promising technique for defining maximum permeability directions and, in turn, the fabric directions.

There is evidence also that orientation of the maximum permeability direction may be determinable from electrical resistivity measurements if substantial directional differences exist. The resistivity is minimal in the direction of least tortuosity of pores and this, in turn, permits rapid determination of the maximum permeability direction (Wyllie and Spangler, 1952).

Capillary Pressure Measurements

For many studies of porous rocks it is useful to record the capillary pressure curves obtained by measuring the pressures required to displace a wetting fluid from the pores of various sizes. The resulting curves (Fig. 6) are generally regarded as indicative of the size distribution of pores in the rock, but they are more closely related to the distribution of pore throat sizes. Values of effective porosity, as well as approximate permeability, may be derived from the curves developed by fluid injection.

One capillary pressure technique is to displace water from a saturated sample by injecting oil or gas under measured pressure. The water escapes through a membrane such as cellophane that will not transmit the other fluid. Moderate displacement pressures of the order of a few hundred psi can be attained on such equipment.

Another rather widely used method for obtaining capillary pressure curves is that of mercury injection against a vacuum (Purcell, 1949).

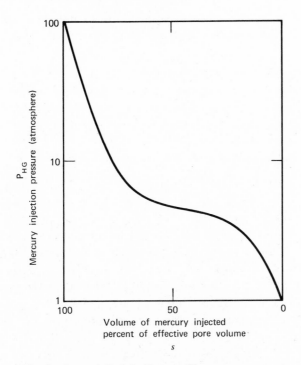

Fig. 6 A mercury-injection capillary pressure curve.

This technique is applicable to irregularly shaped specimens. The sample is placed in a sealed pycnometer cup connected through its base with a volumetric mercury pump and through its top to a gas line. By means of valves the gas line may be used for either creating a vacuum within the cell or for imposing a pressure up to 130 atmospheres. With the sample sealed in the pycnometer cup, the chamber and sample pores are evacuated. A measured volume of mercury is then pumped in to fill the cup to a reference mark and provide the value of cup volume minus sample bulk volume. The pressure on the mercury is next increased in increments by applying gauged pressures in the gas line. With each increase of pressure more mercury is forced into the pores of the sample and the amount is determined at each step by returning the mercury level in the cell to its reference level. Essentially all the available pores are filled eventually, and the total volume of mercury injected represents the effective pore volume. The curve of injected mercury volume versus pressure is of the typical capillary pressure form (Fig. 6).

An estimate of theoretical permeability may be made from mercury injection capillary pressure data by using Purcell's equation

$$k \cong 0.66l \, \phi_\% \int_{0\%}^{100\%} \frac{ds}{P^2_{Hg}} \Bigg|$$

where k is in millidarcys, $\phi_\%$ is the percentage (not decimal) porosity, and s is the percent of the connected pore space filled with mercury at the pressure P_{Hg} expressed in atmospheres. The integral is evaluated graphically from a curve of s versus $1/P^2_{Hg}$, which is derived easily from the capillary pressure curve. A lithologic factor l adjusts the theoretical permeability based on an idealized capillary system to actual values. In practice the l factor is found to vary with permeability, and it must be derived from comparison with other permeability determinations on similar samples. Although this introduces an element of uncertainty into the calculations, the results, even with average l factors, are generally at least of the correct order of magnitude, and the technique provides confirming data.

Capillary pressure curves do not, in fact, represent accurately the pore size distribution as Scheidegger (1960) has noted. Nevertheless they are useful in estimating what kinds of pressures are required to move one fluid against another within single pores of various sizes. This extensive subject has been given a great deal of attention by petroleum engineers and geologists (e.g., Pirson, 1958; Collins, 1961, Amyx and Bass, 1962).

Effective and Relative Permeability

Situations in which rock pores are occupied jointly by two or three fluids have been investigated in connection with ground water and with oil production (e.g., Irmay, 1954; Scheidegger, 1960; Stallman, 1964). In the usual geological case water is present with another sparingly miscible fluid—oil, gas, or air. When there is flow, the fluids interfere with each other and decreased transmission results, just as if the permeability of the rock specimen were lower than its intrinsic value. Values of *effective permeability*, which may then be defined, are functions principally of the nature of the rock matrix, its pores, and of the relative proportions of the two fluids present in the pores.

Expressed in common units of permeability, the effective permeability is the transmitting capacity of the rock for one of the two fluids present when stated proportions of the two fluids are in the pores. The idea may also be extended to the more unusual instances where three fluids are present. Following the convention recommended by American Petroleum Institute (1956), the symbol $k_{o(70,30)}$ denotes effective permeability of the rock to oil when 70% of the pore space is occupied by oil and 30% by water. The symbol $k_{w(70,30)}$ represents the effective permeability to water under the same conditions. A rather easier figure to deal with is *relative permeability* obtained by dividing an effective permeability by the intrinsic permeability of the rock. Thus $k_{ro(70,30)} = k_{o(70,30)}/k$, and relative permeabilities to other fluids are similarly defined. Such a value tells, of course, what fraction of its intrinsic permeability a rock offers to a particular fluid at the stated fluid saturation percentages. The sum of the relative permeabilities to all fluids present is commonly somewhat less than one. If a particular need exists for such information, the rather complicated laboratory equipment required may be called into play to develop diagrams like that shown in Fig. 7. It appears that relative permeability characteristics of rocks may change measurably in response to geostatic load stress (Chierici et al., 1967).

Field Measurement of Permeability

Some empirical relations such as those mentioned by Pirson (1964) have been developed for determining approximate permeabilities and relative permeabilities from well log data. None is claimed to give highly dependable values and none is applicable to rocks other than oil or gas reservoirs.

The methods developed by ground water and petroleum geologists and engineers are rather satisfactory for measuring permeabilities in large volumes of rocks, even though they require special well tests that often involve considerable expense. In ground water work the customary method is the steady pumping of a well accompanied by observations of head drawdown in two wells. In testing wells drilled for petroleum it is common to isolate the face of a fluid-producing formation from the other beds penetrated. The formation is allowed to flow for a period and then closed in while a record of pressure-recovery is made.

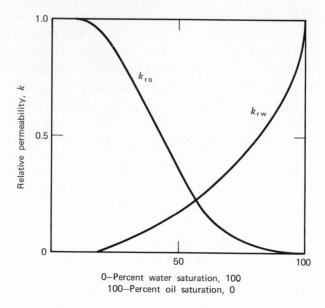

Fig. 7 Typical relative permeability curves for oil (k_{ro}) and water (k_{rw}).

Completed wells or wells temporarily completed with drill-stem testing equipment are suitable sources of these data. A considerable family of equations apply to these pumping and pressure buildup tests. Some allow permeabilities to be calculated either for equilibrium or nonequilibrium drainage of confined or unconfined aquifers (Davis and DeWiest, 1966). Other equations discussed by Theis (1935), Miller et al. (1950), and Horner (1951) permit derivation of effective permeabilities from pressure-recovery records. Articles such as that by Maier (1962) outline modern methods of field permeability analysis.

Although pumping and pressure recovery tests conducted on wells are large and expensive operations, they are made rather frequently because of the important information they provide. They present, especially, a measure of formation permeability based on flow in a very large "sample" of rock. Moreover, the effects of load stresses and of large-scale fracturing are included in the field measurements, but they are not in those made in the laboratory.

RELATION BETWEEN POROSITY AND PERMEABILITY

The scientific literature records many attempts to discern a quantitative relationship between porosity and permeability. There is intuitive expectation of such an affinity, for more connected open spaces certainly would improve a rock's capacity for transmitting fluids. Diagrams of laboratory data from individual formations show that some quite convincing linear trends may be established when the porosity and permeability values are plotted on semilogarithmic coordinates. Archie (1950) found an average of about tenfold increase in permeability for a 3% increase in porosity. Numerous equations, based on theoretical models or on empirical observations, have been proposed to represent such relationships. Although the formulas work rather well under specific conditions or with limited data, they are not very successful when extended to general application. Of the theoretically derived equations, that of Kozeny (1927) and its later modifications are probably the best known. Scheidegger (1960) has presented a good review of such work which shows the subject needs closer study.

One of the important things to be learned from the observational evaluation of permeability developed by Teodorovich (Aschenbrenner and Chilingar, 1960) is that many factors influence permeability. There are no existing means of quantitatively measuring most of these factors. Although porosity is prominent among them and is susceptible to measurement, it is by no means the sole determinant of permeability.

REFERENCES

American Petroleum Institute, 1956, Recommended practice for determining permeability of porous media, A.P.I. R.P. 27, Division of Production.

―――, 1960, A.P.I. recommended practice for core-analysis procedure, A.P.I.R.P. 40, Division of Production.

Amyx, James W., and Daniel M. Bass, Jr., 1962, Properties of reservoir rocks, Chapter 23, in Petroleum production handbook, McGraw-Hill Book Co., 49 chapters.

Archie, G. E., 1950, Introduction to petrophysics of reservoir rocks, *Bull. Am. Assoc. Petrol. Geologists,* **34**, 943—961.

Aschenbrenner, B. C., and G. V. Chilingar, 1960, Teodorovich's method for determining permeability from pore-space characteristics of carbonate rocks, *Bull. Am. Assoc. Petrol. Geologists,* **44**, 1421—1423.

Chierici, G. L., G. M. Ciucci, F. Eva, and G. Long, 1967, Effect of the overburden pressure on some petrophysical parameters of reservoir rocks, *Proc. 7th World Petrol. Congr.,* **2**, 309—338.

Collins, Royal E., 1961, Flow of fluids through porous materials, Reinhold Publ. Co., 270 pp.

Davis, Stanley N., and Roger J. M. DeWiest, 1966, Hydrogeology, John Wiley and Sons, 463 pp.

Evans, Hilton B., 1965, GRAPE*—a device for continuous determination of material density and porosity (*gamma ray attenuation porosity evaluator), Proc. Soc. Prof. Well Log Analysts, 6th Ann. Logging Symposium, B1—B25.

Fatt, I., and D. H. Davis, 1952, Reduction in permeability with overburden pressure; technical note 147, *J. Petrol. Tech.,* **4**, 16.

Harms, J. C., and P. W. Choquette, 1965, Geologic evaluation of a gamma ray porosity device, Proc. Soc. Prof. Well Log Analysts, 6th Ann. Logging Symposium, C1—C37.

Horner, D. R., 1951, Pressure build-up in wells, Proc. 3rd World Petrol. Congr., Section II, 503—521.

Irmay, S., 1954, On the hydraulic conductivity of unsaturated soils, *Trans. Am. Geophysical Union,* **35**, 463—467.

Klinkenberg, L. J., 1941, The permeability of porous media to liquids and gases, Drilling and Production Practice, 200—213.

Kozeny, Josef, 1927, Über kapillare Leitung des Wassers in Boden, *Sitzbericht Akad. Wissenschaften Wien,* **136**—2A, 271—306.

Maier, L. F., 1962, Recent developments in the interpretation and application of DST data, *J. Petrol. Tech.,* **14**, 1213—1222.

Mast, R. F., and P. E. Potter, 1963, Sedimentary structures, sand shape fabrics, and permeability II, *J. Geol.,* **71**, 548—565.

McCulloh, T. H., 1967, Borehole gravimetry: new developments and applications, *Proc. 7th World Petrol. Congr.*, **2**, 736–744.

Miller, C. C., A. B. Dyes, and C. A. Hutchinson, Jr., 1950, The estimation of permeability and reservoir pressure from bottom-hole pressure build-up characteristics, *J. Petrol. Tech.*, **5**, 171–176.

Pirson, S. J., 1958, Oil reservoir engineering, 2nd ed., McGraw-Hill Book Co., 735 pp.

——, 1963, Handbook of well log analysis: Prentice-Hall, 326 pp.

Purcell, W. R., 1949, Capillary pressures—their measurement using mercury and the calculation of permeability therefrom, *Petrol. Trans. Am. Inst. Mining and Metallurgical Eng.*, **186**, 39–48.

Sarmiento, R., 1961, Geological factors influencing porosity estimates from velocity logs, *Bull. Am. Assoc. Petrol. Geologists*, **45**, 633–644.

Scheidegger, A. E., 1960, The physics of flow through porous media, rev. ed., University of Toronto Press, 313 pp.

Stallman, Robert W., 1964, Multiphase fluids in porous media—a review of theories pertinent to hydrologic studies, *U.S. Geol. Surv.* P.P. 411–E, 51 pp.

Theis, C. V., 1935, The relation between the lowering of the piezometric surface and the rate and duration of discharge of a well using ground water storage, *Trans. Am. Geophysical Union*, **16**, 519–524.

Washburn, E. W., and E. N. Bunting, 1922, Determination of porosity by the method of gas expansion, *J. Am. Ceramic Soc.*, **5**, 112–129.

Wyllie, M. R. J., A. R. Gregory, and L. W. Gardner, 1956, Elastic wave velocities in heterogeneous and porous media, *Geophysics*, **21**, 41–70.

Wyllie, M. R. J., and M. B. Spangler, 1952, Application of electrical resistivity measurements to the problem of fluid flow in porous media, *Bull. Am. Assoc. Petrol. Geologists*, **36**, 359–404.

SECTION
V

MINERALOGICAL
ANALYSIS

CHAPTER 15

PREPARATION OF THIN-SECTIONS

HUBERT A. IRELAND

University of Kansas, Lawrence, Kansas

Thin-sections, as used in a geological sense, are very thin slices or fragments of rocks, fossils, or minerals mounted on glass slides for study under a microscope. The study of thin-sections is an important technique for determining mineral content, texture, fabric, structure, fossil content, organic matter, diagenesis, and other aspects of a rock. Thin-sections of igneous and metamorphic rocks are ground to a uniform thickness of 30μ, which provides a standard on which to base interference colors, pleochroism, extinction, and other optical properties. Technique for thin-sectioning of siliceous sedimentary rocks is much the same as for igneous and metamorphic rocks. Sectioning of carbonates and porous, weathered, or unconsolidated sediments or rocks requires essentially the same procedure as for hard rocks, but greater care, more gentle handling, and some other modifications of technique are necessary. Salt, gypsum, and coal require special techniques. Impregnation is required for many samples, and wedge-sections may be desirable for some samples. Thin-sections of fossils may require some sort of special treatment or modification of standard techniques. Discussion of the techniques, equipment, and materials for thin-sections required by paleontol-

367

ogists will not be given here, but some of the discussion applicable to thin-sections of carbonate rocks is also applicable to paleontological work. The technique of thin-section preparation for the special requirements of paleontologists is described in a special section of the *Handbook for Paleontological Techniques* (Kummel and Raup, eds., 1965).

Making thin-sections requires experience, dexterity, and ingenuity. A good section results from using proper cements, grinding to optimum thickness, correct heating, complete impregnation, and other manipulations to suit the sample and type of information desired. It is recommended that an inexperienced person make a few practice thin-sections before attempting work on a valuable specimen.

The basic methods of making thin-sections have changed very little in the last 70 years, but automatic machines, new cements, and various improvements have been developed. Table 1 lists minimum equipment and material required.

TABLE 1　Minimum equipment and material for making thin-sections

Equipment	Supplies
Power grinding lap	Grinding abrasives: coarse to fine
Plate glass lap for hand grinding	Cement for sample and cover glass: Lakeside 70, Canada balsam,
Glass specimen slides	epoxides, or Permount
Cover glasses	Xylene, alcohol, or toluene for cleaning or dilution
Hot plate	Glycol or kerosene for grinding water-soluble specimens
Forceps for handling	Glycerine for temporary covers

A well-equipped, thin-section laboratory should have a diamond saw, multiple grinding and polishing laps, fine to coarse abrasives, a vacuum chamber, and reagents, materials, and gadgets necessary for preparing various types of sections from different types of rock or sediments according to the information desired. The equipment involves substantial cost, but if a large number of thin-sections are needed currently, or anticipated for the future, considerable expense is justified because time, effort, and cost per individual slide can be greatly reduced if the proper equipment is available. Reed and Mergner (1953) give an excellent account of methods, equipment, and supplies used for mass production of thin-sections by the U.S. Geological Survey.

Many types of power saws, grinding machines, and specialized equipment are available from manufacturers and laboratory supply houses. Meyers (1946) describes a handmade vertical saw that cuts a thin-section so thin that it requires only hand smoothing for reduction to the standard 30μ. Microtech Laboratories of Grand Junction, Colorado have placed on the market a newly developed vertical sawing and grinding machine called Micro-Trim (Fig. 1). Seven slides mounted at one time on a vertical wheel may be cut and ground by a second wheel lubricated by water. The slides are held to the vertical wheel by a vacuum. Wheels with slots for several different sizes and shapes of slides are available. Woodbury (1967) discusses grinding methods in general and describes horizontal

Fig. 1 Micro-Trim thin-section machine. (Microtec Development Laboratory, Grand Junction, Colorado)

grinding equipment, cements, and materials available from Buehler, Ltd., of Evanston, Illinois.

Individual taste, experience, purpose, and availability of laboratory equipment play a large part in the methods employed and the cost of thin-sections. Simplicity, low cost, and low expenditure of time and effort are the goals of the methods and recommendations for material and apparatus described in this chapter. Good sections require skill, ingenuity, and time. If time is important or skill is lacking, slides can be made by commercial laboratories for $1.00 to $3.00.

PROCEDURES

Summary of Procedures

The sequence of procedures for making a thin-section is listed below and expanded in the following pages.

1. Prepare the sample for sectioning.
    (a) A hard-rock sample should be chipped, or sawed, to obtain a thin fragment.
    (b) A sample consisting of loose grains, or a porous specimen, or a fractured specimen should be impregnated.
2. Grind one side flat with coarse grit on lap wheel, using water as a lubricant (kerosene or glycol for water soluble samples).
3. Wash thoroughly to remove all coarse grit.
4. Grind on a lap with fine abrasive, or polish with fine abrasive on glass plate.
5. Wash thoroughly to remove all abrasive and dry.
6. Mount specimen, ground side down, on glass slide with Lakeside 70, Canada balsam, or epoxide.
7. Grind specimen with coarse abrasive to 1.0 mm or less thickness on a lap wheel or cut with thin-bladed saw.
8. Grind specimen with coarse abrasive to approximate final thickness, or to a wedge shape, on a glass plate.
9. Wash thoroughly to remove all coarse grit.
10. Complete grinding by hand with a sludge of fine abrasive on a glass plate to a final thickness of 30μ or to a wedge from 30 to 500μ.
11. Wash thoroughly and dry.
12. Stain, etch, or give other treatment as desired.

13. Wash, clean surface, and dry.

14. Apply a cover glass or other protective cover. Use Permount, epoxides, or Lakeside 70 permanent cover cement; glycerine for a temporary cover if subsequent staining, x-ray, luminescence, and so on, are desired.

15. Clean the thin-section and attach a label or inscribe with a diamond pencil.

Preparation of Sample

A sample of the consolidated sedimentary rock to be sectioned is obtained either as a thin chip or as a sawed slab. Sawing is generally done with a diamond saw, but some rocks are sufficiently soft to permit cutting with an ordinary wood saw or a tempered-steel hacksaw blade. In some instances it is necessary or desirable to chip or saw the sample perpendicular to the bedding, or parallel to the bedding, to produce an oriented section. The optimum size for a thin-section is 1 to 2 sq. in., but smaller or larger ones can be used. Thin-sections measuring 75 X 100 mm, required for research work of a special nature, have been successfully made from slabs mounted on a standard lantern slide glass (Hedberg, 1963). The writer has made thin-sections on regular 2-in. square glass slides for projection on a screen without a microscope. Many sedimentologists and paleontologists prefer 25 X 76 mm glass slides, but most prefer the short 27 X 47 mm slides used by igneous and metamorphic petrographers. The long slides, however, allow for a greater surface area, provide needed space for labeling, and can be handled and stored more easily than the short slides.

Impregnation or Aggregation

Porous, poorly cemented samples are generally too weak to permit thin-sectioning without artificial bonding or impregnation of some kind. One method of impregnation is by using a vacuum pump to draw out all interstitial air and allow a cold setting, highly fluid resin or epoxide to fill the pores or fractures and then harden (see Chapter 9). For successful impregnation, a sample must be perfectly dry. Equal parts of Araldite AY105 and Hardener 935P (index 1.55) dissolved in three parts of toluene with subsequent heating for 2 hours at 90°C can be used for impregnation. Woodbury (1967) lists

20-8130 AB with 20-8132 AB hardener as an epoxide, and also 30-8280 AB for Araldite and 30-8282 AB as a hardener, both of which require dilution of one to three parts. The sample should be as thin as possible to ensure maximum capillary filling of voids and bonding of grains.

Although sedimentary rocks can be impregnated with bakelite, it is not as satisfactory as with epoxides or Canada balsam. A specimem to be impregnated with bakelite is placed in a vacuum chamber and impregnated with a fluid consisting of one part finely ground bakelite to three parts acetone, and then placed, for hardening, in an 80°C oven for several hours. Bakelite imparts an undesirable dark color, but it does give contrast to the voids and interstitial spaces. The refractive index is 1.63, which is far above that of other cements.

The author developed a method of impregnation with Canada balsam that is inexpensive, requires only a few minutes time, and has been successful for making a thin-section wafer from loose sand. A solution of high viscosity is made by dissolving one part Canada balsam in three parts xylene. The solution is poured into a small, flat, heat-resistant receptacle 3 to 4 cm in breadth and length and about 1 cm deep. When loose sand is poured into the solution, capillarity draws the solution into the sand, which is then heated to evaporate the solvent and harden the balsam. Avoid overheating the balsam, for it becomes weak and brittle if carbonized. A low oven temperature prevents overheating, but the hardening process is much slower. A sample carefully cured at a relatively low oven temperature can be put on the grinding lap 10 to 15 minutes after cooling.

Dobell and Day (1966) describe a resin called Scotchcast No. 3 (Minnesota Mining and Manufacturing Co.) which has a viscosity of 1000 centipoise at 25°C and 20 centipoise at 120°C. A sample to be impregnated is immersed, or loose grains are poured, into the fluid at room temperature and then baked for 2 hours at 120° to harden the resin. Scotchcast is colorless, isotropic, and has a refractive index of 1.568.

Chatterjee (1966) describes a method of making thin-sections from loose heavy-mineral grains that is much simpler than the methods used by Brison (1951) and by Sanders and Kravitz (1963). Chatterjee used a thermoplastic (no name given) that is soluble in acetone or benzene and at 220°C into a highly viscous fluid with a refractive index of 1.559. The procedure is to drop minerals grains into a drop of melted plastic on a slide, which is covered with another slide that is pressed down firmly. On cooling, the difference in contraction between the glass and the thermoplastic causes separation, which leaves a wafer that can be ground down to one-half the diameter of the enclosed grains, mounted with Canada balsam, Lakeside 70, or an epoxide, and cut to 30μ thickness by standard procedures. It is assumed that any plastic with a high thermal coefficient of expansion could be used. Chatterjee's procedure is valuable because, by other methods, it is very difficult to cement loose grains on a slide, perform the first grind, and then remount the grains on another slide for a second cutting to 30μ. This kind of procedure would be useful in making a thin-section of grains from an unconsolidated sand, such as a dune, beach, or channel sand.

After the sample is impregnated, it can be processed in the same manner as any consolidated chip or slab. However, the first grind, with coarse abrasive, must cut off material at least as thick as one-half the diameter of the largest grain or fragment. Cross sections of the bottom grains must also be obtained, otherwise the section will show only tangential cuts of grains that touch the original bottom surface of the slide.

First Grinding of Specimen

Chips or slabs must be ground to a smooth plane surface on one side prior to mounting on the glass slide. Grinding is normally done on a grooved lap wheel with a coarse abrasive (Carborundum 150 to 300). Water is the normal lubricant, but one should use glycol or kerosene for water-soluble specimens. Kerosene must be removed with a detergent or a solvent before mounting the ground chip and also after the final grinding before the attachment of a cover glass.

Most companies that stock scientific apparatus list grinding and polishing machines and supplies. Machines with one to four lap wheels are commonly available. The use of a battery of lap wheels allows different sizes of abrasive to be used on separate laps and saves the time and effort required to clean a single lap between use of coarse and fine abrasives. One disadvantage of the multiple-wheel setup is the possibility that wheels having finer abrasives may become contaminated by splatter of coarser grit from adjacent wheels, even though partitions are present. Such contamination can be disastrous if a single grain of coarser abrasive rips across a thin, nearly completed section. Automatic grinding and polishing machines are available, but are expensive and justified only on a cost-per-section basis for quantity production of thin-sections. After the coarse grind, the specimen should be thoroughly washed and smoothed with a fine abrasive to eliminate pits and roughness produced in the coarse grind. Carbonate rock sections should always be finished with a fine grind; it is sometimes desirable to polish them. The importance of thorough washing cannot be overemphasized, because a single coarse abrasive can cut grooves or scratches that cause breakage of the slide during the final grind. It is recommended that the fine grind be made by hand on a glass plate with a slurry of abrasive and water. A plane surface can be cut dry for indurated clay, gypsum, or salt by using a coarse and then fine-toothed file or sandpaper.

Mounting of the Slide

Attachment to the desired type of glass slide, generally 27 × 47 or 26 × 76 mm, follows preparation of the primary plane surface. Mounting cements commonly used are Lakeside 70 or Canada balsam with refractive indices of 1.54 to 1.55. Recently, epoxides have come into common use. Kollolith cement is similar to Canada balsam but has a lower index of refraction. It is not recommended for use, since better cements are now available. Permount is a cement frequently used as a cover or for attaching cover glasses. Many different types of cement have been used, some with high indices of refraction and others with features desirable for special purposes. Several are described by Twenhofel and Tyler (1941) and Reed and Mergner (1953).

Lakeside 70 (manufactured by Lakeside Chemical Corp., Chicago) marketed as a long solid stick, is a cement commonly used. It has a refractive index of 1.54, melts at 140°C, and has an advantage of being more viscous than balsam (Von Huene, 1949). The first step in its use is to heat the glass slide on a large iron plate or electric hotplate to a temperature slightly above 140°C. Lakeside is melted onto the glass in quantity sufficient to produce a pool of fluid as large as the surface of the specimen to be mounted. Bubbles of the Lakeside solvent, which may appear at temperatures over 140°C, are undesirable, but stop forming if the temperature is held constant for a time. When solvent bubbles begin to decrease, the slide should be lifted with tweezers and tilted and rotated in various directions so that bubbles float to the outside edge of the fluid and leave the center clear. Reheating may be necessary to obtain a bubble-free fluid. Bubbles will disappear as heating is continued and the solvent is boiled off, but too much heat causes the cement to turn brown and eventually black as the fluid is carbonized. Overheating also alters the index of refraction and causes the cement to become brittle and develop shrinkage cracks on cooling. The worker should make a few slides, experimenting with unimportant specimens, to determine optimum cooking time and to develop dexterity in eliminating bubbles.

The specimen to be mounted must be washed entirely free of abrasive and must be completely dry. The specimen should be warmed for a few seconds on the hot plate, but not warmed to the extent that it is too hot to touch. If not warm, the specimen will chill the heated Lakeside 70, harden it, and prevent a uniform tight seal to the glass slide. When the specimen and mounting cement are ready, the slide should be transferred quickly to a warm, flat surface and the flat side of the specimen pressed down quickly over the slide with a rolling motion, starting at one side or end of the slide. Press down firmly with fingers or a large cork to squeeze out surplus cement, and hold until the cement solidifies. The slide is now ready for the second stage, grinding to the desired thickness.

Although some may prefer Canada balsam as a cement, and it has special value for some purposes, Lakeside 70 is more commonly used. Both have essentially the same index of refraction. Balsam can be purchased as a thick fluid dissolved in xylene or it may be purchased as a solid which has to be dissolved in xylene. The temperature necessary to liquefy solid Canada balsam generally causes overcooking.

Preparation of a specimen mount using Canada balsam follows essentially the same procedure used for Lakeside 70, and the formation of bubbles and danger of overheating also exist. The balsam must be tested for the proper consistency as it is cooked. This is done by using a pin, knife blade, toothpick, or small glass rod to lift a bit of the fluid and touch it to a thumbnail or glass slide for cooling. When the cooled balsam is brittle and does not adhere to a hard surface, it has the right consistency and the warmed specimen must be pressed quickly onto the slide as described above. Canada balsam has been in use for many decades and the 1.55 index of refraction of Canada balsam is a standard for optical examination. Any cement other than Canada balsam or Lakeside 70 should have the index indicated on the slide label.

Several writers recommend applying Lakeside 70 or Canada balsam on both specimen surface and glass slide. This procedure is difficult because both units must be heated to melt the cement, and development of bubbles is likely. Juncture of the two units must be effected quickly at exactly the right moment before the fluid hardens. The procedure is messy, requires special handling and experience, and is not recommended.

Epoxides are now available and are preferred by some workers. Indices of refraction, ranging from 1.54 to 1.58, bracket those of Lakeside 70 and Canada balsam. They are commonly prepared by making a one-to-one mixture of a resin ingredient (A) with a hardener (B). Epoxides normally require 10 or more hours to harden, which is a disadvantage in contrast to the 2 or 3 minutes required for Lakeside 70 or Canada balsam to harden. More rapid hardening of epoxides can be obtained by oven heating for 30 minutes at about 70°C. The use of epoxides permits a cold mount and avoids bubbles if the A and B ingredients are stirred until clear. Differences in indices of various epoxides is not a serious disadvantage if one knows that the specimen is embedded in a epoxide, but the index of refraction should be indicated on the slide label.

Araldite AY205 and Hardener 935F is an available epoxide. Woodbury (1967) lists 20-8130 AB and 20-8132 AB for hardener. It is necessary to select an epoxide that will adhere strongly to a glass surface; otherwise the thin section may peel or become loosened. A frosted surface on the slide will give greater bonding. Ordinarily, thermoplastics should not be used to bond specimens to glass, for the section will commonly separate from the slide. However, use of thermoplastics is desirable in some cases, as described later. After mounting, the final thin-section is cut, stained, or otherwise treated and a cover glass applied with the same epoxide used previously for mounting.

Care must be used in heating gypsum during attachment to a slide and application of a cover glass because dehydration of the gypsum may cause bubbles. Thin-sections of coal require a very special treatment and are described by Thiessen et al. (1938).

Final Grinding

After mounting, the specimen is again put on the grinding lap with coarse abrasive and reduction of the specimen to a thin-section is started. Reduction with coarse abrasive should be down to about 1 mm for soft sedimentary rocks and less for siliceous or hard rocks. If a fine saw-blade is available, the mounted specimen can be sliced to about 1 mm thickness. Chert, sandstone, and other siliceous or hard rocks can be ground much thinner before the grinding is transferred to a glass plate. It is recommended that the final coarse grinding be done by hand with a slurry on a glass plate, and then with a slurry for each finer grade down to the desired thickness. It is most important to remember that a specimen should be thoroughly washed whenever the size of abrasive is changed.

Final grinding on a glass plate should be with a circular motion that covers the corner areas of the glass plate as well as the center portion. Continuous grinding in the center of the plate will gradually cut a depression in the glass that eventually results in production of convex thin-sections. Such a situation becomes apparent when the corners of the glass slide begin to be frosted, indicating that a new glass plate is required. Pressure for grinding is generally applied with three fingers. A mass of plasticine oil-based modeling clay can be pressed to the back of the glass slide to assist in controlling the angle of cutting; or it is also possible to use the flat top of a large cork or a rubber suction cup like those on toy arrows. A slide-holder, which more or less guarantees a uniform thickness if held properly, can be purchased from Buehler, Ltd., or other suppliers. Experience and dexterity must be acquired to cut a section of uniform thickness.

Care should be used in the final grinding of soft rocks, especially carbonates, because a few rotary strokes or a single arm stroke may grind off the entire section. Occasional examination under a petrographic microscope will show, by the interference colors, when the proper thickness of 30μ has been reached.

When ooliths, fossils, fabric, grain orientation, or other aspects of a carbonate or other sedimentary rocks are to be studied, it may be desirable, and many times important, to have a wedged section ranging in thickness from 30 to 500μ instead of the standard 30μ. Wedging can be done on a lap wheel, but hand work on a glass plate is preferable because of the fine control needed when grinding soft rock. Wedging is a common feature resulting from inexperience, but intentional wedge-cutting requires careful control. Wedging can be accomplished by applying greater pressure with one finger on one end of the glass slide. A long slab of the rock is desirable in order to have a low-angle wedge that gives more surface area for each portion of the sample.

Banded or laminated rocks or well-consolidated specimens with partings must be handled carefully to prevent separation in the final stage of grinding. If the specimen is enclosed by a plastic or plaster rim before mounting on the slide, separation can be prevented. Impregnation should be performed as described earlier for samples with voids, open fractures, or for poorly consolidated material.

Etching and Staining

Etching and staining are intermediate steps in completing a thin-section. They are important for the study of many types of sedimentary rock and must be done before application of the cover glass. Acid etching of a carbonate thin-section is generally done with 25%, or less, acetic acid. Very dilute hydrochloric acid may be used, but it works rapidly and is difficult to control. The acid may be applied with a dropper, rod, or by dipping. The writer has dissolved thick sections of carbonate rock, 1 mm thick, in order to study the fabric and orientation of insoluble residues, and has made curved surface sections (Ireland, 1950) for the same purpose. Curved-surface sections are the smooth, or rough, surfaces on chunks of carbonate rock which are withdrawn from an acid bath before complete digestion. They can be used only with reflected light. Use of curved-surface sections is recommended for the study of character, orientation, and distribution of inclusions and clastic grains in sedimentary rocks as well as features of the matrix and cement.

Acid treatment of a carbonate thin-section shows the relations of the insoluble constituents if the thin-section is thinner than the maximum diameter of the included particles. A good representation is possible under any condition, even though some of the grains may be lost because of being enclosed previously by calcite or bonded along only a small portion of the grain perimeter. Acid application allows inspecting the residues without interference from the carbonate contact faces that enclose the particles.

Completion of the Section

When the thin-section has been ground to its final thickness and the etching, staining, and other treatments have been completed, each specimen needs a more or less permanent protective cover. A temporary cover glass applied with glycerine will permit its removal for luminescence or x-ray studies or for some other special treatment. A glass cover is permanent, but other types of covers are available. A clear acrylic plastic spray may be used. It dries quickly, but the thin film is subject to rupture and will not withstand the considerable use and hazard of laboratory classes. A thin coat of epoxide can be spread evenly over the slide, but requires 24 hours to harden at room temperature. Canada balsam applied without a cover glass would have to be cold and would require up to two weeks to harden. The unprotected surface of the balsam might chip with hard usage and also might develop shrinkage cracks after a few months. Heat applied to hasten the hardening of Canada balsam might be disastrous because too much heat could melt the mounting fluid. Lakeside 70 is suitable for attaching a cover glass, but it requires heating and would be difficult to apply without introducing bubbles or melting the mounting cement. Permount is a recently developed cement composed of a 60% solution of synthetic resin in toluene (Fisher Scientific Co., Cat. No. 12-568). It has chemical stability and good adhesion and is simple to use for covering thin-sections and for attaching a cover glass. Permount can be diluted with toluene. Hardening requires one to two days.

A cover glass cemented over the thin-section provides permanent, essentially hazard-free protection. Cover slides are sold in various thicknesses and sizes, generally 25 X 25 or 25 X 45 mm. Epoxides, Lakeside 70, or Permount are the cements most workers prefer.

If Lakeside 70 is used, the procedure is to melt enough of the cement to obtain an excess on the surface of the cover glass, which is then manipulated until bubbles are eliminated. The cover glass is quickly rolled over the thin-section from a side or end and firmly pressed down to squeeze out excess cement. A flat pressure object such as a cork should be used, for the cover glass is very thin and fragile. The slide should be slightly warmed before application of the cover glass, but not enough to melt the bond of the thin section. The cover glass cement hardens in a few minutes and the excess of brittle cement can be scraped off with a knife, razor blade, or edge of a glass slide. After cooling, the entire slide is cleaned with xylene, toluene, or alcohol and is ready for a label or inscribing with a diamond pencil. Hot mounting of a cover glass with Canada balsam is messy and not recommended because the balsam must be cooked on the cover glass. Application of the cover glass is the same as that for Lakeside 70. A cold Canada balsam mount may be done, but a week or two must be allowed for hardening.

An epoxide properly mixed with a hardener can be spread over the slide and a cover glass rolled on to give a cold mount. Some may prefer to apply the epoxide to cover glass surface as well. Hardening will take place in 24 hours, or a very low heat will hasten it. If an epoxide is used for impregnation, the same epoxide should be used for a mounting cement.

Omission of a cover glass permits concurrent study and chemical treatment of a thin-section that may be especially desirable when working with carbonate rocks. Ordinary masking tape (not a plastic tape) can be used to cover all portions of the specimen not to be treated. Unmasked portions can be stained, etched, or otherwise treated chemically as may be desired. A highly adhesive tape such as Scotch tape should be avoided because the removal may pluck out parts of the thin-section as well as soft or friable interstitial material. An exposed thin-section can be protected in storage by a second glass slide, with both slides inserted into an aluminum slide-holder such as used by micropaleontologists. A permanent cover may be applied whenever desired.

SPECIAL TYPES OF THIN-SECTIONS

Thin-sections of small chips from oil-well or water-well samples can be prepared by the methods described by Chatterjee (1966). Another method is to bond the chips into a workable wafer by impregnation with Canada balsam, Lakeside 70, or an epoxide and then process like any other specimen to the standard thickness of 30μ. A simple method, but one that requires dexterity, is to embed the cuttings in Canada balsam or Lakeside 70 on a slide, grind a flat surface, melt the cement just enough to free the particles, turn the slide over, and embed the flat surfaces in prepared hot cement on another glass slide. When cool, the thin-section can be ground to completion and cover glass applied.

REPAIRS TO THIN-SECTIONS

Repairs may be possible for important thin-sections that become broken or have a cover glass broken, become loose, or pop off entirely. Portions of a thin-section on a broken slide cemented with Lakeside 70 or Canada balsam may be heated until the cement melts and then the cover glass slid off with tweezers; never lift. The exposed section on the broken slide is then turned upside down, placed into the hot prepared cement on a new slide and the broken portions of the glass slide slid off laterally; never lift. A new cover glass can then be applied. Dexterity, care, and special attention to temperature are required. One hazard is the development of bubbles that may break the thin rock section. Great care should be used in heating to obtain sufficient fluidity of the original cement to remove the original cover glass and transfer the section to a new slide.

A cover glass that has had insufficient cement may have cavities under the corners or edges which can be filled with Permount, an epoxide, or melted Lakeside 70. These cavities are likely to cause breakage of the cover glass if not filled. If a cover glass becomes loosened or has a poor bond, Newton rings will be observed on the slide. Slight heating and light pressure on the cover glass generally squeeze out a bit more cement and renew the bond, or a small quantity of new cement can be added. Again bubbles can form if too much heat is applied. Too much pressure may break or bend the glass and cause it to pop up later.

Repairs for slides with epoxide cement are difficult. The slide must be soaked in alcohol, toluene, or some other solvent until the section becomes free. Then it is cemented to a new slide. Extreme care must be used because the thin-section of rock is very fragile.

An ingenious person can make many kinds of repairs depending on experience with types of cements, temperature control, character of the thin-section, and proper solvents. Transfer of a broken, impregnated thin-section is essentially impossible because the unit will disperse. However, if a random regrouping of individual grains or fragments is not significant, the thin-section constituents can be slid off, recemented, and covered. The index of refraction and the strong adhesiveness of 3M Scotch transparent tape make it valuable for some types of repair, especially for impregnated sections. (Harbaugh, 1953). If a glass slide is broken into only two or three pieces, the fragments can be matched and secured with Scotch tape. It may be possible for a new glass slide to be taped or cemented onto the back of the broken glass slide.

REFERENCES

Brison, R. J., 1951, A method for polishing thin sections of mineral grains, *Am. Mineralogist,* 36, 732–735.

Chatterjee, B. K., 1966, New technique for preparing polished thin sections of heavy mineral residue, *J. Sed. Pet.,* 36, 268–269.

Dobell, J. P., and D. P. Day, 1966, Preparation of thin sections of unconsolidated or friable sediments, *J. Sed. Pet.,* 36, 254.

Harbaugh, J. W., 1953, Scotch tape aid thin-section studies, *Bull. Am. Assoc. Petrol. Geologists,* 37, 452.

Hedberg, R. M., 1963, Dip Determination in carbonate cores, *J. Sed. Pet.,* 33, 680–693.

Ireland, H. A., 1950, Curved surface sections for microscopic study of calcareous rocks, *Bull. Am. Assoc. Petrol. Geologists,* 34, 1737–1739.

Kummell, B., and David Raup, eds., 1965, Handbook for paleontological techniques, W. H. Freeman, 852 pp.

Meyer, C., 1946, Notes on the cutting and polishing of thin sections, *Econ. Geol.,* 41, 166–172.

Reed, F. S., and J. L., Mergner, 1953, Preparation of thin sections, *Am. Mineralogist,* 38, 1184–1203.

Sanders, J. S., and J. H. Kravitz, 1963, Mounting and polishing mineral grains on a slide for study with reflected and polarized light, *Econ. Geol.,* 59, 291–298.

Thiessen, Reinhardt, et al., 1938, Preparation of thin sections of coal, *U.S. Bureau Mines Inf. Circ. 702l,* 18 pp.

Twenhofel, W. H., and S. A., Tyler, 1941, Methods in the study of sediments, McGraw-Hill Book Co., 183 pp.

Von Huene, R. 1949, Notes on Lakeside No. 70 transparent cement, *Am. Minerologist,* **34,** 125–127.

Woodbury, J. L., 1967, Preparation of petrographic transmitted light sections, *AB Metal Digest,* **13,** No. 1, Pt. 4, 19 pp., Buehler, Ltd., 2120 Greenwood, Evanston, Illinois (See also article on impregnation, **12,** No. 3, Pt. 3.)

The following is a list of selected articles and textbooks not referred to in the text.

Kennedy, G. S., 1945, Preparation of polished thin sections, *Econ. Geol.,* **40,** 355–359.

Krumbein, W. C., and F. J. Pettijohn, 1938, Manual for sedimentary petrography, Appleton-Century, 549 pp.

Rogers, A. F., and P. F. Kerr, 1942, Optical mineralogy, 2nd ed., McGraw–Hill Book Co., 390 pp.

Ross, C. S., 1926, Methods of preparation of sedimentary material for study, *Econ. Geol.,* **21,** 454–468.

Travis, R. B., 1958, Thin section analysis, Chapter 4, in Symposium on subsurface geology in petroleum exploration, LeRoy and Haun, eds., Colorado School of Mines, 1156 pp.

CHAPTER 16

POINT COUNTING

JON S. GALEHOUSE

San Francisco State College, San Francisco, California

The first attempt to determine quantitatively the percentage of various minerals present in a rock was made by Delesse in 1848. His method was commendable but crude and could be applied only to coarse-grained rocks. He used a polished section of rock on which he placed a piece of waxed paper and traced the outlines of the various minerals present. The tracing was then transferred to a piece of tinfoil and the various outlines cut out with tin snips. All pieces representing the same mineral species were collected and weighed. The area percent of any particular mineral was calculated, using the ratio of the weight of the tinfoil representing the particular mineral to the total weight of all the pieces of foil.

As one might guess, Delesse's method never really caught on. The significance of Delesse's pioneering efforts is that he did realize the importance of quantitative mineralogical analysis, and he was the first to suggest that a relationship exists between the area percentages of various minerals in a section and their volume percentages in a rock.

Methods presently used for determining the quantitative mineralogy of thin or polished sections are based on Delesse's area-volume relationship. This relationship is derived mathematically by Chayes (1956, pp. 12—15) and can be stated as follows. In a randomly chosen area of a rock such as may be represented by a thin or polished section, the ratio of the area of a particular mineral to the area of all the minerals is a consistent estimate of the volume percent of the mineral in the rock. In simplified mathematical form, let

X = a particular mineral species in a thin-section,
h = the thickness of the thin-section,
V = volume,
A = area.

Then the volume of X in the thin-section, V_x, equals $A_x \cdot h$ and the total volume, V, equals $A \cdot h$. Therefore

$$\frac{V_x}{V} = \frac{A_x \cdot h}{A \cdot h} = \frac{A_x}{A}$$

The problem of finding the volume percent of various minerals in a rock can therefore be reduced to finding the area percent of the minerals in a section. The area percent can also be converted to weight percent by multiplying by the specific gravity of each mineral species. In addition, a chemical analysis can be derived by using the area percents and the chemical formulas of the minerals (see Friedman, 1960). Two methods described below have been used successfully for determining area percents. Both can be used for studying sedimentary rocks or artificially indurated sediments, but they are by no means unique to sedimentary petrology; in fact, they have been used most often for modal analyses of plutonic rocks.

It was not until 1898, 50 years after Delesse introduced his method, that anyone developed a more practical method of quantitative mineralogical analysis. This delay resulted more from general disinterest in quantitative mineralogical analysis than from widespread acceptance of Delesse's procedure.

ROSIWAL-SHAND METHOD

The Rosiwal-Shand method of determining quantitatively mineral composition is not a "point-counting" technique. A discussion of it is included because it is the precursor of the point-counting techniques presently used and described below.

Rosiwal (1898) developed a linear method of quantitative micrometric mineral analysis (see Table 1). The thin or polished section was moved under the cross-hairs of the microscope in a series of linear traverses (see Fig. 1a), and the length along the traverse that a mineral occupied was noted. After measuring a sample of the entire section, the lengths for a particular mineral were added up and compared with the total length of all the traverses. The ratio of the two was then used to calculate the area percent of the mineral in question.

The basic theoretical difference between Rosiwal's procedure and that of Delesse is that Rosiwal took a sample of the area of the sample, whereas Delesse had measured the entire area of the sample. The validity of parallel lines as estimators of relative area has been mathematically demonstrated by Chayes (1956, pp. 6–9).

The most significant improvement in Rosiwal's procedure was the introduction of a recording micrometer by Shand (1916), hence the designation as the Rosiwal-Shand method. Until 1916 intercept lengths were measured with a micrometer ocular, with each distance for each mineral recorded separately. Shand devised a method in which the section is moved under the cross-hairs by a mechanism,

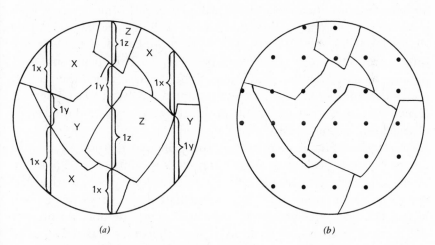

(a)   *(b)*

Fig. 1 Methods of counting thin-sections. *(a)* Rosiwal-Shand method. *(b)* Glagolev-Chayes method.

TABLE 1   Summary of counting methods for mineralogical analysis

Method	What Counted	Based on	What Determined	What Implied	Can Be Converted to
Rosiwal-Shand	Thin or polished section	Length of lines	Area percent	Volume percent, that is, Modal Analysis	Weight percent or chemical analysis
Glagolev-Chayes	Thin or polished section	Grid of points	Area percent	Volume percent, that is, Modal Analysis	Weight percent or chemical analysis
Fleet	Grain mount	All individuals	Number percent	–	–
Ribbon or area	Grain mount	All individuals within band or area	Number percent	–	–
Line	Grain mount	Individuals along lines	Number frequency	–	–

mounted on the stage, which records the length of movement. Subsequent modifications of Shand's device by other investigators resulted in several semiautomatic recording mechanisms (see Heinrich, 1956, pp. 9—10). The most popular of these was the Integration Stage of E. Leitz, Inc. (see Fig. 2). This stage consists of six micrometer spindles, and each of them can advance the section under the microscope. The spindle assigned to a particular mineral is turned whenever that mineral is under the cross-hair, and the length of traverse is automatically recorded.

The Rosiwal-Shand method was the principal method of quantitative mineralogical analysis of thin-sections until the 1950s. By 1956, however, Chayes (p. 36) estimated that its popularity had decreased to the point that only about half the analyses being done in the United States was by the Rosiwal-Shand method. At present the method is used very little. A sign of the times is that Leitz has discontinued manufacturing the Integration Stage. The main reason for the decreasing popularity of the Rosiwal-Shand method was the increasing popularity of the Glagolev-Chayes method.

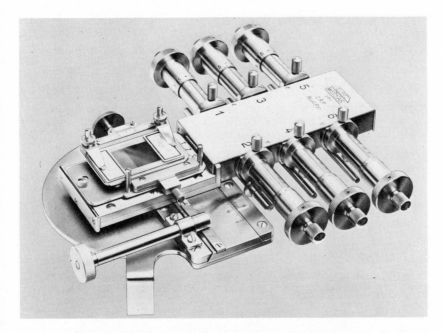

Fig. 2   Integration stage. (Photo courtesy of E. Leitz, Inc.)

GLAGOLEV-CHAYES METHOD

The Glagolev-Chayes method of quantitative mineralogical analysis is presently the most widely used method for studying thin-sections. Glagolev (1933) suggested that instead of measuring continuously along the length of equally spaced lines, one could simplify and expedite procedures by measuring at equally spaced points along the lines. The intersection of the cross-hairs is taken as the point and the mineral beneath it is identified, and one count is recorded for that particular mineral. The section is then moved a given distance and the mineral now beneath the intersection is identified and recorded. Thus the section is sampled by means of a grid of points (see Fig. 1b).

Even though a description of the method was published in English (Glagolev, 1934) it did not immediately become popular in English-speaking countries because of the limited availability of the mechanical counting device he described.

Point counting of thin-sections really started to become popular after Chayes (1949) described a method for modifying existing mechanical stages so that they could be used in the Glagolev method. Chayes (1949) also discussed the precision of the method and stated that the point counter he devised could determine the relative areas at least twice as fast as the most efficient instrument used in the Rosiwal-Shand method. Chayes (1956, pp. 5-6) also established the validity of a grid of points as an estimator of relative areas (see Table 1). Because of Chayes' practical and theoretical contributions to point counting, the procedures described should rightly be known as the Glagolev-Chayes method.

FLEET METHOD

A counting method for determining quantitatively the relative abundance of various minerals in grain mounts was introduced by Fleet (1926). Until then the abundance of minerals in grain mounts was visually estimated and only semiquantitative. Fleet (1926), in dealing with heavy minerals, mounted the entire heavy fraction on the slide and proceeded to identify and count all the grains (see Fig. 3a). The result of this method is a number percent (see Table 1) which cannot be converted to area, volume, or weight percentage without some additional information. This is because the grains are of different sizes and the count implies nothing concerning relative sizes of different mineral species or variations in size within a particular species. Therefore the relative numbers of various species cannot be assumed to represent relative areas of various species. The only way to convert the number percent to area percent is to measure the size of the individual grains (see Chapters 3 to 6) or to mount grains of exactly the same size on the slide. Even then, however, the area percent determined from grain mounts does not directly imply the volume percent. The crystal habit or cleavage form of certain minerals prohibits this. A very thin flake of mica will generally lie with its short axis perpendicular to the slide. An equant grain occupying the same area as the mica and as likely to be counted as the mica would, however, occupy considerably more volume. If the purpose of the investigation is a quantitative mineralogical analysis, there is usually no need to attempt to convert the number percentage to area, volume, or weight percentage. The number percentage has proven to be a very useful parameter in itself.

In using the Fleet method of counting grain mounts or either of the other two methods described below, certain problems arise. The identification of particles in grain mounts is exceedingly more difficult than the identification of constituents in thin-sections. The different thicknesses of the grains give different interference colors. In addition, the larger grains are often opaque. Consequently, in working with heavy minerals it is suggested that no grains larger than 0.5 mm in diameter be mounted. A split of the sample should be kept to be used in immersion media in case there are grains that just cannot be identified on the slide.

One advantage that grain mounts have over thin-sections is that grains usually come to rest with their long and intermediate axes in the plane of the slide. This means that certain minerals commonly show a characteristic shape or a preferred optical orientation; for example, amphiboles almost always appear in elongated rod shapes and mica almost always has its c-axis perpendicular to the plane of the slide.

If the Fleet method is used, it is best to determine the approximate number of grains one desires to count (see discussion below) and then carefully split the sample down to that number and mount it. A difficulty in using the Fleet method is keeping track of which grains have already been counted and which are still to be counted.

Although the Fleet method is not very popular at present, it does have the advantage of yielding a number percent which can be dealt with statistically.

RIBBON OR AREA METHOD

The ribbon or area method of counting grain mounts is similar to the Fleet method except that instead of all the grains on the entire slide being counted, only in certain representative areas of the slide are all the grains counted. Van der Plas (1962, pp. 150–151) describes the ribbon method and advocates its use in granulometric analysis of thin-sections. The method, however, can also be applied to quantitative mineralogical analysis of grain mounts. In the ribbon method (see Fig. 3b), only individuals within certain bands are counted. Actually, the slide is moved along the stage and all individuals between two lines are counted.

The area method is the same as the ribbon method except that the portions of the sample considered are rectangular or square areas rather than elongate bands. Because it is nearly impossible to get a completely random distribution of grains on the slide when preparing grain mounts, it is necessary for the bands or areas to be systematically or randomly arranged over the entire slide. If the band width or the representative areas are considerably larger than the largest grains, the method may be considered to yield a number percent (see Table 1). A serious disadvantage of using the method is that many of the grains to be identified do not appear in the center of the microscopic field. This eliminates the possibility of determining conveniently some of the optical properties and increases the difficulties of identification.

LINE METHOD

The line method is the most popular method now in use for quantitative mineralogical analysis of grain mounts. The method involves counting a sample of the number of the various particles mounted on a slide. This is done by counting individual grains encountered by the intersection of the cross-hairs along linear traverses spaced equadistantly along the slide (see Fig. 3c). As mentioned, the fact that it is almost impossible to prepare a slide with a completely random distribution of grains necessitates setting up traverses that sample the entire area of the section.

The result of the method is a "number frequency" (see Table 1) that simply shows how often particular species were encountered during the count. The number frequency is related to, but distinctly different from, the area, volume, weight, or number percentage. The sample is biased because a larger grain is more likely to be encountered during the analysis than a smaller one.

Even though the area of individual grains does increase their chance of being counted, their different sizes prohibit the number frequency from also being the area percent (for a detailed discussion, see

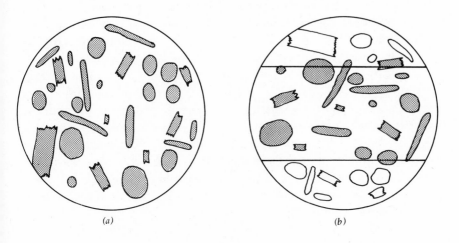

(a)   (b)

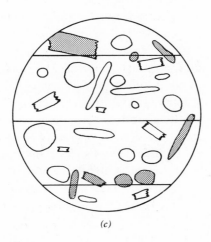

(c)

Fig. 3 Methods of counting grain mounts. (Only the shaded grains are counted. See text for further discussion). (a) Fleet method (all the grains are counted). (b) Ribbon method (all the grains between the two lines are counted). (c) Line method (all the grains that intersect the lines are counted).

Galehouse, 1969). The number frequency underestimates the area of larger grains and overestimates the area of smaller grains; consequently, the number frequency does not imply volume percent and cannot be converted to weight percent or to a chemical analysis.

In addition, the number frequency is different from the number percent (see Galehouse,1969). The number frequency is larger than the number percent for larger grains and smaller than the number percent for smaller grains. Unfortunately, most investigators who use the line method of counting grain mounts refer incorrectly to their results as number percentages instead of number frequencies. The potential danger in doing this is that because of their built-in bias, number frequencies cannot be treated statistically in the same manner as number or area percents (see following discussion).

The discrepancy among number frequency, number percent, and area percent can be minimized by mounting only a limited size interval on the slide. Number frequency, that is, the number of times each species is counted, equals number percentage only when all the grains on the slide are counted as in the Fleet method, all the grains in representative portions of the slide are counted as in the ribbon or area methods, or all the grains have the same cross-sectional area. Number frequency and number percentage equal area percentage only when all the grains have the same cross-sectional area. No method is both practical and theoretically valid for converting number or area percentage in grain mounts to volume or weight percentages. Hunter (1967) proposed a method of converting number percentages of heavy minerals to weight percentages that uses the abundance, density, size, and shape of the grains. His method will yield only an approximate weight percent; however, this is often close enough to be of use in many investigations.

If the results of quantitative mineralogical analyses of grain mounts are to be treated statistically, the number percentage should be determined, that is, the Fleet or ribbon or area method should be used. If the object of the count is only to determine some characterizing aspect of the rock or sediment, the faster line method may be used.

The results of counting grain mounts, both number percents and number frequencies, are commonly used for subsequent interpretations such as correlation, determination of source area and dispersal pattern, or kind and amount of postdepositional diagenesis.

NUMBER OF POINTS OR GRAINS TO BE COUNTED

A significant question confronting the investigator performing quantitative mineralogical analyses of thin-sections or grain mounts is how many points or grains must be counted for a certain amount of confidence in the results. Dryden (1931) was one of the first investigators to deal with the quantitative accuracy of counting grain mounts. Although he was concerned with accuracy in number percentages, his reasoning can also be applied to area percentages.

The basic assumption involved in applying the following statistics is that the encounter of a particular type of mineral for area determinations or a particular individual grain for number determinations is based entirely on abundance and is neither favorably nor unfavorably biased by any other factor during the counting procedure. This assumption is true for the Rosiwal-Shand, Glagolev-Chayes, Fleet, ribbon, and area methods. It is false for the line method since the sample is biased in favor of the larger individuals being counted. If only a limited size range of grains is mounted and counted using the line method, the bias may be small compared to errors of sampling, splitting, and identifying. As the number frequency approaches the number percentage (see above discussion), the following statistics become increasingly appropriate for results of the line method.

The probable errors in percent of individual components at the 50 and 95.4 confidence levels can be calculated as follows:

$$E_{50} = 0.6745 \sqrt{\frac{P(100 - P)}{N}}$$

and

$$E_{95.4} = 2 \sqrt{\frac{P(100 - P)}{N}}$$

where E = probable error in percent,
   N = total number of points or grains counted,
   P = percentage of N of an individual component.

For example, if the total number of heavy-mineral grains counted is 150 and 60 (=40%) of these are sphene, then the probable error

$$E_{50} \quad = 0.6745 \sqrt{\frac{40(60)}{150}}$$

$$= 0.6745 \sqrt{16}$$

$$= (0.6745)(4)$$

$$= 2.698\%$$

and

$$E_{95.4} = 2 \sqrt{\frac{(40)(60)}{150}}$$

$$= (2)(4)$$

$$= 8\%$$

Therefore the chances are 1 in 2 that the real percentage of sphene is between about 37.3 and 42.7 and the chances are 19 out of 20 that the real percentage is between 32 and 48.

Figures 4 and 5 summarize graphically the probable errors at the 50 and 95.4 confidence levels for a component present in various percentages. The graphs show that for general purposes 300 points or grains is a good number to count in order to get the maximum accuracy from a minimum investment of time. Below 300, the probable error increases rapidly, whereas above 300 it decreases slowly. However, the particular type of investigation being done should determine the number of points or grains to be counted. For detailed modal analyses, considerably more than 300 points should be counted, perhaps 1000 or 1500. For classifying heavy-mineral slides into one of several vastly different suites, less than 300 grains may be sufficient; 100 has commonly been used. Table 2 is included to summarize the probable errors in numbers rather than percents.

The great utility of the graphs and table is that the investigator can determine quickly how many grains or points he must count to obtain the desired level of accuracy. In addition, they enable the reader to determine more intelligently the amount of confidence to

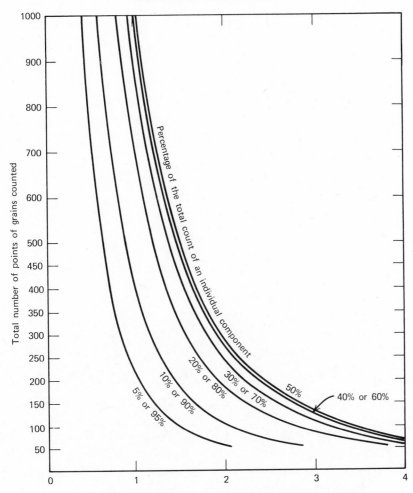

Fig. 4 Probable error at 50 confidence level.

be put in the results of, and the interpretation made from, the counts. For additional discussions, see Chapters 17 and 19.

RECORDING THE DATA

Figure 6 is an example of a data summary sheet that the author uses for point counts of thin-sections or grain mounts. This particular sheet was set up primarily for either the Glagolev-Chayes method of

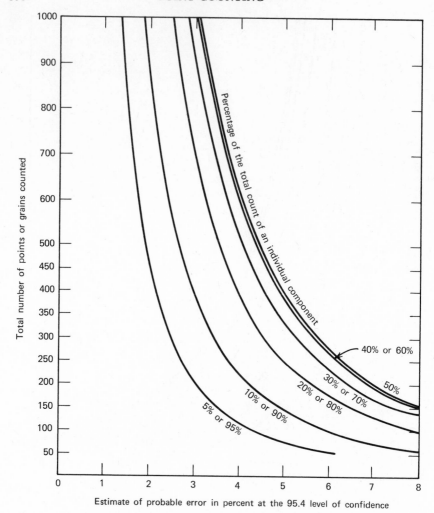

Fig. 5 Probable error at 95.4 confidence level.

counting thin-sections or the line method of counting grain mounts. Similar sheets can easily be designed for the other methods.

It is important to keep a record of the particular traverses made and intervals used. This enables the investigator to return subsequently to the slide and either count additional points or grains or recount the slide to determine the consistency of his identifications. It also enables another operator to count the same points or grains to check for operator variance. Particular points or grains on the slide

TABLE 2 Number of points or grains counted of a particular mineral species and the probable error of the number

Estimated Percent of a Particular Species Present

Total Number of Points or Grains Counted	5	10	20	30	40	50	60	70	80	90	95
100	5±1.5	10±2.0	20±2.7	30±3.1	40±3.3	50±3.4	60±3.3	70±3.1	80±2.7	90±2.0	95±1.5
300	15±2.5	30±3.5	60±4.7	90±5.4	120±5.7	150±5.8	180±5.7	210±5.4	240±4.7	270±3.5	285±2.5
500	25±3.3	50±4.5	100±6.0	150±6.9	200±7.4	250±7.5	300±7.4	350±6.9	400±6.0	450±4.5	475±3.3
1000	50±4.6	100±6.4	200±8.5	300±9.8	400±10.4	500±10.7	600±10.4	700±9.8	800±8.5	900±6.4	950±4.6

50 Level

Estimated Percent of a Particular Species Present

Total Number of Points or Grains Counted	5	10	20	30	40	50	60	70	80	90	95
100	5±4.4	10±6.0	20±8.0	30±9.2	40±9.8	50±10.0	60±9.8	70±9.2	80±8.0	90±6.0	95±4.4
300	15±7.6	30±10.4	60±13.9	90±15.9	120±17.0	150±17.3	180±17.0	210±15.9	240±13.9	270±10.4	285±7.6
500	25±9.7	50±13.4	100±17.9	150±20.5	200±21.9	250±22.4	300±21.9	350±20.5	400±17.9	450±13.4	475±9.7
1000	50±13.8	100±19.0	200±25.3	300±29.0	400±31.0	500±31.6	600±31.0	700±29.0	800±25.3	900±19.0	950±13.8

95.4 Level

Sample Number _____ Rock Name _____

Operator _____ Megascopic Description _____

Date Counted _____ _____

Thin-Section or Grain Mount
Entire Sample
Light Fraction Only
Heavy Fraction Only   Grain Size _____
Other _____

Microscope Number and Type _____

Stage Number and Type _____

Counter Number and Type _____

Slide Position _____

Horizontal Settings   0, 1, 2, 3, 4, 5, 6, 7, 8, 9, 10, 11, 12, 13, 14, 15, 16, 17,

　　　or　　　　　18, 19, 20, 21, 22, 23, 24, 25, 26, 27, 28, 29, 30

Horizontal Interval _____

Vertically Continuous
　　　or
Vertical Interval _____

Mineral Number and Name	Percent	Counter Reading Start	Counter Reading Finish	Number Counted	Comments:
1.					
2.					
3.					
4.					
5.					
6.					
7.					
8.					
9.					
10.					
11.					
12.					
13.					
14. Altered					
15. Opaque					
TOTAL					

Present, but not encountered under cross-hairs _____

Unidentified: Number _____ , Location(s) _____ , Properties _____

　　　　　　Number _____ , Location(s) _____ , Properties _____

Fig. 6 Data summary sheet for point counts of thin-sections or grain mounts

400

can be subsequently relocated if their horizontal and vertical components are read from the graduated stage and recorded. This is especially useful for relocating a particular point or grain to show to someone else, to identify, or to photograph.

In setting up the number of different kinds of classes or minerals to be determined, it is important to be sure that they are all mutually exclusive and exhaustive so that all points or grains can be assigned to one and only one. In working with a large number of slides in a particular study, the author has found it beneficial to look rapidly over all the slides and note which minerals are the most common. These are recorded in the same order on the counters and the data sheets. Consequently, one gets used to the same order on the counter and is able to use it without looking, similar to the way one uses a typewriter. Recording the results in the same order on the data sheets enables more rapid transfer to tables or to IBM cards.

Certain significant mineral species may be rare and may not be encountered under the cross-hair during the counting procedure. These minerals should be noted (see Fig. 6) in order to distinguish them from completely absent minerals in the slide.

EQUIPMENT FOR COUNTING

Microscopes

Most counting is done under the microscope. However, most of the methods and techniques described above can be applied to megascopic mineralogical analysis as well. For point-counting polished sections, a binocular microscope and reflected light are used. For counting thin-sections or grain mounts, a polarizing microscope with a rotating and mechanical stage and a revolving nosepiece with three or more objectives is used. The magnification used depends on how easily the unknowns can be identified. Easily identified or large grains are commonly counted under medium magnification, for example, 80 to 100X, whereas difficult-to-identify or small grains will need higher magnification, that is, 250 to 400X. The author counts heavy mineral grains with a medium power objective and with the substage condenser in. This permits the overall shapes of the grains to be observed, and the additional light is helpful in identifying larger grains which are nearly opaque.

Graduated Mechanical Stages

For the counting of grain mounts, the microscope should be fitted with a mechanical stage that has graduations along both axes. This type of stage enables the slide to be moved a desired amount in either the horizontal or vertical directions (see Fig. 7). American Optical Instrument, Bausch and Lomb, Leitz, and Zeiss are among the companies that make mechanical stages that are popular and completely adequate. Many of these stages are made to attach to several different kinds and makes of microscopes. Before any kind of attachable stage is purchased, one should check with the manufacturer or distributor to make sure it will fit the desired microscope. Most graduated mechanical stages are in the $75 to $150 price range (in 1969).

Point Counting Stages

For the Glagolev-Chayes method of point-counting thin-sections, a graduated mechanical stage that has been modified to a point-counting stage is used. This permits linear traverses to be broken up into equally spaced segments or points. This is commonly accomplished by a device that "clicks" when the section has been moved a given distance. American Optical Instrument, Leitz, and Zeiss are among the companies that make manually operated graduated mechanical stages with the "click" device. These stages are in the $200 to $250 price range (in 1969) and in most cases the "click" device can be quickly disconnected so that the stage can also be used as an ordinary mechanical stage. Figure 8 shows Leitz Stage #40 which comes with seven pairs of interchangeable pinion heads to advance the slide either vertically or horizontally at intervals of 0.1, 0.2, 0.3, 0.4, 0.5, 1.0, or 2.0 mm.

Integrating Eyepiece

An integrating eyepiece has been developed by Carl Zeiss, Inc., which can be used in point-counting thin-sections (see Fig. 9). The method using this eyepiece is a modification of the Glagolev-Chayes method and consists of counting points scattered randomly over the section. The result is an area percent. The integrating eyepiece consists of a plate containing a series of 25 points which is placed in the regular eyepiece of the microscope. When the thin-section is placed on the microscope stage, the field of view appears to be covered by the 25 points. The mineral under each point is identified and recorded. Then, either by moving the section or by rotating the eyepiece, the points overlie new areas and counting continues. The actual distance on the thin-section between adjacent points depends on the magnification used. The manufacturer recommends a magnification such that the average diameter of the grains to be counted is about equal to the distance between adjacent points. Because the distance between points determines how many points will be counted in sampling the entire section, a better method of selecting the proper magnification is to determine the desired number of points to be counted (see above) and pick the magnification that best yields that number.

A serious disadvantage in using the integrating eyepiece is that it is much more difficult to identify minerals in the periphery of a microscope field than those in the center. In the Glagolev-Chayes method, only the minerals in the exact center of the field need to be identified.

Counters

Counting devices are available that tally the number of particular points or grains counted. Those most frequently used are the Clay-Adams and Denominator counters (see Fig. 10). They are available in various sizes, usually with four to eight separate recording keys. In addition, the total of the separate keys is tallied. In most models each separate key can record up to 999 counts and be turned back to zero on completing the count. Most models have space available for writing-in or taping-on the appropriate mineral names above the keys. Prices range from about $75 to $110 (in 1969). Two of the eight-unit counters should be sufficient for almost all analyses.

Automatic Stage and Counter Combination

For counting thin-sections according to the Glagolev-Chayes method, James Swift and Son, Ltd., has developed an automatic

Fig. 7 Mechanical stage. (Photo courtesy of Carl Zeiss, Inc., mechanical stage #57)

point counter (see Fig. 11) that consists of an attachable mechanical stage unit and a counting unit connected to the stage. Each time a key on the counter is punched, the slide is moved automatically a given distance along the traverse. The instrument can be set for either a 1/20, 1/10, 1/6, or 1/3 mm interval between adjacent points on the traverse along one axis. The mechanical stage unit can also be disconnected from the counting unit. The counting unit consists of six separate keys for individual tabulations and a seventh key which can be used to operate the mechanical stage and gives the total number counted. Since the counters cannot be set back to zero, it is essential to take readings at the beginning and end of each count. The instrument that includes the stage and seven-key counting unit

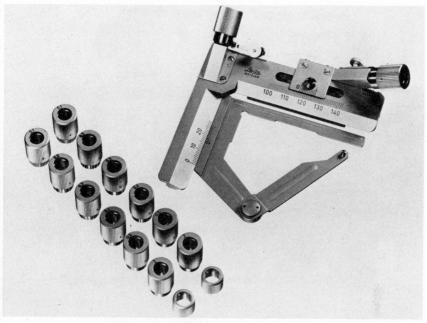

Fig. 8 Point-counting stage. (Photo courtesy of E. Leitz, Inc., stage #40)

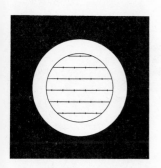

Fig. 9 Integrating eyepiece. (Zeiss integrating eyepiece I)

costs about $650 (in 1969). An additional seven-key counting unit, which can be attached to the other components, costs about $150 and should probably be obtained if one wishes to buy the Swift instrument.

Fig. 10 Counters. (Photo courtesy of Clay-Adams, Inc.)

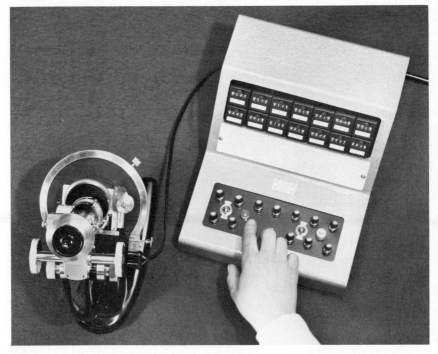

Fig. 11 Automatic point counter. (Photo courtesy of James Swift and Son, Ltd.)

REFERENCES

Chayes, F., 1949, A simple point counter for thin-section analysis; *Am. Mineralogist*, 34, 1-11.

_____, 1956, Petrographic modal analysis; John Wiley and Sons, 113 pp.

Delesse, A., 1848, Procédé méchanique pour déterminer la composition des roches; *Ann. des Mines*, 13, 379-388.

Dryden, A. L., 1931, Accuracy in percentage representation of heavy mineral frequencies; *Proc. National Acad. Sci., U.S.*, 17, 233-238.

Fleet, W. F., 1926, Petrological notes on the Old Red Sandstone of the West Midlands; *Geol. Mag.*, 63, 505-516.

Friedman, G. M., 1960, Chemical analyses of rocks with the petrographic microscope; *Am. Mineralogist*, 45, 69-78.

Galehouse, J. S., 1969, Counting grain mounts: number percentage vs. number frequency; *J. Sed. Pet.*, 39, 812-815.

Glagolev, A. A., 1933, On the geometrical methods of quantitative mineralogical analysis of rocks (in Russian); *Trans. Inst. Econ. Min., Moscow*, 59, 1-47.

_____, 1934, Quantitative analysis with the microscope by the point method; *Engineering Mining J.*, 135, 399.

Heinrich, E. W., 1956, Microscopic petrography; McGraw-Hill Book Co., 296 pp.

Hunter, R. E., 1967, A rapid method for determining weight percentages of unsieved heavy minerals; *J. Sed. Pet.*, 37, 521-529.

Rosiwal, A., 1898, Uber geometrische Gesteinsanalysen; *Verh. der Kaiserlich-Koniglichen Geol. Reichsanstalt*, 143-175.

Shand, S. J., 1916, A recording micrometer for geometrical rock analysis; *J. Geol.*, 24, 394-404.

Van der Plas, L., 1962, Preliminary note on the granulometric analysis of sedimentary rocks; *Sedimentology*, 1, 145-157.

CHAPTER 17

MATHEMATICAL ANALYSIS OF POINT COUNT DATA

JAMES C. KELLEY

University of Washington, Seattle, Washington

THE POINT-COUNTING PROCEDURE

In Chapter 16, a number of techniques for point counting were introduced. For the purposes of this chapter we shall assume that a *sample* of size N has been drawn from the *population* of interest. For example, suppose that 100 points have been counted in each of 10 grain mounts prepared from a beach sand. The sample is the set of points counted of size $N = 1000$, and the population of interest is all of the sand on the beach. As the sample is drawn, each of the points is classified into one and only one of k categories. Examples are clastic and nonclastic grains ($k = 2$); detrital grains, biogenic grains, ash, wood, and opaque minerals ($k = 5$); and a complete bulk-mineral analysis, including quartz, plagioclase and so on, biotite, and mafic minerals (e.g., Table 1). We shall assume that the categories are mutually exclusive and together include all elements in the population.

In this framework point-count data are frequencies of elements in the different categories. The basic aim of point counting is to estimate the proportions of the elements in different categories of the population from a knowledge of the frequencies in the same categories in the sample.

Point-counting analysis is based on the assumption that the frequencies observed in the sample are representative of the proportions in the population and the various techniques of point counting are designed principally to ensure that this assumption will be correct (Chayes, 1956, Chapter 1). Statistically, essentially the same assumption is made, but with the additional requirement that the counting process must sample the population *randomly*. In order to ensure the validity of this assumption, steps must be taken to introduce a randomization procedure into the collection of the original field sample, the preparation of the section, and the point-counting process itself. Sampling procedures are treated elsewhere (Chapter 1; Griffiths, 1967, Chapter 2; Krumbein and Graybill, 1965, Chapter 7). Assume here that the sample has been taken randomly from the population and that the basic assumptions of point counting are not violated.

TABLE 1   Ten bulk-mineral analyses of sediment samples from the Columbia River

Sample No.	Quartz	Plagioclase	Potassium Feldspar	Lithic Fragments	Opaques	Biotite	Mafics
c—c7	41	13	8	26	2	1	10
c—c36	34	11	11	33	1	3	6
c—c50	30	13	7	35	2	0	14
c—c62	15	14	5	60	0	0	5
c—c64	30	11	10	35	2	2	9
c—c73	31	17	14	25	2	1	11
c—c75	34	12	8	37	2	7	0
c—c81	32	18	8	33	2	1	6
c—c82	38	13	9	22	2	1	14
c—c85	39	9	11	30	10	1	0

If the elements in the population categories are randomly distributed within the population (i.e., if grains of the same class are not clustered together) and if a fixed total count procedure is used, the population may be assumed to have the *multinomial* density. (Bennett and Franklin, 1954, p. 114).

ESTIMATION AND CONFIDENCE LIMITS

In random samples from a multinomial distribution, the sample frequencies can be used to estimate the population proportions by the maximum-likelihood criterion. That is, given a random sample from a population whose elements can be classified into one of k categories, in which the elements exist in proportions p_1, p_2, ..., p_j, ..., p_k, one can estimate the proportion p_j for the jth category by the frequency with which members of that category appear in the sample. This frequency is

$$\hat{p}_j = \frac{n_j}{N} \tag{1}$$

The caret notation indicates that $\hat{p}_j$ is an estimate of p_j; n_j is the number of counts recorded for category j, and N is the total number of counts made.

Once an estimate has been obtained it is useful to have a measure of confidence in the estimate. Appropriate simultaneous confidence intervals (Miller, 1966, p. 216) for the $\hat{p}_j$ when N is large (>100) are given by

$$p_j \pm \left[\chi^2_{k-1,\alpha}\right]^{1/2} \left(\frac{\hat{\sigma}^2_j}{N}\right)^{1/2} \qquad j = 1, 2, \ldots, k \tag{2}$$

where $\chi^2_{k-1,\alpha}$ is the upper-tail ($P(\chi^2) < 0.5$) of the χ^2 distribution with $k-1$ degrees of freedom. For example, for $k = 10$ and $\alpha = 0.05$ (95% confidence) the value of χ^2 is 8.343 (Table 2). The statistic $\hat{\sigma}^2_j$

TABLE 2 Upper percentage points of the χ^2 distribution

$$P(\chi^2) = \int_{\chi^2}^{\infty} \frac{1}{\left(\frac{f-2}{2}\right)! \, 2^{f/2}} (\chi^2)^{(f-2)/2} e^{-\chi^2/2} d(\chi^2)$$

f \\ $P(\chi^2)$	0.995	0.990	0.975	0.950	9.900	0.750	0.500	0.250	0.100	0.050	0.025	0.010	0.005
1	3927×10^{-8}	1571×10^{-7}	9821×10^{-7}	3932×10^{-6}	0.01579	0.1015	0.4549	1.323	2.706	3.841	5.024	6.635	7.879
2	0.01003	0.02010	0.05064	0.1026	0.2107	0.5754	1.386	2.773	4.605	5.991	7.378	9.210	10.60
3	0.07172	0.1148	0.2158	0.3518	0.5844	1.213	2.366	4.108	6.251	7.815	9.348	11.34	12.84
4	0.2070	0.2971	0.4844	0.7107	1.064	1.923	3.357	5.385	7.779	9.488	11.14	13.28	14.86
5	0.4117	0.5543	0.8312	1.145	1.610	2.675	4.351	6.626	9.236	11.07	12.83	15.09	16.75
6	0.6757	0.8721	1.237	1.635	2.204	3.455	5.348	7.841	10.64	12.59	14.45	16.81	18.55
7	0.9893	1.209	1.690	2.167	2.833	4.255	6.346	9.037	12.02	14.07	16.01	18.48	20.28
8	1.344	1.646	1.180	2.733	3.499	5.071	7.344	10.22	13.36	15.51	17.53	20.09	21.96
9	1.735	2.088	2.700	3.325	4.168	5.899	8.343	11.39	14.68	16.92	19.02	21.67	23.59
10	2.156	2.588	3.247	3.940	4.865	6.737	9.342	12.55	15.99	18.31	20.48	23.21	25.19
11	2.603	3.053	3.816	4.575	5.578	7.584	10.34	13.70	17.28	19.68	21.92	24.72	26.76
12	2.074	3.571	4.404	5.226	6.304	8.438	11.34	14.85	18.55	21.03	23.34	26.22	28.30
13	3.565	4.107	5.009	5.892	7.042	9.299	12.34	15.98	19.81	22.36	24.74	27.69	29.82
14	4.075	4.660	5.629	6.571	7.790	10.17	13.34	17.12	21.06	23.68	26.12	29.14	31.32
15	4.601	5.229	6.262	7.261	8.547	11.04	14.34	18.25	22.31	25.00	27.49	30.58	32.80
16	5.142	5.812	6.908	7.962	9.312	11.91	15.34	19.37	23.54	26.30	28.85	32.00	34.27
17	5.697	6.408	7.564	8.672	10.09	12.79	16.34	20.49	24.77	27.59	30.19	33.41	35.72
18	6.265	7.051	8.231	9.390	10.86	13.68	17.34	21.60	25.99	28.87	31.53	34.81	37.16
19	6.844	7.633	8.907	10.12	11.65	14.56	18.34	22.72	27.20	30.14	32.85	36.19	38.58
20	7.434	8.260	9.591	10.85	12.44	15.45	19.34	23.83	28.41	31.41	34.17	37.57	40.00

f													
21	8.034	8.897	10.28	11.59	13.24	16.34	20.34	24.93	29.62	32.67	35.48	38.93	41.40
22	8.643	9.542	10.98	12.34	14.04	17.24	21.34	26.04	30.81	33.92	36.78	40.29	42.80
23	9.260	10.20	11.69	13.09	14.85	18.14	22.34	27.14	32.01	35.17	38.08	41.64	44.18
24	9.886	10.86	12.40	13.85	15.66	19.04	23.34	28.24	33.20	36.42	39.36	42.98	45.56
25	10.52	11.52	13.12	14.61	16.47	19.94	24.34	29.34	34.38	37.65	40.65	44.31	46.93
26	11.16	12.20	13.84	15.38	17.29	20.84	25.34	30.43	35.56	38.89	41.92	45.64	48.29
27	11.81	12.88	14.57	16.15	18.11	21.75	26.34	31.53	36.74	40.11	43.19	46.96	49.64
28	12.46	13.56	15.31	16.93	18.94	22.66	27.34	32.62	37.92	41.34	44.46	48.28	50.99
29	13.12	14.26	16.05	17.71	19.77	23.57	28.34	33.71	39.09	42.56	45.72	49.59	52.34
30	13.79	14.95	16.79	18.49	20.60	24.48	29.34	34.80	40.26	43.77	46.98	50.89	53.67
40	20.71	22.16	24.43	26.51	29.05	33.66	39.34	45.62	51.80	55.76	59.34	63.69	66.77
50	27.99	29.71	32.36	34.76	37.69	42.49	49.33	56.33	63.17	67.50	71.42	76.15	79.49
60	35.53	37.48	40.48	43.19	46.46	52.29	59.33	66.98	74.40	79.08	83.30	88.38	91.95
70	43.28	45.44	48.76	51.74	55.33	61.70	69.33	77.58	85.53	90.53	95.02	100.42	104.22
80	51.17	53.54	57.15	60.39	64.28	71.14	79.33	88.13	96.58	101.88	106.63	112.33	116.63
90	59.20	61.75	65.65	69.13	73.29	80.62	89.33	98.65	107.56	113.14	118.14	124.12	128.30
100	67.33	70.06	74.22	77.93	82.36	90.13	99.33	109.14	118.50	123.34	129.56	135.81	140.17
z_p	-2.576	-2.326	-1.960	-1.645	-1.282	-0.6745	0.0000	$+0.6745$	$+1.282$	$+1.645$	$+1.960$	$+2.326$	$+2.576$

For a number of degrees of freedom $f > 100$, take

$$\chi^2(P) = f\left(1 - \frac{2}{9f} + z_p\ \sqrt{\frac{2}{9f}}\right)^2.$$

where z_p is negative for $P < 0.5$. This table was adapted, with the permission of the author and the editor, from C. M. Thompson, *Biometrika*, 32 (1941–1942), pp. 188–89.

is the estimated variance of the proportion. For multinomially distributed variables, the variance is given by

$$\sigma^2_j = p_j(1 - p_j) \tag{3}$$

and the appropriate estimate is

$$\hat{\sigma}^2_j = \hat{p}_j(1 - \hat{p}_j) \tag{4}$$

Thus the confidence intervals may be written as

$$\hat{p}_j \pm \left(\chi^2_{k-1,\alpha}\right)^{1/2} \left(\frac{\hat{p}_j(1 - \hat{p}_j)}{N}\right)^{1/2} \qquad j = 1, 2, \ldots, k \tag{5}$$

Usually, in point counting, N is reasonably large. If N is small (<50) the confidence intervals given in (4) may be inappropriate. For a discussion of these and other confidence intervals, see Goodman (1965).

In order to test whether a set of observed counts $n_1, n_2, \ldots, n_k$, ($\Sigma_j n_j = N$) are drawn from a particular multinomial distribution with probabilities $p_1, p_2, \ldots, p_k$, we can compute the statistic

$$\chi^2 = \sum_{j=1}^{k} \frac{(n_j - Npj)^2}{Npj} \tag{6}$$

which has approximately the χ^2 distribution with $k-1$ degrees of freedom if the p_j are all specified, rather than estimated from the sample (Bennett and Franklin, 1954, p. 620). If the statistic calculated in this manner exceeds the tabled value of χ^2, the null hypothesis which the sample was drawn from the specified multinomial population density is rejected.

AN EXAMPLE

To illustrate the preceding procedure, consider the results of the first analysis in Table 1. The estimate of the proportion of quartz in this sediment is 0.41 (=41%). The variance of this proportion is

$$\hat{\sigma}_1^2 = 0.41(1 - 0.41) = 0.2419 \tag{7}$$

The confidence interval for the quartz percentage with $\alpha = 0.05$ is computed as

$$0.41 \pm (12.59)^{1/2} \frac{0.2419^{1/2}}{300}$$

or

$$0.41 \pm 0.10 \tag{8}$$

Thus, with $100(1 - \alpha) = 95\%$ confidence, the limits 0.31 and 0.51 contain the true proportion of quartz in the population.

In the case of K-feldspar, where $\hat{p}_3 = 0.08$, the confidence limits are

$$\hat{p}_3 \pm (12.59)^{1/2} \left(\frac{0.08 \times 0.92}{300} \right)^{1/2}$$

or

$$0.08 \pm .055 \tag{9}$$

Again, with 95% confidence (which means this statement is incorrect, on the average one time in twenty) the limits 0.025 and 0.135 contain the true proportion of K-feldspar in the population.

THE EFFECT OF CLOSURE

It has been shown that point-count data can be used to estimate the k proportions of the population of interest. Frequently, it is desirable to do more than simply estimate the composition of a sediment or sedimentary rock. Perhaps the most common extension is the study of compositional variation within the population by sampling from different areas and estimating local compositions. In general, these extensions involve studying the relationships between variables. When such a study is contemplated, special care is required when dealing with variables of the type encountered in point counting (i.e., frequencies). Variables of this type require special treatment because they are subject to a linear constraint and thus are not independent because the variables must always have a constant sum

$$\sum_{j=1}^{k} p_j = 1 \tag{10}$$

Knowledge of any $k - 1$ of the frequencies fixes the value of the kth frequency, since

$$p_k = 1 - \sum_{j=1}^{k-1} p_j \tag{11}$$

Another way of expressing this is to say that there are only $k - 1$ degrees of freedom available in our choice of the p_j's. Variables that have a constant sum are said to be *closed* and Equation (10) is called the *closure constraint* on the variables p_j. The variables are not independent, and the relationship between them is solely an artifact of the way in which they are computed, and in this sense is spurious.

The lack of independence among variables is an important consideration when relationships between the proportions are to be investigated. The usual measure of relationships between two variables is the Pearson product-moment correlation coefficient, $\rho_{jj'}$ which measures the degree of association between variables p_j and $p_{j'}$. The correlation is defined as

$$\rho_{jj'} = \frac{\sigma_{jj'}}{\sigma_j \sigma_{j'}} \tag{12}$$

where $\sigma_{jj'}$ is the *covariance* of p_j and $p_{j'}$. The value of $\rho_{jj'}$ can range from -1 to $+1$. A value of $+1$ indicates that high values of p_j are perfectly correlated with high values of $p_{j'}$, and a value of zero indicates that there is no correlation between them.

An important result of the closure constraint is that the covariances of the frequencies are not zero, as is the case for independent variables. This result is important because many commonly used statistical techniques are based on the assumption that the covariances of the variables are equal to zero, that is,

$$\text{cov}\,(p_j p_{j'}) = 0 \tag{13}$$

Instead, for the multinomial distribution

$$\text{cov}\,(p_j p_{j'}) = \sigma_{jj'} = -p_j p_{j'} \tag{14}$$

The value of $\sigma_{jj'}$ can be estimated by replacing the p_j's by their estimates, $\hat{p}_j$, giving the relationship

$$\hat{\sigma}_{jj'} = -\hat{p}_j \hat{p}_{j'} \tag{15}$$

Since both p_j and $p_{j'}$ are nonnegative, the value of $\sigma_{jj'}$ is negative. This negative value for the covariance indicates that, at least to some extent, high values of $\hat{p}_j$ tend to occur with low values of $\hat{p}_{j'}$ and vice versa. Obviously if $k = 2$, as when a sediment is analyzed for the proportion of clastic to nonclastic grains, any increase in clastic grains must be accompanied by an equal decrease in nonclastic grains. As the number of classes is increased, say by separating nonclastic grains into biogenic grains, chemical precipitates, and so on, the effect of closure is decreased, but it cannot be eliminated. Clearly, the effect is most severe on those constituents that are present in large amounts.

Since σ_j and $\sigma_{j'}$ in equation (12) are the positive square roots of σ_j^2 and $\sigma_{j'}^2$, the expected value of $\rho_{jj'}$ is negative when computed for variables from a multinomial distribution in which there is no real relationship between the variables. Furthermore, the value it will take can be determined if the proportions p_j and $p_{j'}$ are known. Substituting (3) and (15) in (17), we have

$$\rho_{jj'} = -\left[\frac{p_j p_{j'}}{(1-p_j)\,(1-p_{j'})}\right]^{1/2} \tag{16}$$

For a more general treatment of this subject see Mosimann (1962), Chayes and Kruskal (1966), and Chayes (1967), and Darroch and Ratcliff (1970).

The value of $\rho_{jj}{}'$ can be estimated from the sample by

$$\rho_{jj}{}' = - \left[\frac{\hat{p}_j \hat{p}_{j}{}'}{(1-\hat{p}_j)\ (1-\hat{p}_{j}{}')} \right]^{\frac{1}{2}} \tag{17}$$

The effect of the closure is most apparent on those variables that contribute the most to the total sum. If the correlation coefficient between two proportions is computed directly from the point-count data, its value must be compared to the null value given by Equation (17) rather than against a null value to zero. Chayes (1966, p. 696) suggests the use of the Fisher Z—test for making this comparison.

AN EXAMPLE

Suppose it is desired to determine whether there is any relationship between the amount of quartz and the proportion of lithic fragments in the samples of Table 1. Computing the correlation coefficient from data in the table yields a value of -0.88.

The proportions of the two constituents are estimated as 0.32 and 0.34, respectively. According to Equation (17) if these constituents are completely uncorrelated, the value would be expected to be

$$\rho^*{}_{jj}{}' = -0.49 \tag{18}$$

Given this null value, the calculated value of the correlation coefficient may be tested for significance by computing the statistic

$$\xi = \frac{(Z\hat{\rho} - Z\rho^*)}{(n - 3)^{\frac{1}{2}}} \tag{19}$$

where $Z\hat{\rho}$ is the value of the Fisher Z—statistic corresponding to the coefficient $\hat{\rho}$, and $Z\rho^*$ is the corresponding Z—value for ρ^*. Here n is the sample size—that is, the number of independent estimates on which the correlation coefficient is based; it might represent, say, a number of replicate analyses of a slide or replicate slides. Values for Z in terms of the correlation coefficients are given in Table 3. It is suggested that n be rather large (>15) if this test is to be sufficiently powerful (Chayes, 1966). The test statistic, ξ, is distributed as $N(0, 1)$, and thus the table for the standard normal deviate (Table 4) can be used to determine confidence intervals for ξ. For example, the 95% confidence interval is

$$-1.96 < \xi < 1.96 \tag{20}$$

If·the value of ξ falls outside this interval, the value of the correlation coefficient computed from the sample is considered to be different from the null value, but it cannot be concluded that there is a significant degree of correlation between the variables.

TABLE 3   Table of $Z = \frac{1}{2} \log_e (1 + r)/(1 - r)$ to transform the correlation coefficient

r	0.00	0.01	0.02	0.03	0.04	0.05	0.06	0.07	0.08	0.09
0.0	0.000	0.010	0.020	0.030	0.040	0.050	0.060	0.070	0.080	0.090
0.1	0.100	0.110	0.121	0.131	0.141	0.151	0.161	0.172	0.182	0.192
0.2	0.203	0.213	0.224	0.234	0.245	0.255	0.266	0.277	0.288	0.299
0.3	0.310	0.321	0.332	0.343	0.354	0.365	0.377	0.388	0.400	0.412
0.4	0.424	0.436	0.448	0.460	0.472	0.485	0.497	0.510	0.523	0.536
0.5	0.549	0.563	0.576	0.590	0.604	0.618	0.633	0.648	0.662	0.678
0.6	0.693	0.709	0.725	0.741	0.758	0.775	0.793	0.811	0.829	0.848
0.7	0.867	0.887	0.908	0.929	0.950	0.973	0.996	1.020	1.045	1.071
0.8	1.099	1.127	1.157	1.188	1.221	1.256	1.293	1.333	1.376	1.422

r	0.000	0.001	0.002	0.003	0.004	0.005	0.006	0.007	0.008	0.009
0.90	1.472	1.478	1.483	1.488	1.494	1.499	1.505	1.510	1.516	1.522
0.91	1.528	1.533	1.539	1.545	1.551	1.557	1.564	1.570	1.576	1.583
0.92	1.589	1.596	1.602	1.609	1.616	1.623	1.630	1.637	1.644	1.651
0.93	1.658	1.666	1.673	1.681	1.689	1.697	1.705	1.713	1.721	1.730
0.94	1.738	1.747	1.756	1.764	1.774	1.783	1.792	1.802	1.812	1.822
0.95	1.832	1.842	1.853	1.863	1.874	1.886	1.897	1.909	1.921	1.933
0.96	1.946	1.959	1.972	1.986	2.000	2.014	2.029	2.044	2.060	2.076
0.97	2.092	2.109	2.127	2.146	2.165	2.185	2.205	2.227	2.249	2.273
0.98	2.298	2.323	2.351	2.380	2.410	2.443	2.477	2.515	2.555	2.599
0.99	2.646	2.700	2.759	2.826	2.903	2.994	3.106	3.250	3.453	3.800

TABLE 4 Cumulative normal distribution

$$F(Z) = \int_{-\infty}^{z} \frac{1}{\sqrt{2\pi}}\, e^{-z^2/2}\, dz$$

r	.00	.01	.02	.03	.04	.05	.06	.07	.08	.09
0	.5000	.5040	.5080	.5120	.5160	.5199	.5239	.5279	.5319	.5359
1	.5398	.5438	.5478	.5517	.5557	.5596	.5636	.5675	.5714	.5753
2	.5793	.5832	.5871	.5910	.5948	.5987	.6026	.6064	.6103	.6141
3	.6179	.6217	.6255	.6293	.6331	.6368	.6406	.6443	.6480	.6517
4	.6554	.6591	.6628	.6664	.6700	.6736	.6772	.6808	.6844	.6879
5	.6915	.6950	.6985	.7019	.7054	.7088	.7123	.7157	.7190	.7224
6	.7257	.7291	.7324	.7357	.7389	.7422	.7454	.7486	.7517	.7549
7	.7580	.7611	.7642	.7673	.7704	.7734	.7764	.7794	.7823	.7852
8	.7881	.7910	.7939	.7967	.7995	.8023	.8051	.8078	.8106	.8133
9	.8159	.8186	.8212	.8238	.8264	.8289	.8315	.8340	.8365	.8389
10	.8413	.8438	.8461	.8485	.8508	.8531	.8554	.8577	.8599	.8621
11	.8643	.8665	.8686	.8708	.8729	.8749	.8770	.8790	.8810	.8830
12	.8849	.8869	.8888	.8907	.8925	.8944	.8962	.8980	.8997	.9015
13	.9032	.9049	.9066	.9082	.9099	.9115	.9131	.9147	.9162	.9177
14	.9192	.9207	.9222	.9236	.9251	.9265	.9279	.9292	.9306	.9319
15	.9332	.9345	.9357	.9370	.9382	.9394	.9406	.9418	.9429	.9441
16	.9452	.9463	.9474	.9484	.9495	.9505	.9515	.9525	.9535	.9545
17	.9554	.9564	.9573	.9582	.9591	.9599	.9608	.9616	.9625	.9633
18	.9641	.9649	.9656	.9664	.9671	.9678	.9686	.9693	.9699	.9706
19	.9713	.9719	.9726	.9732	.9738	.9744	.9750	.9756	.9761	.9767
20	.9772	.9778	.9783	.9788	.9793	.9798	.9803	.9808	.9812	.9817
21	.9821	.9826	.9830	.9834	.9838	.9842	.9846	.9850	.9854	.9857
22	.9861	.9864	.9868	.9871	.9875	.9878	.9881	.9884	.9887	.9890
23	.9893	.9896	.9898	.9901	.9904	.9906	.9909	.9911	.9913	.9916
24	.9918	.9920	.9922	.9925	.9927	.9929	.9931	.9932	.9934	.9936
25	.9938	.9940	.9941	.9943	.9945	.9946	.9948	.9949	.9951	.9952
26	.9953	.9955	.9956	.9957	.9959	.9960	.9961	.9962	.9963	.9964
27	.9965	.9966	.9967	.9968	.9969	.9970	.9971	.9972	.9973	.9974
28	.9974	.9975	.9976	.9977	.9977	.9978	.9979	.9979	.9980	.9981
29	.9981	.9982	.9982	.9983	.9984	.9984	.9985	.9985	.9986	.9986
30	.9987	.9987	.9987	.9988	.9988	.9989	.9989	.9989	.9990	.9990
31	9990	.9991	.9991	.9991	.9992	.9992	.9992	.9992	.9993	.9993
32	.9993	.9993	.9994	.9994	.9994	.9994	.9994	.9995	.9995	.9995
33	.9995	.9995	.9995	.9996	.9996	.9996	.9996	.9996	.9996	.9997
34	.9997	.9997	.9997	.9997	.9997	.9997	.9997	.9997	.9997	.9998

For more extensive tables, see National Bureau of Standards, *Tables of Normal Probability Functions,* Washington, U. S. Government Printing Office, 1953 Applied Mathematics Series 23. Note that they show

$$\int_{-z}^{z} f(z)\,dz, \quad \text{not} \quad \int_{-\infty}^{z} f(z)\,dz$$

It can be stated with 95% confidence that the hypothesis that the samples are drawn from a single multinomial population can be rejected. Departures from the expected value of the correlation coefficient can be the result of either true correlations or sampling from different populations, or any of a number of other subtle effects. Interpretation is difficult under these circumstances and must be undertaken with great care. For further information on the complexities involved, see Mosimann (1962, 1963).

OTHER EFFECTS OF CLOSURE

Although the effect of closure on correlations between frequencies cannot be eliminated, at least the magnitude of the effect can be estimated. With relationships between frequencies and other variables not affected by the closure, a more difficult problem arises. Suppose that the problem is to determine whether the percentage of clastic grains in a sediment is related to mean grain size, or depth of water, or distance offshore. The usual procedure is to calculate a regression equation, for example, one clastic fraction as a function of water depth. If the regression proves to be significant, a casual relationship between two variables is suggested, but cannot be accepted on the basis of the regression alone. In this case the regression relationship itself is suspect, because if one of the variables in a closed system is related to an external variable, all the other variables in the closed system reflect this relationship. There is no way to separate these effects without additional information. For example, suppose a hypothetical continental shelf sediment is composed entirely of land-derived detrial grains and microfauna tests. A study of this sediment might show that the percentage of detrial grains decreases with increased distance offshore. On the basis of these data alone, however, it is impossible to choose among the following alternative interpretations of the regression: (a) more detrital grains are deposited nearshore and test deposition is constant over the area nearshore, (b) more tests are deposited offshore and detrial sedimentation is constant over the area, or (c) any interpretation intermediate between (a) and (b).

The same arguments apply to more complicated regressions, as in trend-surface analysis and other polynomial-regression analyses.

MULTIVARIATE STATISTICAL TECHNIQUES

Since the techniques discussed in this section are complex, no attempt at a detailed and necessarily lengthy explanation is made. Some suggestions are offered that may be helpful to those who contemplate the use of these techniques.

Many multivariate techniques are postulated on the assumption of independence of the variables. Where this assumption holds, the matrix of sums of the squares and sums of the cross products of the deviations from the means is of rank k for the case of k variables, and the off-diagonal elements (the sums of cross products) are all expected to be zero. For closed variables it can be shown that this matrix has rank at most equal to $k - 1$ (Mosimann, 1962).

Because

$$\sum_{j=1}^{k} p_j = 1 \tag{21}$$

and because by the definition of the mean given a sample of size n on which are measured the k proportions, p_j,

$$\sum_{j=1}^{k} (p_{ij} - \bar{p}_j) = 0 \tag{22}$$

the matrix $\mathbf{D} = \{p_{ij} - \bar{p}_j\}$, which is $n \times k$, has rank at most equal to $k - 1$ given that $n \geq k$. Furthermore, the matrix $\mathbf{C} = \mathbf{D}'\mathbf{D}$, where $\mathbf{D}'$ is the transpose of $\mathbf{D}$ is of rank at most equal to $k - 1$. The matrix $\mathbf{C}$ is extremely important in multivariate statistics where it often appears after multiplication by a constant $\frac{1}{n-1}$ as the covariance matrix Σ. But this matrix must usually be inverted and thus it is advantageous to assure that $\mathbf{D}$ is of full rank at the outset.

Two suggestions have been made to circumvent this difficulty. The first is simply to eliminate one of the variables and carry out the calculations using only $k - 1$ variables. This procedure is most useful in procedures such as discriminant function analysis where the interpretation of the solution requires that the variables retain their significance in the subject matter context. The alternative suggestion is to transform the variables, for example, by the arcsine transformation

$$x^* = \sin^{-1} (\sqrt{x})$$
(23)

where x is the untransformed variate and x^* is the transformed variate (Bartlett, 1947). This transformation, which is appropriate for proportions, eliminates the algebraic restrictions but affects the distribution of the variable, because it is not of the form

$$x^* = a + bx$$
(24)

and the transformed variables are frequently difficult to interpret. For more information on transformation see Bartlett (1947). For other multivariate techniques, such as the procedures factor analysis, closure does not introduce great difficulties, but it must be remembered, for example, that the R–matrix in R–mode analysis has rank at most equal to $k - 1$. The results of the analysis must be interpreted in the light of these restrictions.

COMPONENTS PRESENT IN SMALL AMOUNT

If one or more of the components of a sediment occur in very small amounts, say when $p_j < 0.05$, two aspects of the treatment of the variables change.

First, the effect of the closure constraint is quite small and for most purposes can be ignored. Remember, however, that the p_j refers to the component's proportion in the population under analysis, so that although the proportion of, say, garnet in a sediment is small (e.g., $\sim 1\%$), in dealing with the heavy minerals alone, p_j refers to the proportion of garnet in the population of heavy minerals and this proportion may be somewhat larger (e.g., $\sim 10\%$).

Second, individual proportions in a multinomial distribution are binomially distributed. For components present in small amount the binomial distribution approximates the Poisson distribution and the Poisson approximations may be used (Griffiths, 1967, p. 288). In the

Poisson distribution, the population proportion is estimated by the frequency, just as in the binomial distribution

$$\hat{p}_j = \frac{n_j}{N} \qquad (25)$$

The variance estimate is different, however, because in the Poisson distribution the variance is equal to the mean

$$\hat{\sigma}_j^2 = \hat{p}_j \qquad (26)$$

CONCLUSIONS

Given certain assumptions concerning the manner in which the point-counting procedure draws a sample from the population, which is the sediment or sedimentary rock of interest, a number of parameters that characterize the population can be estimated from the sample. Variables measured from the sample can be analyzed by a number of standard statistical techniques, but care must be exercised to ensure that the assumptions implicit in the method are not violated. Of special concern is the assessment of relationships between variables, where the effect of closure on point-count data must be recognized and where the results of the analyses must be interpreted accordingly.

REFERENCES

Bartlett, W., 1947, The use of transformations, *Biometrics,* 3, 39–52.
Bennett, C. A., and N. L. Franklin, 1954, Statistical analysis in chemistry and the chemical industry, John Wiley and Sons, 724 pp.
Chayes, F., 1956, Petrographic modal analysis, John Wiley and Sons, 113 pp.

_____, 1967, Statistical petrography, Ann. Rept. to the Director, Carnegie Institution, 372–379.

_____, and W. Kruskal, 1966, An approximate statistical test for correlation between proportions, *J. Geol.* 74, 692–702.

Darroch, J. N., and D. Ratcliff, 1970, Null correlation for proportions – II, J. Inter. Assoc. Mathematical Geology, 2, 307–312.

Goodman, L. A., 1965, On simultaneous confidence intervals for multinomial populations, *Technometrics*, 7, 247–254.

Griffiths, J. C., 1967, Scientific method in the analysis of sediments, McGraw-Hill Book Co., 509 pp.

Krumbein, W. C., and F. A. Graybill, 1965, An introduction to statistical models in geology, McGraw-Hill Book Co., 475 pp.

Miller, R. G., 1966, Simultaneous statistical inference, McGraw-Hill Book Co., 272 pp.

Mosimann, J. E., 1962, On the compound multinomial distribution, the multivariate β–distribution, and correlations among proportions, *Biometrika*, 49, 65–82.

_____, 1963, On the compound negative multinomial distribution and correlations among inversely sampled pollen counts, *Biometrika*, 50, 47–54.

CHAPTER 18

HEAVY-MINERAL SEPARATION

ROBERT E. CARVER

University of Georgia, Athens, Georgia

OBJECTIVES OF HEAVY-MINERAL SEPARATION

The heavy-mineral component of a terrigenous sediment consists of all clastic grains with specific gravities greater than about 2.9. The heavy-mineral suite is commonly a significant feature of a sediment or sedimentary rock, especially in mature sands that lack rock fragments or light-mineral grains diagnostic of provenance. Although some heavy-mineral species may be selectively destroyed during transport and diagenesis, the remaining heavy minerals are commonly the best, and sometimes the only, guide to provenance of sandstones. The combined percentage of zircon, tourmaline, and rutile (ZTR index, Hubert, 1962, p. 443–444) is an index of mineralogical maturity. ZTR index is independent of the maturity index based on feldspars and it is applicable to sandstones that contain very little feldspar. Solution or corrosion of heavy minerals may indicate intrastratal solution and provide information about conditions of diagenesis.

Unfortunately, intrastratal solution may completely destroy minerals that are important provenance indicators (i.e., pyroxenes, hornblende) or change the ZTR index so that it does not indicate the degree of maturity at the time of deposition. Additional interpretation problems are caused by differential size distribution of heavy minerals. Two sediment samples from the same source, with identical histories, will have different heavy-mineral suites if they are of different grain size, primarily because all sizes of a heavy-mineral species are not equally abundant. Large zircon grains, for example, are rare, because the original grains tend to be small. Consequently, percentages of zircon in coarse-grained sediments are lower than percentages in fine-grained sediments from the same source (van Andel and Poole, 1960). Blatt and Sutherland (1969) have shown that the rate of intrastratal solution is higher in coarse-grained sediments than in less permeable, fine-grained sediments, and this differential solution may increase differences between heavy-mineral suites in fine- and coarse-grained sediments from identical sources.

There is little doubt that the negative aspects of size distribution and chemical alteration of heavy minerals have been overemphasized. For many years the usefulness of heavy-mineral studies was discounted, but about ten years ago, beginning perhaps with van Andel and Poole's (1960) work on Holocene sediments of the northern Gulf of Mexico, sedimentologists began to look at heavy minerals with a renewed interest based on a greater understanding of the processes which produce diversity in heavy-mineral suites.

Walker (1967), employing heavy-mineral studies in part, attacked the red-bed problem, which had been nearly dormant since publication of Krynine's (1949) paper on the tropical origin of many red beds. They convincingly demonstrated that Pliocene red beds in California have formed anthigenically through intrastratal alteration of hornblende and biotite. Hand (1967), working with the concept of hydraulic equivalence introduced by Rubey (1933) and expanded by Rittenhouse (1943), used differences in the settling velocities of light- and heavy-mineral fractions to distinguish eolian sands from beach sands, extending earlier work by Rukhin (1937), who compared the size distributions of light- and heavy-mineral fractions for the same purpose. White and Williams (1967) found that differences in settling velocities in quartz and tourmaline in various parts of a cross-bed set revealed differences in the ratios of traction to suspension sedimentation in different elements of the structure.

Heavy-mineral studies have been applied to an increasingly broad field of problems in sedimentary petrology, but there are a number of other applications of interest to sedimentary petrologists.

Metamorphic isograds in the Appalachian Piedmont have been mapped by panning alluvium from streams with small drainage areas and noting occurrences of minerals which are indices for metamorphic grade (Overstreet, 1962). Panning in the field, or standard separation in the laboratory, are commonly used in exploration and evaluation of sedimentary ore deposits (placer deposits, heavy-mineral sands) and upstream tracing of hydrothermal and plutonic ore minerals. Plutonic ore minerals include diamonds, by their association with garnet and chrome diopside. Heavy-mineral separation techniques may be useful in routine analysis of ore tenor, for example, percentage of kyanite in quartz gangue. Gravity and magnetic separation are employed in accessory mineral studies in igneous petrology, and are especially useful in isolating single mineral species for radiometric or mineralogical studies. In paleontology, heavy-liquid separation techniques are used to separate calcareous microfossils from a quartz-sand matrix.

Techniques for gravity or magnetic separation of individual mineral species or groups of minerals are widely applicable to geological problems, although they are most frequently applied, and most useful, in studies of sedimentary petrology and sedimentation.

THEORY

All standard laboratory techniques for heavy-mineral separation are based on mass separation in a liquid with a specific gravity between the specific gravities of the minerals or groups of minerals to be separated. Ordinarily, the separation is accomplished by gravity settling or centrifuging of the sample in a brominated hydrocarbon liquid. There is no essential theoretical difference between gravity settling and centrifuge methods. In the centrifuge method, settling velocities of grains heavier than the surrounding liquid are increased by increasing the acceleration-of-gravity (g) factor in any of the equations for fall of a small body in a fluid. The greater part of the voluminous literature concerned with the centrifuge method deals with designs of centrifuge tubes that permit clean separation of the light and heavy fractions. Since development of the partial freezing method, these elaborations are not necessary and the centrifuge method has, perhaps, become the most popular and practical method of heavy-mineral separation.

Other techniques of separation, more commonly applied to industrial processes or mineral prospecting, depend on gravity settling of heavy minerals in a suspension of solids in a fluid medium. Solid-liquid suspensions act as liquids of high specific gravity in panning, jigging, and in several commercial heavy-media beneficiation processes. These methods have not been widely used in the laboratory, but panning can be a valuable field tool, and solid-liquid suspensions have been marketed for laboratory use (see Browning, 1961, pp. 12–13).

Differences in laboratory procedure for heavy-mineral separation center on pretreatment of samples, selection of a heavy liquid and diluent, use of a centrifuge to accelerate and improve separation, and the extent of separation of heavy minerals into single species or groups of species. In the following discussion, the emphasis is on standard methods most frequently employed; variations in procedure are implied or only briefly discussed.

PRETREATMENT OF SAMPLES

Disaggregation

The first step in pretreatment of samples for heavy-mineral separation is disaggregation. Sediments and rocks vary so widely in the degree of compaction and cementation, and mineralogy of the cement, that there is no standard treatment. Methods of disaggregation are outlined by Ingram in Chapter 3, Griffiths (1967, pp. 47–54), and Müller (1967, pp. 35–41) and are discussed by Milner (1962), Twenhofel and Tyler (1941), Krumbein and Pettijohn (1938) and others. The available methods are myriad and a thorough knowledge of methods, and some experimentation, will be probably required to determine the best method for a given refractory rock type.

Unconsolidated samples of sand-sized sediment should be thoroughly dried, aggregates crushed with the fingers, and the sample washed to remove fines, wood fragments, and soluble salts. If the sample is too well cemented to be crushed with the fingers, crushing in a jaw crusher or light crushing (not grinding) with a hammer and anvil, to a size coarser than 1 cm, will, in all but the most tightly cemented rocks, do little harm to the constituent grains. When the cement is weaker than the grains, the main effect of coarse crushing is to produce fines from the crushed cement. If the cement is not too strong, coarse aggregates remaining after jaw crushing can be reduced by rolling the sample with a rubber-covered rolling pin on a thick rubber sheet. If the sample cannot be completely disaggregated by the methods above, without undue damage to detrital grains, the cement should be attacked chemically.

Calcite-cemented rocks are easily disaggregated with 10% hydrochloric or acetic acid (one part standard reagent to nine parts distilled water). Iron-oxide cements are normally soluble in 10% hydrochloric acid, at room temperature, on 24-hour exposure. Stronger hydrochloric acid solutions, or hot hydrochloric acid, can be used, but with increasing liability to other mineral components of the rock, to attack iron-oxide, dolomite, gypsum, and other cements. This, however, is normally neither necessary nor desirable. Given an adequate volume of cold, dilute hydrochloric acid, periodic stirring, and enough time, all of the common cements, except silica, will yield to treatment in cold, dilute hydrochloric acid. Treatment with hydrochloric acid will destroy apatite, and hydrochloric acid treatment should therefore be avoided unless a preliminary investigation indicates that the occurrence of apatite is not a critical factor in the study.

Unfortunately, little can be done with rocks which are firmly cemented by silica; reagents attacking the silica cement also attack, to an unacceptable degree, silicates in the heavy-mineral suite. Fine crushing frees heavy-mineral fragments from the siliceous matrix, but tends to fragment heavy-mineral grains. The degree of fragmentation depends on cleavage, hardness, grain size, and adhesion to silica cement, rather than mineral abundance. Grain counts of crushed heavy minerals are therefore dependent on a number of uncontrollable variables and are not directly comparable among tightly cemented samples with differing amounts of silica cement or different proportions of heavy minerals.

Unconsolidated fine-grained sediments should be washed in distilled water to remove soluble salts. Compact clays and shales may disaggregate in cold water, boiling water, or dilute acid solutions. If these methods are not effective, the rock may yield to prolonged boiling in sodium bicarbonate solution, treatment with kerosene and water, or treatment with hydrogen peroxide. In the kerosene method, the sample is dried thoroughly, saturated with kerosene, and covered with water. The water has a much higher surface tension than kerosene and destructively replaces kerosene in the smaller pores of the rock. The treatment should be repeated until disaggregation is complete. Saturation with kerosene is best accomplished by pouring kerosene slowly into a tray containing fragments of the rock. The upper parts of the fragments should remain exposed so that air is freely displaced from pore spaces as kerosene saturation proceeds by capillary action. Most rocks can be completely saturated with fluids, by capillary action, without resort to vacuum, if the rock is very thoroughly dried and is only partly immersed in the fluid. Thin slabs of rock can be saturated with fluids of high viscosity (i.e., melted Canada balsam) by this method. In the hydrogen peroxide method, attributed by Müller (1967) to Wick (1947), the sample is fragmented, dried thoroughly, and covered with 15% hydrogen peroxide solution (one part fresh, commercial, 30% H_2O_2 solution to one part water). Release of oxygen, accompanied by a vigorous boiling action, should begin within 15 minutes. If the reaction has not started after 15 minutes, it should be catalyzed by adding a few milliliters of potassium hydroxide solution. As in the kerosene method, the treatment should be repeated until disaggregation is complete.

Surface stains and undesirable matrix components should be removed by chemical treatment regardless of whether the rock is rigidly cemented. Dilute hydrochloric acid, oxalic acid, or a mixture of stannous chloride and hydrochloric acid (5 gm $SnCl_2$ to 10 ml HCl diluted to 10% solution with nine parts distilled water) effectively removes iron-oxide stains. Sodium hypochlorite (5% commercial solution) will oxidize organic matter, which may react deleteriously with the heavy liquid, and bitumens can be removed by washing with petroleum distillates, typically kerosene.

Samples treated with water-soluble chemicals should be thoroughly washed before drying at low temperature. Hydrochloric acid fumes are dangerous to humans and highly corrosive to ferrous metals. All operations involving hydrochloric acid should be performed in a fume hood. Samples treated with hydrochloric acid should be washed until a drop of 5% silver nitrate solution added to a final, distilled-water wash yields no visible precipitate of silver chloride. Ingestion of oxalic acid solutions will cause severe burns of the mouth, throat, and stomach lining and may result in death (Sax, 1963). Oxalic acid should be handled with extreme caution. Kerosene evaporates slowly at room temperature, leaving very little residue, but it may be desirable to wash kerosene-treated samples with acetone.

Size Separation

Rubey (1933) showed that grains of heavy minerals are hydraulicly equivalent to light-mineral grains of some larger size, depending on the specific gravity of the heavy mineral, and heavy-mineral suites therefore vary in composition with grain size of the sample. An extended discussion of the hydraulic equivalence problem is presented by Hubert in Chapter 19. A method of partially negating the grain-size effect, first suggested by Rubey (1933), is to examine a single, small size fraction of all samples. Although this practice has been widely adopted, a standard size fraction has not been established.

Young (1966), in a reexamination of Rittenhouse's (1944) Rio Grande River samples, found little difference in the average reproducibility of grain counts between the 61 to 88 μ and 124 to 175 μ size fractions, but recommended the smaller fraction on the basis of tests of reproducibility of heavy-mineral ratios. Krumbein and Rasmussen (1941), in a study of sources of error in heavy-mineral studies, examined the 124 to 175 μ and 175 to 246 μ fractions separately and found that sampling and laboratory errors were lower for the coarser fraction. Müller (1967) recommends study of the 63 to 200 μ fraction.

Some recent studies illustrate the range of practice in this respect: Giles and Pilkey (1965) studied the 125 to 250 μ fraction of all samples; Galehouse (1967), the 63 to 500 μ fraction; and Stanley (1965), the 100 to 450 μ fraction. Van Andel and Poole (1960) studied the 63 to 500 μ fraction and demonstrated that heavy-mineral distribution patterns in Holocene sediments of the northern Gulf of Mexico are independent of grain size. However, in a study of Holocene sediments of the Gulf of California, van Andel (1964) found that percentages of some minerals, in the suite larger than 62 μ correlated with median size of the sample.

As Griffiths states (1967, p. 204), the identification of heavy minerals with the petrographic microscope is a difficult task and (Griffiths and Rosenfeld, 1954) operator variation is probably a major source of error in all scientific work involving subjective assessment, as in the identification of heavy-mineral grains. The difficulty of identifying heavy minerals in grain mounts, and consequent operator variation, is very much reduced if all grains are of approximately the same size, because color, birefringence, and other optical properties tend to be more nearly constant for a given mineral species or variety (Rubey 1933). All optical effects, especially pleochroism and color, tend to be more pronounced in larger grains, and large grains are easier to identify consistently and correctly than small grains.

It is therefore desirable to restrict heavy-mineral studies to a single size class narrow enough to produce uniform optical effects and reduce or eliminate variation in heavy-mineral proportions caused by differences in grain size of the samples. The size class should be coarse enough that characteristic optical properties are easily observed. However, the size class should not be so restricted that excessively large bulk samples are required to obtain 300 to 400 heavy-mineral grains of the required size, nor should the size class be so different from the modal class of most samples that very large bulk samples must be processed to obtain a reasonable crop of heavy minerals in the desired size range. For most studies, the 125 to 250 μ fraction seems an excellent compromise. After samples are disaggregated and cleaned, they should be sieved and the heavy minerals from a single sieve fraction, that is, the 2ϕ to 3ϕ fraction, separated for analysis.

HEAVY LIQUIDS

Either tetrabromoethane or tribromomethane (bromoform) are used in standard heavy-mineral separation procedures. Their properties are similar (Table 1) and the decision as to which to use is normally based on cost, ready availability, or familiarity with a particular procedure. At full strength, the specific gravity of tetrabromoethane is 2.97 and bromoform 2.89. Originally "heavy minerals" were defined as all detrital components with specific gravities greater than about 2.8 (Howell, 1957, light minerals).

TABLE 1 Properties of commonly used heavy liquids

Data from O'Connel (1963), Sax (1963), Browning (1961), and Jahns (1939)

Heavy Liquid Common Synonyms	Chemical Formula	Specific Gravity 20°/4° C–Δ/°C	M.P., °C B.P., °C	Vapor Pressure, mm, 25°C	Sol. in H$_2$O gm/100 gm	Toxicity Hazard Vapor Inhal.	Toxicity Hazard Skin Cont.	Acetone	Alcohol	CCl$_4$	Benzene
Dibromomethane Methylene bromide MB	CH$_2$Br$_2$	2.48–0.0037	−52.6 97.	45.3	1.1	Moderate	Moderate	M	M	M	M
Tribromomethane Bromoform	CHBr$_3$	2.89–0.0023	8. 149.	5.8	0.311	Severe	Moderate	M	M	M	M
Tetrabromoethane 1,1,2,2-tetrabromoethane Acetylene tetrabromide Tetrabromethane TBR, TBE	C$_2$H$_2$Br$_4$	2.97–0.0023	0.1 243.	1.02	0.065	Severe	Moderate	M	M	M	M
Diiodomethane Methylene iodide Iodoform MI	CH$_2$I$_2$	3.32–0.0021	6.1 182.	1.25	0.124	Unknown	Unknown	M	M	M	M
Thallous formate Soln.	HCO$_2$Tl	3.4 (1)			800+	Extreme	Extreme				
Thallous formate-malonate Soln. Clerici's solution	HCO$_2$Tl CH$_2$(COOTl)$_2$	4.3 (1)				Extreme	Extreme				

(1) Specific gravity varies with degree of saturation of solution and saturation varies with solution temperature. Maximum specific gravities for saturated solutions at 90°C are approximately 5.0 for both solutions. Specific gravity figures given are for saturated solutions at room temperature.

Toxicities listed are, in general, the worst of the combinations of chronic and acute hazards listed by Sax (1963).

435

If further separation of the heavy-mineral suite, or a greater selectivity in the specific gravity of minerals to be separated is required, heavy liquids ranging in specific gravity from about 0.8 to 5.0 can be prepared from materials listed in Tables 1 and 2. Precise gravity separations can be acomplished in two successive float-sink operations. Table 1 presents the more commonly used heavy liquids; Twenhofel and Tyler (1941, pp. 75—76) and O'Connel (1963) list a few others which are rarely used. The major problem associated with producing a heavy liquid of any desired specific gravity is the selection of a diluent with a vapor pressure low enough that specific gravity of the mixture will remain constant over a reasonable length of time. A simple method for complete recovery of the heavy liquid, after the required seprations have been made, is also desirable. For halogenated hydrocarbons, the ideal diluent should have a vapor pressure equal to the vapor pressure of the heavy liquid. The diluent should be either miscible with water or have a freezing point significantly different from that of the heavy liquid, so that separation of the heavy liquid and solvent can be accomplished by solvent extraction or freezing (see section on recovery of heavy liquid). Twenhofel and Tyler (1941) reject acetone and ethanol as diluents, because of their volatility, and recommend carbon tetrachloride or benzene, even though the heavy liquid must be recovered by distillation.

Dimethylsulfoxide has been recommended as a diluent for bromoform (Meyrowitz et al., 1959) and methylene iodide (Cuttitta et al., 1960). Dimethylsulfoxide causes oxidation of methylene iodide, with release of free iodine and darkening of the solution to opacity within a few weeks. It is not therefore recommended as a diluent for methylene iodine. N, N-dimethylformamide has been recommended as a diluent for both methylene iodide and bromoform (Meyrowitz et al., 1960; Hickling et al., 1961). Like dimethylsulfoxide, it has a low vapor pressure and is miscible with water, acetone and alcohol, but it does not oxidize and discolor methylene iodide. Sax (1963), however, suggests that mixtures of N, N-dimethylformamide and halogenated hydrocarbons are hazardous (possibly explosive) and the compound is not recommended for use with any of the common heavy liquids.

For separations involving specific gravities above 3.3, thallous formate in water, or a mixture of equal parts of thallous formate and

TABLE 2 Properties of solvents and diluents

Data are from Weast (1967) unless otherwise noted

Liquid Common Synonyms	Chemical Formula	Specific Gravity	M.P., °C B.P., °C	Vapor Pressure, mm, 20°C[b]	Sol. in H_2O[c]	Toxicity Hazard Vapor Inhal.	Skin Cont.
Methanol Methyl alcohol Wood alcohol	CH_3OH	0.79	−97.8 65.0	93	M	Severe	Severe
Ethanol Ethyl alcohol Alcohol Grain alcohol	C_2H_5OH	0.79	−117.3 78.5	44	M	Low	Low
Acetone 2-propanone Dimethyl ketone	CH_3COCH_3	0.79	−95.3 56.2	185	M	Low	Low
Benzene Benzol Phene	C_6H_6	0.88	5.5 80.1	79	SLT.	Severe	Severe
Tetrachloromethane Carbon tetrachloride	CCl_4	1.59	−23.0 76.7	91	INS.	Severe	Severe

(continued)

TABLE 2 Properties of solvents and diluents (Continued)

Data are from Weast (1967) unless otherwise noted

Liquid Common Synonyms	Chemical Formula	Specific Gravity	M.P., °C B.P., °C	Vapor Pressure, mm, 20° C[b]	Sol. in H_2O[c]	Toxicity Hazard Vapor Inhal.	Skin Cont.
Dimethylsulfoxide[a] Methylsulfoxide DMSO	$(CH_3)_2SO$	1.10	18.4 189	0.37	M	Unknown	Unknown
N,N-Dimethylformamide[a] DMF	$HCON(CH_3)_2$	0.95	−61 153	2.7	M	Severe	Severe

Toxicity ratings are, in general, the worst of the combinations of acute and chronic hazards listed by Sax (1963).

[a]Data from Meyrowitz et al. (1959).

[b]Data from Washburn (1928).

[c]M indicates miscible in all, or nearly all proportions; SLT., slightly soluble; INS., insoluble.

thallous malonate in water (Clerici's solution), are commonly used (7 to 10 gm of each of the thallous compounds per milliliter of water). At room temperature, saturated Clerici's solution has a specific gravity of about 4.25, and the specific gravity can be accurately determined from the refractive index (Vassar, 1925; Jahns, 1939, Table 3). For high specific gravities a thermostatically controlled heating-jacket around a container holding the Clerici's solution is recommended. The specific gravity approaches 5.0 at 90°C (Browning, 1961). Hot water is used to wash samples after immersion in Clerici's solution, and slow evaporation at controlled temperature restores the specific gravity of diluted solutions. The thallium compounds are recovered from washings by evaporation. Thallous formate solution and Clerici's solution are absorbed readily through the skin and the vapors absorbed through the lungs. Both are extremely poisonous, producing symptoms similar to arsenic poisoning and worse. Browning (1961) recommends the use of rubber gloves, protective clothing, goggles, and a respirator when working with thallium compounds. More than 0.1 mg of vapor per cubic meter of air is considered damaging (Sax, 1963) and care should be taken to see that all the work is done in an efficient hood and below the decomposition temperature of 130°C.

SEPARATION BY GRAVITY METHODS

Separation of heavy minerals by settling is normally accomplished in glass funnels. Funnels without ribs or flutes are very desirable for this process because ribs or flutes tend to interfere with the thorough stirring required to obtain complete separation. Separatory funnels are sometimes used, but their disadvantage is that sand grains tend to stick in the glass stopcocks, thereby making it impossible to control the flow of liquid and sediment.

In the most common arrangement, a funnel that will hold the heavy liquid and sample is supported above a filtering funnel and receiving flask as shown in Fig. 1. It is of some advantage to square-off the lower end of the holding funnel before use, because a beveled-end sometimes tends to trap heavy minerals in the tube. A short piece of rubber tubing is fitted over the end of the holding funnel and a pinchcock placed on the rubber tubing. A clamp of the

TABLE 3   Refractive index and specific gravity of Clerici solution at
19.5°C (Vassar, 1925) and 21°C (Jahns, 1939). For a given index
of refraction, the specific gravity increases about 0.001 per degree
above 21°C and decreases about 0.001 per degree below 21°C (Jahns,
1939). Because the degree of saturation may vary with the procedure
used, or fractional crystallization of thallium malonate may occur,
each investigator should construct his own working curve and
check it against the data given below

19.5°C		21°C	
Refractive Index	Specific Gravity	Refractive Index	Specific Gravity
1.6761	4.076	1.6954	4.233
1.6296	3.695	1.6769	4.069
1.6154	3.580	1.6571	3.889
1.5990	3.434	1.6307	3.665
1.5815	3.280	1.6165	3.562
1.5693	3.184	1.5917	3.341
1.5620	3.114	1.5727	3.157
1.5515	3.024	1.5467	2.950
1.5363	2.884	1.5317	2.815
1.5156	2.692	1.5100	2.625
		1.4832	2.378
		1.4561	2.157

Day or Fisher design, which can be removed without passing the
clamp across the end of the tubing, is best for this purpose. If a large
holding funnel is used, pressure at the bottom of the funnel stem will
be high and a clamp with a relatively strong spring will be required.
When the equipment is properly assembled, the holding flask is
partly filled with heavy liquid and the first sediment sample is
poured into the heavy liquid. The sample is stirred periodically until
heavy minerals can no longer be observed to settle into the stem of
the holding funnel. If the sample is large, some of the sample will
adhere to the wetted area above the liquid line in the funnel and
some heavy minerals may be trapped in this portion of the sample. It
is therefore advisable to wash the funnel walls with heavy liquid from
a plastic squeeze bottle, after most of the heavy minerals have been
separated, remixing the sample, and allowing the remaining heavy
minerals to settle out.

A small funnel with a filter paper is placed under the holding
funnel and a flask reserved for pure heavy-liquid is placed under the

lower funnel. The accumulated heavy minerals are dropped from the stem of the holding funnel by opening briefly the clamp on the tubing of the upper funnel. Again, pressure at the base of the funnel stem may be high, and considerable care must be used to avoid splashing the heavy minerals and heavy liquid out of the filter paper. The heavy liquid is allowed to filter into a clearly marked flask reserved for undiluted heavy liquid. The small funnel containing the

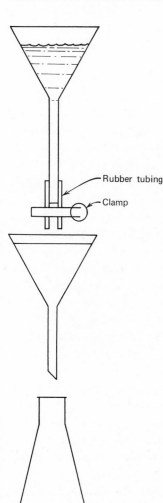

Rubber tubing

Clamp

Fig. 1 Arrangement of equipment for gravity settling of heavy minerals.

heavy minerals is transferred to a different rack, or the flask containing the heavy liquid is removed and replaced by a flask reserved for washings. The heavy minerals are washed several times by directing a stream of solvent from a squeeze bottle into the filter paper, and are set aside to dry.

A second funnel with a filter paper, and a flask reserved for undiluted heavy liquid, are placed under the funnel containing the heavy liquid and light minerals. The heavy liquid, carrying part of the sample with it, is drained into the filter paper and then allowed to filter into the flask. The pinchcock on the holding funnel is completely removed during this operation, allowing as much as possible of the heavy liquid to filter to the heavy-liquid flask. When the light fraction of the sample has drained, the flask for undiluted heavy liquid is removed and replaced by the flask reserved for washings. The remaining part of the sample, adhering to the sides and tube of the upper funnel, is then washed down through the upper funnel and into the filter paper with a jet of solvent from a plastic squeeze bottle. The light-mineral fraction, caught in the filter paper, is washed several times with small quantities of solvent and set aside to dry.

The most common error in procedure is contamination of the pure heavy-liquid with solvent, either by failure to remove the flask of heavy liquid before washing the sample with solvent, or by filtering undiluted heavy liquid into a flask of washings. Cleanup of diluted heavy liquid is, in most cases, a simple process and the error is not serious as long as the operator is aware that dilution of the heavy liquid has occurred. The best way to avoid this type of accident is to mark clearly the solvent and heavy liquid flasks with large labels of contrasting color.

In the settling method the greater part of the heavy liquid is recycled. The only temporary loss is that part of the heavy liquid which wets the sand and filter paper and is collected in the washings. It is therefore possible to run a large number of samples with a minimum of cleanup time and a relatively small volume of heavy liquid.

If a large number of samples is to be run, it is very desirable to set up a multiple funnel rack which permits a number of samples to be handled simultaneously. For 90-mm diameter funnels, boards with 2-in. diameter holes on 5-in. centers, with slots cut to the edge of the board, make very satisfactory funnel racks. The racks are supported by finger clamps attached to ring stands. Because the time required for settling and filtering operations to come to completion is the major time factor, running eight to ten samples simultaneously reduces greatly the average separation time.

SEPARATION BY CENTRIFUGE METHODS

Centrifuge methods of heavy-mineral separation are invariably used with sediments finer than 63 μ and are commonly used with sand-sized samples. On the whole, centrifuging methods are probably superior to settling methods, but neither method has an overwhelming advantage in terms of completeness of separation, cost, or average separation time. The centrifuge technique does, however, permit a little greater flexibility of operation. Centrifuge tubes can be tightly corked and conveniently set aside for a few hours or days at several points in the procedure. It is usually impractical to halt settling procedures, except at the end of a complete separation, because of the bulky equipment setup and evaporation of heavy liquid from open glassware.

Barsdate (1962) used a hypodermic syringe to remove heavy minerals from centrifuge tubes. In the method suggested by Barsdate the sample is immersed in heavy liquid in a 15-ml conical-bottom tube and centrifuged. Heavy minerals are removed with the aid of a 1.7-mm diameter hypodermic needle attached to a 2-ml syringe. The plunger of the syringe is depressed slightly as the needle tip presses through the light-mineral aggregate at the top of the heavy liquid to expel air or liquid and prevent light minerals from entering the needle. With the needle at the bottom of the tube, the plunger is retracted and the heavy minerals are drawn into the needle and barrel of the syringe. The needle is withdrawn from the centrifuge tube and the heavy minerals discharged into a funnel fitted with a filter paper. The process is repeated until all heavy minerals have been removed from the centrifuge tube, and the syringe is rinsed with heavy liquid several times to transfer all heavy minerals to the filter paper. The remaining heavy liquid and light minerals are then poured and washed into a second funnel and filter paper. The procedure for washing the sample and collecting the undiluted heavy liquid is similar to the procedure for gravity methods. Barsdate's method does not require freezing of all, or part, of the heavy liquid, nor does it require centrifuge tubes of special design. However, manipulation of the hypodermic syringe may prove tedious if large numbers of samples are to be run, and the partial-freezing method is recommended for investigations requiring many heavy-mineral separations.

Straight-walled, Pyrex, 50-ml capacity (or larger) centrifuge tubes with conical bottoms are recommended for the partial-freezing procedure described below. Two or more centrifuge tubes, normally a multiple of the number of tubes required to fill the centrifuge head, are set out and labeled. Sample splits weighing 10 to 20 gm are placed in the tubes. The tubes are partially filled with heavy liquid so that the top of the sediment and heavy-liquid mixture is at approximately the same level in all the tubes. Each tube is weighed on a platform balance and the weight brought to within 0.1 gm of some common weight by adding small quantities of heavy liquid. Rubber cushions designed to receive the conical tube ends are placed in the bottom of the metal shields (tube holders) and the balanced tubes are put in the shields and centrifuged.

Müller (1967) recommends centrifuging for 15 minutes at 3000 to 4000 rpm for samples in the 63 to 200 μ size range, but shorter periods and lower speeds are probably acceptable, especially with the coarser grain sizes recommended here (125 to 250 μ). Rittenhouse and Bertholf (1942) obtained slightly greater weights of heavy minerals by centrifuging at 1800 rpm for 5 minutes than by settling with continuous stirring for 2 hours, based on experiments with samples in the size range 88 to 124 μ. Centrifuging for 5 minutes at 2000 rpm should be adequate for most sand samples, but the time and speed should be greatly increased for sediments smaller than 63 μ.

The frozen heavy-liquid method of separation of light- and heavy-mineral fractions is commonly attributed to Mackay of the Royal School of Mines, London (Griffiths, 1967). The much superior method of partial freezing is described by Fessenden (1959) and the use of liquid nitrogen for freezing by Scull (1960). In the partial freezing method of separation, a test-tube rack with openings large enough to hold the centrifuge tubes is placed in a flat pan or tray and the base of the rack covered with crushed solid carbon dioxide (dry ice) or a liquefied gas (liquid air or liquid nitrogen). The centrifuge tubes are placed in the rack so the tube bottoms rest on, and in, the cold medium. Freezing of the heavy liquid at the base of the tube begins in a few minutes and is allowed to continue until the heavy liquid is frozen to 2 to 4 cm above the heavy minerals concentrated in the bottom of the tube.

With the heavy minerals solidly frozen at the bottom of the centrifuge tube, the unfrozen heavy liquid and light-mineral fraction is poured into a funnel fitted with a filter paper, and the heavy liquid is filtered into a clearly marked flask reserved for undiluted heavy liquid. The centrifuge tube is inverted over the filter paper and any light-fraction sand remaining in the tube is removed by washing with a jet of heavy liquid, preferably cold, from a squeeze bottle. In this step the squeeze bottle must be operated in an inverted position and the bottle should be modified by tightly sealing the annulus around the delivery tube and cutting the tube off near the top of the bottle. Pollack (1962) recommends that the frozen plug of heavy liquid be held in the tube with a rod and small stopper while the light fraction is washed out, but this is probably not necessary if cold heavy liquid is used for washing. After the heavy liquid has filtered from the light-mineral fraction, the receiving flask is changed and the light-mineral fraction is thoroughly washed with solvent before being set aside to dry.

The centrifuge tube is placed in a rack at ambient temperature long enough for the frozen heavy liquid to melt, and the heavy minerals are poured and washed into a funnel fitted with a filter paper. When the heavy liquid has filtered off, the heavy mineral fraction is thoroughly washed with solvent and dried.

As in the settling method, most of the heavy liquid is recovered after each separation, and a large number of samples can be run with a small volume of heavy liquid. The equipment requirements of the centrifuge methods are somewhat greater than the settling method, the centrifuge and a platform balance being the major items. The average time per sample is small, especially if a large number of filter funnels can be set up.

RECOVERY OF HEAVY LIQUIDS

Bromoform or tetrabromethane mixed with the solvent used to wash samples and clean glassware can be recovered by solvent extraction. Bromoform and tetrabromethane are almost immiscible with water (Table 1), whereas acetone and alcohols are miscible in all, or nearly all, proportions. The heavy liquid can therefore be recovered by bubbling water slowly through the mixture or washing the mixture with water several times.

If the mixture of heavy liquid and diluent is to be washed, it is poured into a large bottle and the bottle is nearly filled with water. The bottle is vigorously shaken and the liquid mixtures are allowed to separate before the water, now containing some of the solvent, is poured, or preferably siphoned, from the bottle. The process is repeated until mineral fragments of high specific gravity will float in the purified heavy liquid. Calcite, specific gravity 2.72; aragonite, 2.95; dolomite, 2.85; and muscovite selected for a specific gravity near that of the heavy liquid are good choices for indicator minerals. When washing is complete, a separatory funnel is used to separate the heavy liquid from water that cannot be poured off. The heavy liquid should be filtered before reuse, primarily to remove any remaining water, and clarified, if required, as described below.

Initial separation of water-miscible solvent and heavy liquid can also be accomplished by bubbling water slowly through the mixture, but the flowing water tends to entrain, or float, heavy liquid and losses may be greater than in the washing method. If water-immiscible compounds are used as solvents or diluents, the light liquid is removed by distillation. Clerici's solution is restored to saturation by evaporation of water in a water bath.

Halogenated hydrocarbons may break down, to some extent, under the influence of heat or light and become very dark in color due to the presence of free halogens. They can be clarified by shaking with a dilute solution of NaOH.

TOXICITY OF HEAVY LIQUIDS, DILUTENTS AND SOLVENTS

All halogenated hydrocarbons are toxic to some extent, and low-carbon-number chlorinated hydrocarbons are particularly dangerous. Short exposure to high concentrations of vapor may cause unconsciousness, respiratory failure, and death in some cases. Prolonged or repeated inhalation of low concentrations of vapor, accidental ingestion, or prolonged or repeated skin contact can cause irreversible damage to the liver, kidneys, or lungs. Carbon tetrachloride is especially toxic in all these respects, whereas bromoform and tetrabromoethane are only moderately toxic and have low vapor pressures which further reduce the hazard.

Heating of halogenated hydrocarbons should be avoided or done only under the most rigorously controlled conditions. Carbon tetrachloride releases phosgene vapor if overheated, and bromoform and tetrabromoethane release highly toxic carbonyl bromide vapor (Sax, 1963).

All operations with halogenated hydrocarbons should be carried out in a hood with the front closed as much of the time as possible. Skin contact should be avoided and, if it occurs, the skin should be immediately and thoroughly washed.

With the exception of methyl alcohol, the alcohols and ketones (acetone) likely to be used in heavy-mineral separation are not very dangerous, but excessive inhalation or skin contact should be avoided. Ingestion of methyl alcohol or exposure to high concentrations of vapor is very dangerous. Methyl-alcohol poisoning may be cumulative on a day to day basis, and may result in unconsciousness and blindness (Sax, 1963). Benzene is also highly toxic, concentrations less than 3000 ppm may cause unconsciousness and eventual death, and exposures to low vapor concentrations over long periods of time may cause severe damage to the blood-forming tissues with consequent deterioration of the blood.

MAGNETIC SEPARATION METHODS

Magnetic separation of mineral grains depends on the magnetic susceptibility of the minerals to be separated. Magnetic susceptibility is a complex phenomenon depending on chemical composition, particularly minor amounts of iron or manganese, and lattice structure. Magnetic susceptibility varies widely within some mineral species (Table 4). Magnetic-separation methods therefore tend to be most valuable and most precise when all grains of a given mineral species have identical, or nearly identical, compositions and structures, as in igneous rocks. Sedimentary rocks normally include several varieties of a given mineral species, all with different magnetic susceptibilities, so that precise separations of single species are more difficult. However, the method is sometimes useful in heavy-mineral

TABLE 4 Ranges of magnetic susceptibilities for some common heavy minerals and siderite

Mineral	Susceptibility Range $\times 10^{-6}$ emu	Number of Samples	Source
Ilmenite	15.45– 70.00	3	Powell & Ballard, 1968
Rutile	0.85– 4.78	5	Powell & Ballard, 1968
Sphene	4.43– 5.84	3	Powell & Ballard, 1968
Zircon	0.16– 0.26	4	Powell & Ballard, 1968
Siderite	66.09–103.81	15	Powell and Miller, 1963
Garnets	11.5 – 75.6	23	Frost, 1960

studies. White and Williams (1967) magnetically separated tourmaline from other minerals in samples from a large trough cross-bed set and compared settling velocities of quartz and tourmaline to assess the dominant sediment transport mechanism in different parts of the structure.

Magnetite is the only common detrital material which can be removed from sediments with a hand magnet. Other minerals are commonly separated with a Frantz isodynamic separator, after removal of magnetite which tends to clog the machine. The Frantz separator consists of an electromagnet with two elongated pole pieces shaped so that the gap between poles is considerably wider on one side than the other. A vibrating metal chute runs between the pole pieces, parallel to their length. The magnet is mounted at an angle which permits mineral grains to be fed into the upper end of the chute and, with the aid of the vibrator, slide toward the lower end. Grains with higher magnetic susceptibility are pulled toward the side of the chute where the pole gap is narrow and the magnetic flux greatest. A dividing edge near the lower end of the chute separates the grains into two streams, one consisting of grains of higher susceptibility, which are caught in separate hoppers.

The Frantz isodynamic separator is mounted on a universal mount which allows it to be rotated in the direction of grain movement (slope) and, transversely, in a direction normal to the grain flow (tilt). Separation of minerals in the chute depends on the tilt of the chute, the amperage applied to the electromagnet, and, to some extent, on the slope and rate of feed to the machine. The current and tilt required to separate minerals of known magnetic susceptibility can be calculated (Müller, 1967). However, because of the wide variation in magnetic susceptibility of minerals of a given species (Table 4), the common practice is to set the slope at 10 to 30° (20° is more or less standard), the tilt at 5 to 20° (15° for minerals of moderate to high susceptibility, 5° for minerals of low susceptibility), and determine the most efficient amperage by trial and error. More than one pass through the separator is normally required for a clean separation.

REFERENCES

Barsdate, R.J., 1962, Rapid heavy mineral separation, *J. Sed. Pet.,* **32**, 608–620.

Blatt, Harvey, and Berry Sutherland, 1969, Intrastratal solution and non-opaque heavy minerals in shales, *J. Sed. Pet.,* **39**, 591–600.

Browning, J. S., 1961, Heavy liquids and procedures for laboratory separation of minerals, U.S. Bureau of Mines Inf. Circ. 8007, 14 pp.

Cuttitta, F., R. Meyrowitz, and B. Levin, 1960, Dimethylsulfoxide, a new diluent for methylene iodide heavy liquid, *Am. Mineralogist,* **45**, 726–728.

Fessenden, F. W., 1959, Removal of heavy liquid separates from glass centrifuge tubes, *J. Sed. Pet.,* **29**, 621.

Frost, M. J., 1960, Magnetic susceptibility of garnet: Mineralogical Magazine, **32**, 573–576.

Galehouse, J. S., 1967, Provenance and paleoccurrents of the Paso Robles Formation, California, *Bull. Geol. Soc. America* **78**, 951–978.

Giles, R. T., and O. H. Pilkey, 1965, Atlantic beach and dune sediments of the southern United States, *J. Sed. Pet.,* **35**, 900–910.

Griffiths, J. C., 1967, Scientific method in the analysis of sediments, McGraw-Hill Book Co., 508 pp.

——, and M. A. Rosenfeld, 1954, Operator variation in experimental research, *J. Geol.,* **62**, 74–91.

Hand, B. M., 1967, Differentiation of beach and dune sands, using settling velocities of light and heavy minerals, *J. Sed. Pet.,* **37**, 514–520.

Hickling, N., F. Cuttitta, and R. Meyrowitz, 1961, N,N-dimethylformamide, a new diluent for bromoform used as a heavy liquid, *Am. Mineralogist,* 46, 1502-1503.

Howell, J. V., 1957, Glossary of geology and related sciences, Am. Geol. Inst., Washington, D. C., 325 pp.

Hubert, J. F., 1962, A zircon-tourmaline-rutile maturity index and the interdependence of the composition of heavy mineral assemblages with the gross composition and texture of sandstones, *J. Sed. Pet.,* 32, 440–450.

———, and W. J. Neal, 1967, Mineral composition and dispersal patterns of deep-sea sands in the western North Atlantic Petrologic province, *Bull. Geol. Soc. America,* 78, 749–772.

Jahns, R. H., 1939, Clerici solution for the specific gravity determination of small mineral grains, *Am. Mineralogist,* 24, 116–122.

Krumbein, W. C. and F. J. Pettijohn, 1938, Manual of sedimentary petrography, Appleton-Century-Crofts, 549 pp.

———, and W.C. Rasmussen, 1941, The probable error of sampling beach sand for heavy mineral analysis, *J. Sed. Pet.,* 11, 10–20.

Krynine, P. D., 1949, The origin of red beds, *New York Acad. Sci. Trans., Series 2,* 11, 60–68.

Meyrowitz, R., F. Cuttitta, and N. Hickling, 1959, A new diluent for bromoform in heavy liquid separation of minerals, *Am. Mineralogist,* 44, 884–885.

———, ———, and B. Levin, 1960, N,N-dimethylformamide, a new diluent for methylene iodide heavy liquid, *Am. Mineralogist,* 45, 1278–1280.

Milner, H. B., ed., 1962, Sedimentary petrography, 4th revised ed., MacMillen, Vol. I, 643 pp.

Müller, German, 1967, Sedimentary petrology, Part I, Methods in sedimentary petrology, translated by Hans—Ulrich Schmincke, Hafner Publishing Co., 283 pp.

O'Connel, W. L., 1963, Properties of heavy liquids, *Trans. Am. Inst. Mining, Metallurgical and Petrol. Eng.,* 226, 126–132.

Overstreet, W. C., 1962, A review of regional heavy-mineral reconnaissance and its application in the southeastern Piedmont, *Southeastern Geol.,* 3, 133–173.

Pollack, J. M., 1962, Removal of heavy liquid separates from glass centrifuge tubes—additional suggestions, *J. Sed. Pet.,* 32, 607.

Powell, H. E., and C. K. Miller, 1963, Magnetic susceptibility of siderite, U.S. Bureau of Mines Rept. Inv. 6224, 19 pp.

_____, and L. M. Ballard, 1968, Magnetic susceptibility of group IV B, V B, and VI B metal-bearing minerals, U.S. Bureau of Mines Inf. Circ. 8360, 9 pp.

Rittenhouse, Gordon, 1943, Transportation and deposition of heavy minerals, *Bull. Geol. Soc. America* **54**, 1725—1780.

_____, 1944, Sources of modern sands in the middle Rio Grande Valley, New Mexico, *J. Geol.* **52**, 145—183.

_____, and W. E. Bertholf, Jr., 1942, Gravity versus centrifuge separation of heavy minerals from sand, *J. Sed. Pet.,* **12**, 85—89.

Rubey, W. W., 1933, The size-distribution of heavy minerals within a water-laid sandstone, *J. Sed. Pet.,* **3**, 3—29.

Rukhin, L. B., 1937, A new method for the determination of the conditions of deposition of ancient sands, Problems of Soviet Geology, **7**, 953—959.

Sax, N. I., 1963, Dangerous properties of industrial materials, 2nd ed., Reinhold, 1343 pp.

Scull, B. J., 1960, Removal of heavy liquid separates from glass centrifuge tubes—alternate method, *J. Sed. Pet.,* **30**, 626.

Stanley, D. J., 1965, Heavy minerals and provenance of sands in flysch of central and southern French Alps, *Bull. Am. Assoc. Petrol. Geologists,* **49**, 22—40.

Twenhofel, W. H., and S. A. Tyler, 1941, Methods of study of sediments, McGraw-Hill Book Co., 183 pp.

Van Andel, T. H., 1964, Recent marine sediments of Gulf of California, pp. 216—310, *in* T. H. van Andel and G. G. Shor, Jr., eds., Marine Geology of the Gulf of California, Am. Assoc. Petrol. Geologists Memoir 3, 408 pp.

_____, and D. M. Poole, 1960, Sources of recent sediments in the northern Gulf of Mexico, *J. Sed. Pet.,* **30** 91—122.

Vassar, H. E., 1925, Clerici solution for mineral separation by gravity, *Am. Mineralogist,* **10**, 123—125.

Walker, T. R., 1967, Formation of red beds in modern and ancient deserts, *Bull. Geol. Soc. America,* **78**, 353—368.

Washburn, E. W., ed., 1928, International critical tables of numerical data, physics, chemistry and technology, 3, McGraw-Hill Book Co.

Weast, R. C., 1967, Handbook of chemistry and physics, 48th ed., Chemical Rubber Co.

White, J. R., and E. G. Williams 1967, The nature of a fluvial process as defined by settling velocities of heavy and light minerals, *J. Sed. Pet.*, 37, 530—539.

Wick, Werner, 1947, Aufbereitungsmethoden in der Mikropalaontologie, *Naturalhistorische Ges. zu Hannover, Jahrsbericht*, 94—98, 35—41, ns.

Young, E. J., 1966, A critique of methods for comparing heavy mineral suites, *J. Sed. Pet.*, 36, 57—65.

CHAPTER 19

ANALYSIS OF HEAVY-MINERAL ASSEMBLAGES

JOHN F. HUBERT

University of Massachusetts, Amherst, Massachusetts

Sedimentary petrology was founded in 1851 when Sorby described a cherty limestone that he examined with a petrographic microscope. Studies of accessory heavy-mineral assemblages in sands and sandstones began about 1870. The papers published from 1800 to 1900 by Retgers in Germany, Artini in Italy, Cayeux in France, and Hume and Mackie in Britain remain as models of thoroughness for the description of heavy minerals. Quantitative treatment of heavy-mineral assemblages began with Artini's classic paper (1898) on heavy minerals in River Po sediments in Italy. In the 30 to 40 years that followed, heavy minerals were commonly the only components of sands and sandstones studied in detail. This no doubt stemmed from their obvious usefulness in interpreting source areas, and perhaps also because of their inherent beauty under the microscope.

Early in this century, the study of heavy minerals flourished, especially in Britain, resulting in 1922 in the pioneering book by Milner, *Introduction to Sedimentary Petrology*. Descriptions and illustrations in this book, which is in its fourth edition (1962), are still the best available for students learning to identify heavy minerals. The magnitude of the early work on heavy minerals is reflected in Boswell's 1933 volume, *Mineralogy of Sedimentary Rocks,* which included annotated references to more than 1000 papers. His review of the history of heavy-mineral studies (pp. 15—28) can be read with profit and enjoyment.

During the period from 1920 to 1940, petroleum geologists commonly used heavy-mineral studies in problems of provenance and correlation of sandstones, but this approach was soon replaced by micropaleontology and electric-logging techniques. Investigations of heavy minerals declined primarily because of difficulties encountered with wide fluctuations in the proportions of heavy-mineral species due to size sorting. In addition, it was claimed that heavy minerals undergo intensive and widespread modification by intrastratal solution. Opposing evidence, however, suggests that the degree of selective sorting can be assessed, and the importance of intrastratal solution has been overemphasized.

Numerous studies demonstrate that heavy-mineral assemblages in most sandstones can be used successfully to help clarify their geologic history. The interpretation of heavy-mineral assemblages is discussed in the following sections in the sequence: (a) reliability of heavy-mineral proportions in modal analyses, (b) assessment of the effects and volumetric importance of intrastratal solution, (c) provenance of heavy minerals, (d) evaluation of selective sorting and hydraulic ratios, and (e) mapping of dispersal patterns of sediments in sedimentary petrologic provinces.

MODAL ANALYSIS OF HEAVY MINERALS

Laboratory Procedures

Proportions of heavy minerals can be determined utilizing an entire sample, or the sediment can be sieve-fractioned before the heavy minerals are isolated with heavy liquids. It is advisable to microsplit the bulk sample, or subfractions of it, before heavy-liquid separation and then to include the entire heavy-mineral fraction on the slide. A common procedure is to analyze the size range only from about 0.5 to 0.03 mm. Only a few minerals, such as garnet, are commonly larger than 0.5 mm. The lower limit of 0.03 mm reduces the amount of zircon that so commonly floods the silt grades and excludes many grains too small for practical study. Narrow size fractions must be used if hydraulic ratios are to be calculated, or if comparisons based on a restricted size interval are to be made among samples. If the heavy minerals are cleaned with stannous chloride or hydrochloric acid solutions, care must be taken not to destroy the more soluble minerals such as apatite.

Heavy minerals are commonly counted in "ribbon" traverses randomly selected on a mechanical stage where each grain that falls within the field of view is identified and tabulated. This procedure yields mineral percentages based on number of grains. The number percentages can be converted to weight percentages if desired (Hunter, 1967).

The nonopaque suite of minerals is most useful in making genetic interpretations, and it is usually recalculated to total 100% by omitting micas and opaque grains. The nonopaque minerals comprise a fairly homogenous hydraulic suite, thereby reducing problems of selective sorting that may be severe for the total assemblage. Furthermore, quantification of micas is difficult because their densities straddle that of common heavy liquids (bromoform and tetrabromoethane) which result in incomplete or irregular separation. Opaque grains are commonly not identified according to mineral species, although they might reveal more information than is generally thought.

Doeglas (1940, pp. 115–116) suggested a useful procedure to increase the precision with which the nonopaque suite is determined in each sample. First, the ratio of opaques to all other minerals is counted in, say, 200 points by ribbon traverses. The nonopaque count is then increased to 200 grains, thus substantially increasing the reliability of the proportions of these minerals. For example, if there are 80% opaque grains in the total assemblage and 200 nonopaques are counted, this is equivalent to counting 1000 grains of the total assemblage.

Reliability of Heavy-Mineral Proportions

Sources of laboratory errors in modal analysis are (a) splitting the bulk sample, (b) sieving the subsample into size grades, (c) separating, microsplitting, and mounting the heavy minerals, and (d) identifying and counting the minerals. The most important source of error arises from counting a limited number of grains. This type of error is not to be confused with errors in identification or tabulating the percentages (Krumbein and Rasmussen, 1941, pp. 17—18). The other errors are considerably less important.

The counting error is a function of the number of grains counted in a sample of grains randomly drawn from the total number, or population, of grains on the heavy-mineral mount. The population has a fixed number of grains, but it is so large that in practice it can be considered infinite. If the sample size is fixed at, say, 200 grains, and the real value of probability p for the occurrence of the event, that is, one grain of the mineral, lies between 0.10 and 0.90, then the binomial (Bernoulli) frequency distribution is an appropriate model for describing the sampling distribution of the mineral percentages (Griffiths, 1960, p. 354). If p is much less than 0.10 or greater than 0.90, then the Poisson frequency distribution is a suitable approximation to the binomial. The Poisson frequency distribution need not concern us here, however, because little interpretative value can be placed on small percentages that involve sampling errors that are large by comparison.

The binomial is a probability density function of the form

$$f(x) = (p + q)^n,$$

where x is the number of grains of the specific mineral encountered in the ribbon traverses, p is the probability of encountering the mineral, q is the probability of not encountering the mineral, and n is the total number of grains in the ribbon traverses. The sum of $p + q$ is 1.0. The population mean of a binomial distribution is np and the variance is npq.

Using the mean, $\bar{x}$, and size, n, of a sample, the 95% confidence interval around the population mean of a binomial distribution can be read from Fig. 1. For example, if a mineral forms 20% of the heavy-mineral assemblage based on a count of 50 grains, the interval from 10 to 34% has a 95% probability of including the fixed, real population mean. The population mean will lie outside the confidence interval in only 5% of the cases on the average in repeated sampling, which is the usual acceptable margin of error. If the number of grains is increased to 250, the confidence interval is reduced to between 15 and 26%. As would be expected, the confidence interval narrows with increased information, that is, an increase in number of grains counted.

A further example shows that when the sample mean is 70%, the confidence interval with 50 grains is between 56 and 82%, whereas with 250 grains it is reduced to between 64 and 76%. These examples of the increase in precision with an increase from 50 to 250 grains illustrate why many investigators count 200 or 300 grains per sample. Examination of the chart shows that increasing the count to 1000 grains does not decrease the confidence interval enough to

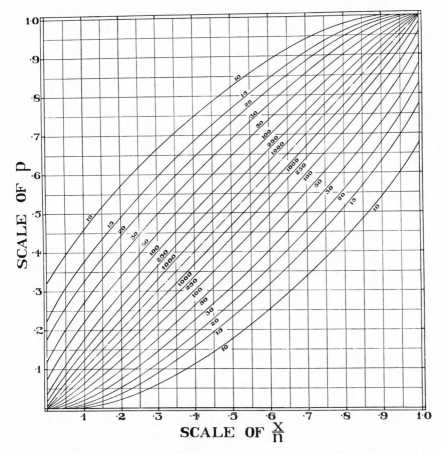

Fig. 1 Ninety-five percent confidence limits on means drawn from the binomial distribution. Sample $\overline{x} = x/n$.

justify the additional time spent making the analysis. It is preferable to count only 200 to 300 grains and use the remaining time to analyze additional samples.

INTRASTRATAL SOLUTION

Whether or not intrastratal solution of heavy minerals is a major geologic process is currently being debated. The possible effect of intrastratal solution on any particular heavy-mineral assemblage must be assessed before interpreting source-rock lithology, weathering in the source area, destruction by abrasion during transport, and selective sorting. Table 1 shows the relative chemical stability of some common heavy minerals.

Evidence of intrastratal solution of heavy minerals in sandstones is of two types. The first is the occurrence of higher proportions of unstable heavy minerals in impermeable calcareous concretions than in the surrounding porous sandstone (Bramlette, 1941, pp. 34–35; Weyl, 1952; Todd and Folk, 1957, p. 2560). The minerals most affected by solution in the formations studied are hornblende, epidote, and garnet.

A second evidence of intrastratal solution is the gradual decrease in unstable heavy minerals downward in the stratigraphic column or, more rarely, within specific stratigraphic intervals. Partial solution of heavy minerals is reflected by corroded grain surfaces and skeletal grains (Edelman and Doeglas, 1932). A noteworthy example is corroded hornblende in several hundred feet of Pliocene-Pleistocene-Recent intertidal arkosic sediments in a desert region of Baja, California (Walker, 1967, pp. 359–360). Solution of hornblende in these sediments has been more thorough in the lower parts of the section. Another example is partially dissolved heavy-mineral grains, notably augite, hornblende, kyanite, and staurolite, in Tertiary geosynclinal sandstone of the Northern Appennines, Italy (Gazzi, 1965, pp. 111–113).

On the other hand, heavy–mineral zones and dispersal patterns have been successfully delineated in geosynclinal flysch sandstones of Upper Cretaceous to Lower Tertiary age in the Central and Southern French Alps without detecting intrastratal solution (Stanley, 1965, pp. 32–33). Van Andel (1959, p. 159) searched without success for partially dissolved, corroded grains of heavy minerals in thick sedimentary sequences from various parts of the world. Furthermore, the relative absence of unstable heavy minerals in Late Paleozoic slate-phyllite litharenites of the Appalachians and their abundance in Triassic arkosic sandstones is clearly controlled by source area lithology, not intrastratal solution (Krynine, 1950, pp. 89—90).

Some caution is necessary before interpreting all raggedly angular, "hacksaw" grains of garnet, staurolite, augite, and hornblende as having been produced by intrastratal solution. In some cases these grains reflect their original shapes in the source rocks. Heavy-mineral grains with fragile shapes can survive hundreds of miles of fluvial transport. Some irregular shapes are produced by breakage during transport.

Many orthoquartzite sandstones, especially of Early Paleozoic but also of Cretaceous and Early Tertiary age, are rich in ultrastable

TABLE 1   Approximate relative chemical stability
of some common heavy minerals.
Modified from Smithson (1941) and Pettijohn (1941, 1957)

Ultrastable	Rutile
	Zircon
	Tourmaline
Stable	Leucoxene
	Muscovite
	Chlorite
	Hematite (unstable under reducing conditions)
Semistable	Apatite
	Monazite
	Staurolite
	Sillimanite
	Kyanite
	Epidote group
Unstable	Biotite
	Garnet
	Magnetite (unstable under oxidizing conditions)
	Ilmenite
	Hornblende
	Augite
	Olivine

zircon, tourmaline, and rutile and impoverished in relatively unstable heavy minerals. Several authors interpret this to reflect pervasive intrastratal solution that increases with geologic age (Bramlette, 1941, p. 35; Pettijohn, 1941, p. 621, 1957, pp. 514–520). Krynine (1942, pp. 545–546) pointed out, however, that light- and heavy-mineral assemblages commonly show similar degrees of mineralogical maturity. In orthoquartzite sandstones the grains are almost exclusively ultrastable quartz, chert, zircon, tourmaline, and rutile, concentrated during very long time intervals under conditions of regional tectonic stability. The mineralogically immature lithic-bearing and arkosic sandstones of orogenic belts of any age, by contrast, commonly contain large proportions of unstable heavy minerals.

Limited data indicate that in feldspathic and lithic sandstones that grade into quartz-chert arenites the combined proportion of zircon, tourmaline, and rutile in the nonopaque suite (ZTR index, Hubert, 1962, p. 445) increases, whereas feldspar and igneous and metamorphic fragments decrease in the light fraction. The parallel increase in mineralogical maturity of the light- and heavy-mineral assemblages suggests that intrastratal solution is probably not a widespread geological process. Inversions in ZTR index do occur in some lithic sandstones eroded from source rocks that inherently contain mostly zircon, tourmaline, and/or rutile. Examples are (a) the Lamotte Sandstone (Cambrian of Missouri) where the quartz porphyry arkoses have a zircon-dominated nonopaque assemblage (Ojakangas, 1963, pp. 864–866), and (b) many slate-phyllite litharenites (Krynine, 1942, pp. 548–549).

Modification of nonopaque heavy minerals to an ultrastable zircon-tourmaline-rutile assemblage is usually slower than alteration of the light fraction to an ultrastable quartz-chert composition (Hubert, 1962, p. 447). Because of their slower modification rate, the heavy minerals in many quartzose sandstones are a better indicator of source-rock lithology, whereas the light minerals reflect better the abrasion history of the sand. For example, the faster modification rate of the light minerals in the western North Atlantic shelf and deep-sea sands is apparent in the feldspathic quartzite and orthoquartzite sands whose nonopaque heavy-mineral assemblages are immature and contain low percentages of zircon-tourmaline-rutile (Hubert and Neal, 1967, p. 751). The quartzose sands were modified to a quartz-feldspar-mica assemblage without distinctive rock fragments, but the heavy minerals were retained, including some that reflect changes in source area lithology. Durability of heavy minerals makes possible a subdivision of the western North Atlantic province into five sedimentary subprovinces.

PROVENANCE

To aid in interpreting source rock lithology, heavy minerals are usually grouped into genetic suites such as reworked sedimentary, low- and high-rank metamorphic, sialic and mafic igneous, pegmatitic, and authigenic suites (Table 2). The relative importance of each suite is then evaluated. Another possible subdivision is supracrustal (sediments and low-rank metamorphic) versus crustal (high-rank metamorphic and plutonic) suites which yields clues on the magnitude of uplift in the source area (Pettijohn, 1957, p. 512). Although severe weathering can modify profoundly heavy-mineral assemblages in soil profiles, its effect on the composition of sands in basins with a moderate to rapid rate of erosion and sedimentation is negligible.

Even small amounts of debris from ash falls or erosion of nearby ashes and tuffs can be recognized by a characteristic suite of minerals. Most common are euhedral crystals of brown biotite, apatite, and zircon and, less commonly, augite and hornblende. Associated light minerals may include euhedral, clear quartz crystals of β-quartz form, and glass shards, or zeolite pseudomorphs after glass shards. Examples are the Cretaceous Mowry bentonites in Wyoming (Slaughter and Earley, 1965, pp. 25–28) and Cretaceous Mesaverde sandstones, Wyoming (Cole, 1960, pp. 39, 50).

Distinctive mineral varieties defined by color, inclusions, morphology, overgrowths, composition, and so on, can be used to trace detrital grains of sandstones to specific source rocks. This approach is most successful when applied to lithic and arkosic sandstones deposited in fluvial environments relatively near their source. Once mixing of grains from several sources has occurred in a marine environment it is more difficult to identify specific sources. A few examples will clarify this method.

Broken pieces of large, idiomorphic blue tourmaline (indicolite) crystals in the fluvial Triassic arkoses of Connecticut were traced to

TABLE 2   Provenance of some common heavy minerals

Reworked sediments	Well-rounded grains of rutile, tourmaline, zircon.
Low-rank metamorphic	Biotite, chlorite, spessartite garnet, tourmaline (especially small, euhedral, brown crystals with graphite inclusions).
High-rank metamorphic	Actinolite, andalusite, apatite, almandine garnet, biotite diopside, epidote, clinozoisite, glaucophane, hornblende (including blue-green varieties), ilmenite, kyanite, magnetite, sillimanite, sphene, staurolite, tourmaline, tremolite, zircon.
Sialic igneous	Apatite, biotite, hornblende, ilmenite, monazite, muscovite, rutile, sphene, tourmaline, zircon.
Mafic igneous	Augite, diopside, epidote, hornblende, hypersthene, ilmenite, magnetite, olivine, oxyhornblende, pyrope garnet, serpentine.
Pegmatites	Apatite, biotite, cassiterite, garnet, monazite, muscovite, rutile, tourmaline (especially indicolite).
Ash falls	Euhedral crystals of apatite, augite, biotite, hornblende, and zircon.
Authigenic	Hematite, leucoxene, limonite, tourmaline, zircon; euhedral crystals of anatase, brookite, pyrite, rutile, and sphene.

pegmatites in eastern Connecticut (Krynine, 1950, pp. 192–193, 211). Apatite grains with graphite inclusions arranged subparallel to the c-axes in the fluvial Fountain arkose of Permo-Pennsylvanian age along the Front Range, Colorado, were derived from the Precambrian Idaho Springs Schist that crops out to the west in the Front Range (Hubert, 1960, pp. 193–194). The abraided "stumps" of colorless authigenic overgrowths on tourmaline grains in the Middle Ordovician Bellefonte orthoquartzite sandstone in central Pennsylvania were eroded from Upper Cambrian Gatesburg orthoquartzites known to contain tourmaline colorless overgrowths (Krynine, 1946, pp. 72–73). Recycling of tourmaline grains in these sandstones was demonstrated by use of 13 varieties of tourmaline based on color and inclusions (Krynine, 1946, p. 82).

In some instances assemblages of minerals, none individually diagnostic, are characteristic of specific source rocks. A classic example was discovered ,in the study of the unroofing of the Dartmoor granites as reflected by the Permian to Pliocene succession in southern England (Groves, 1931). The Dartmoor assemblage consists of distinctively zoned zircon, andalusite (including chiasto-lite), indicolite, manganiferous garnet, anatase, rutile, brookite, sphene, monazite, cassiterite, topaz, cordierite, sillimanite, dumortierite, and corundum. A necessary preliminary to Grove's interpretations was that the accessory minerals of the Dartmoor granites were known in great detail (Brammall, 1928).

A unique search for the source of a sand based on its heavy-mineral assemblage occurred in World War II. During 1944 to 1945, the Japanese released 9000 incendiary balloons to drift toward North America in the high-velocity air currents of the upper atmosphere, hoping to set the forests on fire during the summer dry season. Each balloon carried thirty 6-pound bags of ballast sand which dropped successively as the balloon fell below 30,000 feet altitude. The composition of the sand was determined as 52% hypersthene, 10% magnetite, 8% volcanic quartz phenocrysts, 8% common quartz, 8% hornblende, 7% augite, 6% plagioclase, 1% garnet, and traces of biotite, hornblende schist, sphene, volcanic glass, and zircon, together with shell fragments (Ross, 1950, p. 908). Based on a knowledge of the geology of Japan, it was determined that there were only five sites where beach sand of this composition could have been obtained. Air force planes photographed these sites and located a manufacturing plant with partially inflated gas bags nearby. The threat from incendiary balloons was subsequently eliminated from the war.

SELECTIVE SORTING AND HYDRAULIC RATIOS

Selective sorting of heavy minerals has been a persistent problem in studies of correlation, provenance, and dispersal patterns. Even closely spaced samples of varying grain size yield markedly different heavy-mineral assemblages because heavy minerals are deposited with light minerals of equivalent hydraulic size. Hydraulic equivalent size is defined as the difference in size between a given heavy-mineral species and the size of a quartz sphere with the same settling velocity in water (Rubey, 1933). The hydraulic size of a heavy mineral may be measured operationally as the difference in ϕ units between the size of the mineral and the size of the light minerals with which it is deposited. The size of a grain on the ϕ scale equals $-\log_2 X$ where X is the grain diameter in millimeters (Krumbein, 1936, p. 38).

A convenient way to determine hydraulic size is to measure the separation of the modal classes of grain-size distributions of light and heavy minerals (Briggs, 1965, p. 944). The size distribution of light minerals is determined by standard sieving techniques. The size distribution curve for each heavy mineral is obtained by converting the number frequency in each narrow sieve interval to weight percentage and then plotting the distribution as a frequency curve. The hydraulic sizes of heavy minerals show a positive correlation with density (Table 3; Briggs, 1965, p. 945). A necessary precaution when using the method of hydraulic ratios is that each investigator redetermine hydraulic sizes of the heavy minerals in his samples because some of the heavy minerals may have truncated size-frequency distributions.

The hydraulic size measured empirically includes the effects of size, density, shape, availability of the mineral in various sizes, and nature of the depositing medium. The shape factor is especially important when differences in shape among heavy minerals of about the same specific gravity are marked (Briggs et al., 1962, p. 653).

Rittenhouse (1943, p. 1743) proposed the concept of hydraulic ratios to resolve the problem of selective sorting. The hydraulic ratio of a heavy mineral is defined as "100 times the weight of that mineral in a known range of sizes divided by the weight of light minerals of equivalent hydraulic size." The weight of the heavy mineral is determined in an index size interval and divided by the weight of the hydraulically equivalent lights in the size interval that is coarser by the hydraulic size of the mineral. In a regional study, samples with larger hydraulic ratios for a given mineral indicate areas where the mineral was relatively more available for deposition; the mineral was presumably deposited in the proportion that it had in the transport load of the current.

An example of the calculations is given in Table 4 and Fig. 2. In hydraulic equivalence studies it is important that a homogenous sedimentation unit be sampled so that each sample was deposited within a narrow range of hydraulic conditions.

The method of hydraulic ratios is based on two fundamental assumptions, in addition to the absence of intrastratal solution. First, the hydraulic size of each heavy mineral is assumed to be constant in the samples studied. For Tertiary sandstones in Hyerfano Park, Colorado, Briggs (1965, p. 941) found that the hydraulic sizes for the nonmagnetic heavy minerals were constant in the finer sands with a median diameter less than 0.125 mm (3ϕ) but inversely proportional to the median diameter in coarser samples. Even in the finer sands, however, the hydraulic sizes of the heavy minerals were larger than theoretical predictions based on settling velocity equations (Rubey, 1933, p. 332).

Second, use of hydraulic ratios assumes that the heavy minerals are supplied by the source area in sizes appropriate for deposition in the sieve intervals analyzed. Some minerals, however, tend to occur in small crystals in the source rocks, notably zircon, sphene, rutile,

TABLE 3 Approximate hydraulic size and density
of some common heavy minerals
(Rittenhouse, 1944, p. 165)

Mineral	Density	Hydraulic Size in ϕ Units (Best Value Over All Size Intervals and All Samples)
Magnetite	5.2	1.0
Ilmenite	4.7	1.0
Zircon	4.6	0.9
Barite	4.5	0.5
Garnet	3.8	0.6
Kyanite	3.6	0.3
Sphene	3.5	0.5
Clinopyroxene	3.5	0.3
Hypersthene	3.4	0.4
Diopside	3.3	0.4
Hornblende	3.2	0.2
Apatite	3.2	0.4
Tourmaline	3.1	0.2

Note: 0.8 may be a better value for magnetite and ilmenite (Rittenhouse, 1944, p. 160).

TABLE 4 Sample calculation of hydraulic ratios of minerals in the 0.125 to 0.088 mm (3.0 to 3.5 ϕ) size interval (Rittenhouse, 1944, p. 166)

Mineral	I Number Frequency, %	II Density	III (I X II)	IV (III/404.0)	V Weight (IV X 0.175[a]), gm	Hydraulic Ratio (V/wt. of Equivalent Lights from Fig. 2)
Garnet	15.0	3.8	57.0	14.1	0.0247	0.0247/24.1 = 0.102
Hornblende	35.0	3.2	112.0	27.7	0.0485	0.0485/13.0 = 0.374
Ilmenite	50.0	4.7	235.0	58.2	0.1018	0.1018/28.5 = 0.357
Sum	100.0%		404.0	100.0	0.1750	

Hydraulic sizes: garnet (0.6), hornblende (0.2), ilmenite (0.8).

[a] Total weight of heavies in this size interval was 0.175 gm.

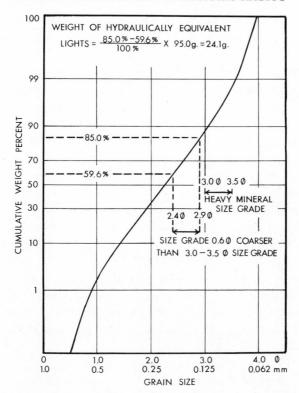

Fig. 2 Sample calculation of the weight of the light minerals in the size grade hydraulically equivalent to garnet in the 0.125 to 0.088 mm (3.0 to 3.5φ) size interval (Rittenhouse, 1944, p. 166). Total weight of the sample is 95.0 gm. The grain-size distribution is plotted on arithmetic probability paper.

epidote, ilmenite, and monazite. A low hydraulic ratio for a coarse sieve interval would be misleading if a mineral were abundant in the transport load of the current, but of a size inherently too small to be available in that size interval. Because of this factor, plus the relative constancy of hydraulic sizes in finer sands, hydraulic ratios seem to work best in fine and very fine sands. Briggs (1965, pp. 941–946) illustrates a method to correct for the effects of specific gravity and shape of heavy minerals in sands coarser than fine sand.

Despite the time-consuming laboratory work and calculations necessary to determine hydraulic ratios, they show somewhat greater sampling variability on replicated samples than the simpler, more direct methods of weight percentages of minerals in sieved-size intervals or ratios of number percentages of minerals of similar density in a fine sieve-size interval, for example, 0.061 to 0.88 mm (Young, 1966, pp. 63—65). In general, published studies show that mineralogically complex provinces and dispersal patterns can be successfully mapped if (a) minerals or ratios of minerals that show either weak correlation or no correlation with mean grain size of the samples are used, (b) large numbers of samples are used for an "averaging" effect, and (c) attention is given to ratios of the proportions of color varieties of minerals where the varieties are of similar specific gravity.

The hydraulic equivalence concept has also been used to interpret sedimentary processes on the basis of departures from the theoretical hydraulic sizes. Based on sampling individual laminae, McIntyre (1959, pp. 295—296) showed that foreshore beach sands at Lorain on Lake Erie have smaller hydraulic sizes than those predicted for garnet, hornblende, clinopyroxene, and hypersthere. Evidently, these minerals lodge among the light minerals and some of the coarser light grains are preferentially swept away by the swash and backwash currents.

Rukhin (1937) proposed a "coefficient of shift," that is, the difference in grain diameter of quartz and a given heavy mineral X 100, to differentiate a wide variety of dunes with values less than +2 from rivers and marine sands with values from +3 to +10. Hand (1967, p. 518), using settling velocities rather than grain diameters, proposed a "delta value" (i.e., the log median velocity of heavy mineral less the log median velocity of quartz). He reasoned that delta values are zero if the grain were deposited in water by settling alone, and either positive or negative if factors other than settling were involved. For New Jersey coastal sands, Hand found that dune sands always showed relatively more positive delta values than did marine beach sands. In most of the sands heavy minerals were smaller than those predicted by settling equations because, once deposited, heavy minerals are evidently more difficult to entrain than quartz (Hand, 1967, p. 516). White and Williams (1967, p. 533) found that in a cross-bedded Pleistocene fluvial sand in Pennsylvania, the difference in settling velocities of quartz and tourmaline increased systematically from about zero in bottomset laminations to a maximum in the upper part of the foreset laminations. The bottomset laminations were inferred to form from suspension, and the foresets from both suspension and traction. This corroborates the claim by Hand (1967) that heavy minerals in bed-load sands commonly have smaller settling velocities than quartz because they are more difficult to entrain. As would be expected, the grain-size distributions for quartz and tourmaline in foreset laminations were each bimodal, with the relatively coarser modes reflecting traction and the finer modes reflecting suspension (White and Williams, 1967, p. 533).

SEDIMENTARY PETROLOGIC PROVINCES

The concept of heavy-mineral and light-mineral petrologic provinces allows mapping of dispersal patterns of sediments. A petrologic province is defined as a complex of distinctive, homogenous sediments that constitute a natural unit in terms of age, origin, and distribution (Baturin, 1931, p. 5; Edelman, 1933). Mapping of provinces requires a regional and stratigraphic approach because a province is a three-dimensional body of rock characterized by a distinctive suite of minerals. Some provinces cut time lines at high angles and interfinger or grade transitionally into adjoining provinces; an example occurs in the Cenozoic of the Gulf Coast (Cogen, 1940, pp. 2070–2071).

Boundaries between heavy-mineral provinces have been mapped using several criteria. Ideally, a province is distinguished by the occurrence of volumetrically important heavy minerals that are absent or rare in adjoining provinces. Practically, however, provinces are often differentiated by variations in the relative proportions of minerals. Commonly, some samples do not "fit" the province in which they occur because of selective sorting (Rittenhouse, 1943, p. 1728), mixing of sediment from two provinces, or sampling, counting, and laboratory errors (van Andel and Poole, 1960, p. 96).

In order to evaluate partially the degree of control of selective sorting on nonopaque heavy-mineral proportions, the mean grain size of samples and the percentages of specific minerals are normally plotted on scatter diagrams (e.g., Galehouse, 1967, p. 959). Minerals that show little or no correlation with grain size can then be mapped with some confidence to draw province boundaries. Although not commonly done, plots can also be made of size sorting, shape sorting, and roundness versus the proportions of selected nonopaque minerals. The difficulties of selective sorting in the total heavy-mineral assemblage are reduced if micas are omitted from calculation of mineral percentages and in some cases also the opaque grains.

An example of a reasonably sharp province boundary occurs in the Gulf of Mexico between the shelf sands in the Eastern Gulf Province, rich in kyanite and staurolite, and the Mississippi Province dominated by amphiboles and pyroxenes (Fig. 3; Table 5; Goldstein, 1942, p. 78). Approximately the same boundary is obtained using isopleths drawn on the ratios of hornblende/kyanite plus staurolite (van Andel and Poole, 1960, p. 101). This boundary also illustrates the importance of the absolute amount of heavy minerals supplied by each province. Mississippi sediment is nearly pure mud with a small percentage of sand and consequently only 0.15% total heavy minerals by weight, whereas sandy deposits of the Eastern Gulf contain 0.40% heavy minerals. Thus even a small amount of Eastern Gulf sand will greatly modify the heavy-mineral assemblage of Mississippi sediment; for example, when the hornblende/kyanite plus staurolite ratio is 1.0, the Eastern Gulf contribution to the sediment is only about 5% of the total volume of the sediment (van Andel and Poole, 1960, p. 101).

An approach that allows using a mineral that shows strong correlation with mean grain size is to map isopleths based on ratios of color varieties of similar specific gravity. The ratio of blue-green to common-brown hornblende was of diagnostic value in the shelf sands of the Gulf of Mexico (van Andel and Poole, 1960, p. 105).

The cost and time necessary to calculate hydraulic equivalent ratios may be justified when the problem of mineral segregation by selective sorting is severe. McMaster (1954, p. 115) sorted out heavy-mineral provinces in Recent beach sands along the 135-mile-long coast of New Jersey using hydraulic equivalent ratios for 11 minerals and mineral groups.

The size of a sedimentary petrologic province depends on (a) the extent of the area covered by the current dispersal system, (b) the volume of sediment and the rate at which it is supplied to the province, and (c) the degree of mixing of sediment from adjacent provinces which in extreme cases can obscure province boundaries. The mineral composition, areal extent, and sediment dispersal patterns of Wisconsin and postglacial heavy-mineral provinces in the Gulf of Mexico and the western North Atlantic are shown in Fig. 3

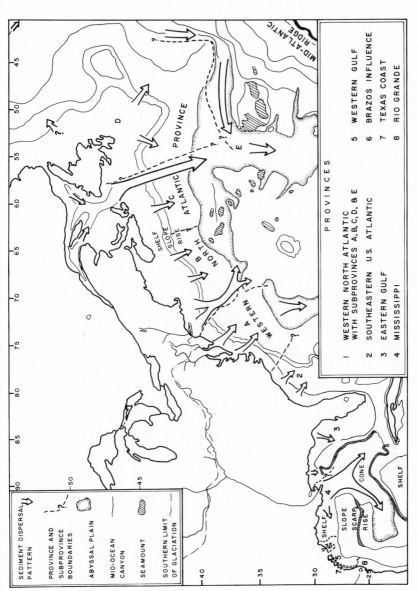

Fig. 3 Heavy-mineral provinces in the Gulf of Mexico and western North Atlantic.

471

TABLE 5 Composition of heavy-mineral provinces in the Gulf of Mexico and western North Atlantic

Province	Source Area	Amphiboles	Epidote	Garnet	Kyanite	Pyroxenes	Staurolite	Tourmaline	Zircon	Other	Number of Samples
Western North Atlantic	Glacial area of eastern North America	44	6	19	Tr.	5	2	2	6	16	102
Southeastern U.S. Atlantic	Piedmont, southern Appalachians, Tertiary-Quaternary	31	23	8	6	Tr.	13	4	6	9	59
Eastern Gulf	Cretaceous through Quaternary and southern Appalachians	13	16	2	16	3	16	12	12	10	31
Mississippi	Glacial area of northern U.S.	42	16	9	—	25	Tr.	2	2	3	116
Texas Coast	Rivers between Nueces and Brazos; reworked Pleistocene	7	11	9	5	3	8	28	23	6	54
Western Gulf	Complex	58	17	7	1	3	1	5	4	4	127
Brazos Group	Cretaceous through Quaternary of Central Texas	33	15	8	3	3	3	8	20	7	19
Rio Grande	Upper Rio Grande, Pecos; volcanics of Chihuahua	30	15	10	2	24	1	3	6	9	41

Data for Southeastern U.S. Atlantic Province from Pilkey, 1963, p. 646.

Data for the Eastern Gulf, Mississippi, Texas Coast, Western Gulf, Brazos Group, and Rio Grande Provinces from van Andel and Poole, 1960, pp. 98–99.

Data for the Western North Atlantic Province from Hubert and Neal, 1967, p. 767.

The amphiboles in the Mississippi Province include 7% basaltic Hornblende.

and Table 5. Sand dispersal patterns on the shelf are controlled by longshore and shelf bottom currents. Submarine canyons act as sediment traps that funnel sand to the deep-sea fans; the sand is then transported across the abyssal plains. The Mississippi and western North Atlantic Provinces are exceptionally large because the source areas are covered by extensive, thick deposits of unconsolidated glacial debris and involve integrated, subcontinental drainage basins. The five western North Atlantic subprovinces are mapped on their proportions of epidote, total kyanite-sillimanite-staurolite, tourmaline, and pyroxene, and their ratios of garnet to kyanite-staurolite. None of these indices is correlated with mean grain size of the sands and silts.

The sequences of mineral zones (provinces) in many thick sequences of sediments, especially in geosynclines, commonly reflect progressive "unroofing" of the source area as erosion proceeds to deeper levels of the earth's crust (Lapworth, cited in Boswell, 1933, p. 48). Grains eroded from the layers of the crust appear in reverse order upwards in the sedimentary sequence: weathered regolith, sedimentary rocks, low-rank metamorphics, and high-rank metamorphics and plutonics. Griffiths (1967, p. 215), van Andel (1959, p. 160), and Pettijohn (1957, p. 517) discuss seven tectonic sequences varying in age from Late Paleozoic through Tertiary where heavy-mineral zones are well established. The composition of lithic fragments in the light fraction and heavy-mineral zones both reflect progressive changes in source rock lithology, thus providing an argument against formation of the mineral zones by intrastratal solution.

The use of eigenvector analysis is an important advance in the technique of handling complex heavy-mineral assemblages that produces objective and reproducable maps of mineral distribution patterns. The vector analysis method is explained in Harmon (1960). This approach is perhaps the only one available when source areas are remote and petrographically complex and the sands are extensively mixed during long distances of transport. Such a case is the Orinoco-Guayana Shelf where a Q-mode vector analysis after varimax rotation yielded maps of factor loadings on each of six vectors (samples with end-member or diagnostic mineral associations) which are more easily interpreted than visual inspection of the raw data (Imbrie and van Andel, 1964, pp. 1146–1152). The vector projection contour maps show clearly that zones of littoral transport shifted landward during the post-Pleistocene rise in sea level.

The extensive calculations necessary for vector analysis require a computer program (e.g., The COVAP program of Manson and Imbrie, 1964). The Q-mode analysis uses the samples as vectors and does not require the absence of a linear restraint on the data; that is, the mineral percentages in each sample can add to 100%. An approximate test is available for the statistical significance of a parametric correlation coefficient between mineral proportions in samples where the percentages sum to 100% (Chayes and Kruskal, 1966).

The R-mode type of vector analysis, however, uses minerals as vectors and a linear restraint must not be present in the data. A method of avoiding a linear restraint on the data is to record heavy-mineral frequencies as mineral weights. When this is done, an R-mode vector analysis can be used, such as that run on heavy minerals in four Tertiary formations in Hyerfano Park, Colorado, by Briggs (1965). The minerals having high correlations with each other are grouped to define factor axes. The factor axes are thus a reduced number of variables that explain most of the variance in the original mineral frequency matrix. The most important factor axes represent major mineral associations. The interpretative problem is to locate source rocks that upon erosion would yield the mineral associations (factors) observed in each of the formations.

ACKNOWLEDGMENTS

Figure 1 is reproduced by permission of the editor of *Biometrika* from Clopper and Pearson (1934). The author thanks L. I. Briggs, A. B. Carpenter, R. E. Carver, Tom Freeman, J. S. Galehouse, G. M. Griffin, Tj. H. van Andel, and E. G. Williams for critical reading of the manuscript and helpful suggestions for its improvement.

REFERENCES

Artini, E., 1898, Intorno alla composizione mineralogica della sabbie di alcuni fiumi del Veneto, con applicazione terreni di trasporto, *Rivista Mineralogia e Cristallographia Italiana, Padova,* **19,** 33–94.

Baturin, V. P., 1931, Petrography of the sands and sandstones of the productive series, *Bull. Trans. Azerbaidjan Petrol. Inst.*, 1, 95 pp.

Boswell, P. H. G., 1933, Mineralogy of sedimentary rocks, Thomas Murby Publishing Co., London, 393 pp.

Bramlette, M. M., 1941, The stability of minerals in sandstone, *J. Sed. Pet.*, 11, 32–36.

Brammall, A., 1928, Dartmoor detritals: a study in provenance, *Proc. Geol. Assoc.*, 39, 17–48.

Briggs, L. I., 1965, Heavy mineral correlations and provenances, *J. Sed. Pet.*, 35, 939–955.

_____, D. S. McCulloch, and Frank Moser, 1962, The hydraulic shape of sand particles, *J. Sed. Pet.*, 32, 645–656.

Chayes, Felix, and William Kruskal, 1966, An approximate statistical test for correlations between proportions, *J. Geol.*, 74, 962–702.

Clopper, C. J., and E. S. Pearson, 1934, The use of confidence or fiducial limits illustrated in the case of the binomial, *Biometrika*, 26, 404–413.

Cogen, W. M., 1940, Heavy mineral zones of Louisiana and Texas Gulf coast sediments, *Bull. Am. Assoc. Petrol. Geologists*, 24, 2069–2101.

Cole, D. L., 1960, Petrology of the Mesaverde Sandstone, Big Horn Basin, Wyoming, unpublished M.A. thesis, Univ. Missouri at Columbia, 126 pp.

Doeglas, D. J., 1940, The importance of heavy mineral analysis for regional sedimentary petrology, National Res. Council, Comm. on Sedimentation Rept. 1939–1940, Exhibit G, 102–121.

Edelman, C. H., 1933, Petrologische provinces in het Nederlandse Kwartair, Centen Publishing Co., *Amsterdam*, 104 pp.

_____, and D. J. Doeglas, 1932, Relikstrurkturen detritischer Pyroxenen und Amphibolen, *Tschermak's Mineralogische u. Petrographische Mittheilungen*, 42, 482–489.

Galehouse, J. S., 1967, Provenance and paleocurrents of the Paso Robles Formation, California, *Bull Geol. Soc. Am.*, 78, 951–978.

Gazzi, Paolo, 1965, On the heavy mineral zones in the geosyncline series, recent studies in the Northern Appennines, Italy, *J. Sed. Pet.*, 35, 109–115.

Goldstein, A., 1942, Sedimentary petrologic provinces of the northern Gulf of Mexico, *J. Sed. Pet.*, 12, 77–84.

Griffiths, J. C., 1967, Measurement of mineral composition: accessory minerals, *in* Scientific method of analysis of sediments, McGraw-Hill Book Co., pp. 200–219.

———, 1960, Frequency distributions in accessory mineral analysis, *J. Geol.* 68, 353–365.

Groves, A. W., 1931, The unroofing of the Dartmoor Granite and the distribution of its detritus in the sediments of southern England, *Quart. J. Geol. Soc. London*, 87, 62–96.

Hand, B. M., 1967, Differentiation of beach and dune sands, using settling velocities of light and heavy minerals, *J. Sed. Pet.*, 37, 514–520.

Harman, H. H., 1960, Modern factor analysis, Univ. Chicago Press, 462 pp.

Hubert, J. F., 1962, A zircon-tourmaline-rutile maturity index and the interdependence of the composition of heavy mineral assemblages with the gross composition and texture of sandstones, *J. Sed. Pet.*, 32, 440–450.

———, 1960, Petrology of the Fountain and Lyons formations, Front Range, Colorado, *Colorado School of Mines Quart.*, 55, 242 pp.

———, and W. J. Neal, 1967, Mineral composition and dispersal patterns of deep-sea sands in the western North Atlantic petrologic province, *Bull. Geol. Soc. Am.*, 78, 749–772.

Hunter, R. E., 1967, A rapid method for determining weight percentages of unsieved heavy minerals, *J. Sed. Pet.*, 37, 521–529.

Imbrie, John, and T. H. van Andel, 1964. Vector analysis of heavy-mineral data, *Bull. Geol. Soc. Am.*, 75, 1131–1156.

Krumbein, W. C., 1936, Application of logarithmic moments to size frequency distributions of sediments, *J. Sed. Pet.*, 6, 35–47.

———, and W. C. Rasmussen, 1941, The probable error of sampling beach sand for heavy mineral analysis, *J. Sed. Pet.*, 11, 10–20.

Krynine, P. D., 1950, Petrology, stratigraphy, and origin of the Triassic sedimentary rocks of Connecticut, *Bull. Connecticut Geol. and Nat. History Surv.*, 73, 239 pp.

———, 1946, The tourmaline group in sediments, *J. Geol.*, 54, 65–87.

———, 1942, Differential sedimentation and its products during one complete geosynclinal cycle, *1st Panamerican Congr. Mining Engineering and Geology Proc.*, 2, 537–582.

Manson, Vincent, and John Imbrie, 1964, Fortran program for factor and vector analysis of geologic data using an IBM 7090 or 7090/1401 computer system, Kansas Geol. Surv. Computer Contribution No. 13, 47 pp.

McIntyre, D. D., 1959, The hydraulic equivalence and size distributions of some mineral grains from a beach, *J. Geol.*, 67, 278–301.

McMaster, R. L., 1954, Petrography and genesis of the New Jersey beach sands, New Jersey Dept. Conservation and Econ. Devel., Geol. Series, Bull. 63, 239 pp.

Milner, H. B., 1962, Sedimentary petrography, Macmillan, Vol. II, 715 pp.

Ojakangas, R. W., 1963, Petrology and sedimentation of the Upper Cambrian Lamotte sandstone in Missouri, *J. Sed. Pet.*, 33, 860–873.

Pettijohn, F. J., 1957, Sedimentary rocks, 2nd ed., Harper and Brothers, 718 pp.

———, 1941, Persistence of heavy minerals and geologic age, *J. Geol.*, 49, 610–625.

Pilkey, O. H., 1963, Heavy minerals of the U.S. South Atlantic shelf and slope, *Bull. Geol. Soc. Am.*, 74, 641–648.

Rittenhouse, G., 1944, Sources of modern sands in the Middle Rio Grande Valley, New Mexico, *J. Geol.*, 52, 145–183.

———, 1943, Transportation and deposition of heavy minerals, *Bull. Geol. Soc. Am.*, 54, 1725–1780.

Ross, C. S., 1950, The dark-field stereoscopic microscope for mineralogic studies, *Am. Mineralogist*, 35, 906–910.

Rubey, W. W., 1933, Settling velocities of gravel, sand, and silt particles, *Am. J. Sci., 5th Series*, 25, 325–338.

Rukhin, L. B., 1937, A new method for the determination of the conditions of deposition of ancient sands, *Problems of Soviet Geology* 7, 953–959.

Slaughter, M., and J. W. Earley, 1965, Mineralogy and geological significance of the Mowry bentonites, Wyoming, Geol. Soc. America Spec. Paper 83, 116 pp.

Smithson, Frank, 1941, The alteration of detrital minerals in the Mesozoic rocks of Yorkshire, *Geol. Mag.*, **78**, 97—112.

Sorby, H. C., 1851, On the microscopical structure of the calcareous grit of the Yorkshire Coast, *Quart. J. Geol. Soc. London*, **7**, 1—6.

Stanley, D. J., 1965, Heavy minerals and provenance of sands in flysch of central and southern French Alps, *Bull. Am. Assoc. Petrol. Geologists*, **49**, 22—40.

Todd, T. W., and R. L. Folk, 1957, Basal Claiborne of Texas, record of Appalachian tectonism during Eocene, *Bull. Am. Assoc. Petrol. Geologists*, **41**, 2545—2566.

van Andel, T. H., 1959, Reflections on the interpretation of heavy mineral analyses, *J. Sed. Pet.*, **29**, 153—163.

———, and D. M. Poole, 1960, Sources of recent sediments in the northern Gulf of Mexico, *J. Sed. Pet.*, **30**, 91—122.

Walker, T. R., 1967, Formation of red beds in modern and ancient deserts, *Bull. Geol. Soc. Am.*, **78**, 353—368.

Weyl, R., 1952, Zur Frage der Schermineralverwitterung in Sedimenten, *Erdol und Kohle*, **5**, 29—33.

White, J. R., and E. G. Williams, 1967, The nature of a fluvial process as defined by settling velocities of heavy and light minerals, *J. Sed. Pet.*, **37**, 530—539.

Young, E. J., 1966, A critique of methods for comparing heavy mineral suites, *J. Sed. Pet.*, **36**, 57—65.

CHAPTER 20

INSOLUBLE RESIDUES

HUBERT A. IRELAND

University of Kansas, Lawrence, Kansas

Insoluble residues are materials remaining from digestion of a sample in a fluid solvent, normally an acid. Acids commonly used are hydrochloric, acetic, and formic, but other solvents are used for special purposes. Carbonates are the most commonly dissolved rocks, but silicates can be treated with hydrofluoric acid, and some evaporites are soluble in water. Residues made during the early development of the technique were called "siliceous residues" because limestones and dolomites were most commonly studied and chert and quartzose material were the chief remnants. The principal insoluble residues are chert, quartz, pyrite, siliceous fossils, and aggregates of clay, silt, or sand. Gypsum, anhydrite, glauconite, mica, iron oxides, and many other minerals are accessories, and some can be used for correlation of strata.

The discovery of abundant arenaceous Foraminifera, conodonts, graptolites, chitinozoans, spores, and other microfossils in residue of carbonate rocks has led to a spectacular increase in the study of insoluble residues during the last two decades. It is estimated that 70% or more of insoluble residue work now done is applied to paleontology. This book is directed chiefly toward petrographic studies, but the techniques, choice of solvents, general procedure, and discussion apply equally as well to paleontology. From one point of view, fossils are as important as detrital constituents, and in some cases are more important, because fossils give information on salinity, depth of water, currents, and other aspects of environments of deposition. The *Handbook of Paleontological Techniques* (Kummel and Raup, 1965) covers application of acid residues to various aspects of paleontology.

Features of insoluble residues can be used for correlation, subdivision, identification, and description of thick stratigraphic sections of carbonates where fossils are absent. The noncarbonate constituents of limestones are liberated and concentrated by removal of carbonate matrix, making it possible to examine and search for diagnostic minerals, siliceous oolites, chert, and other rock fragments that make possible identification and correlation from place to place, whether on the surface or the subsurface. Some characteristics can be traced over extensive areas, but it is unwise to attempt correlation on physical or petrologic characteristics for more than several miles unless knowledge of various controls, such as facies changes, or alteration of constituents is available. Locally, in subsurface work, residues are good for determining structure when other criteria are not conclusive.

Acid residues are especially valuable for study of well samples. Identification of strata using lithologic characteristics as seen under a binocular microscope is difficult when cuttings are from a thick carbonate section that lacks differentiating lithologic features. Thin-sections of cuttings or cores may be made, but preparation of insoluble residues has been found to be easier and, in many cases, more diagnostic. Unless cores are available, chips from well cuttings are generally so small that diagnostic megafossils are destroyed. Acid residues liberate both diagnostic petrologic constituents and any insoluble microfossils present. Well samples commonly include cavings from above the sample interval, or other foreign material, and cores of carbonate rocks are preferable for detailed study because acid residues, thin-sections, and acetate peels can be made and studied.

Insoluble residues were used only sporadically until the late 1920s when McQueen (1932) began using them extensively for correlation and subdivision of the thick, nonfossiliferous Cambro-Ordovician section penetrated by water wells in Missouri. The success in Missouri led Ireland (1936) to apply residue studies on regional scale, using over 8000 samples to correlate Lower and Middle Paleozoic rocks from outcrops in the Arbuckle Mountains, through the subsurface, with outcrops in northeastern Oklahoma. Correlations were based chiefly on petrographic criteria, but the discovery of abundant arenaceous Foraminifera in samples from Siluro-Devonian beds (Hunton Group) helped considerably. The study led to descriptions of the Foraminifera by Moreman (1930) and by Ireland (1939). Utilization of lithologic character in combination with microfossil content and constituents of acid insoluble residues is an important aid in studying subsurface geology (Ireland, 1967b).

Discussions of insoluble residue technique and application have been published by Ireland (1936, 1958, 1963), Groskopf and McCracken (1949), McCracken (1955), and others. A national conference of residue workers resulted in a standardization of terminology (Ireland, 1946). The use of diagnostic petrologic residues made it possible to determine the original thickness of the Cambro-Ordovician beds of northeastern Oklahoma and parts of adjacent states and thus outline the topography of the Precambrian surface (Ireland, 1955).

COLLECTION AND PREPARATION OF SAMPLES

Most samples to be treated for insoluble residues are calcareous material from outcrops, well cuttings, cores, or dredge samples. All or part of the sample may be used, depending on the amount available or accessible for collection and the purpose of the study. Generally, part of the sample should be saved for lithologic examination. Well cuttings are generally limited in quantity and represent a continuous or channel sample for an interval of several feet, generally 5 or 10 ft in rotary-drilled wells and 2 or 3 ft in old cable-tool wells. Rotary-well samples commonly contain material caved from above, expecially shale, which reduces the carbonate percentage of the sample, but normally can be differentiated from indigenous shale partings of a carbonate sequence. A continuous or channel sample from an outcrop or a core from a well is desirable for either petrographic or paleontologic study. Channel sampling on an outcrop is difficult and often impossible. On outcrops the writer selects representative pieces of about 100 gm at 6-in. or smaller intervals to make a composite sample of a 5-ft interval. This gives a much greater possibility of finding diagnostic features which might be missed if a single large sample was taken for the 5-ft interval. A single sample is a point or spot sample and not entirely representative of a specific interval. A single sample taken 2 ft from the base of a 5-ft interval might actually represent the same stratigraphic level as a sample 4 ft from the base of a 5-ft interval taken several miles away. A composite sample would overlap, but a single sample would not. An essentially continuous channel sample is obviously the optimum. Core samples from well borings are continuous samples and most desirable, if available.

Chert on an outcrop, or in the subsurface, may range from a solid band to a few scattered nodules within a limestone bed. For an acid residue to have chert representative of the interval or as part of a spot sample on an outcrop, it may be necessary to move right or left of the line of section to collect samples of adjacent chert nodules for mixture with the sample. The chert must be included because it is a significant part of the lithology and may be the only residue after the sample is acidized. The presence of even a small quantity of a specific type of chert is sometimes more diagnostic than a large volume of some other type of chert. It is possible for a well boring to pass through a bed containing scattered chert nodules and not encounter any nodule which otherwise would supply a diagnostic residue.

Many samples that appear to be pure carbonates and effervesce readily when tested in the field may contain 50 to 80% clay which is not apparent until the samples are processed in the laboratory. This obscure clay content is typical of Silurian limestones in England, Scandinavia, and other places. Sublithographic, light-colored Hunton Limestone from Oklahoma, apparently with a high carbonate content, contains 40 to 50% clay and silt but only 0.5% coarse-fraction residue. Undoubtedly many other beds have similar constituents.

Well cuttings are limited in amount to the envelope or sack coming from the sample split at the distribution center of commercial service companies, and generally weigh less than 200 gm. If portions are obtained directly from the original bag of cuttings from a company or state survey file, the quantity may be much greater. Much of each sample from a rotary-drilled well may be cavings from overlying shale. Percentages of the residue in the original rock cannot be computed for rotary-rig samples unless indigenous chips from the original rock are laboriously picked out with tweezers. A concentration of significant constituents from a sample containing large quantities of cavings can be accomplished by sieving the residues through a 1.0 to 2.0 mm screen. Most of the cavings will be too coarse to pass the sieve and large fragments of chert or other indigenous matter can be picked out with tweezers.

Cores used for insoluble residues should be sawed longitudinally to obtain a continuous sample, either as a slab from the side or as a quarter or half of the core. The segment of the sawed core should be subdivided into lengths of 1 ft or less as determined by the nature of the project or the size of acid receptacle.

A small piece or chip of the original core or outcrop sample should be preserved for reference to the exact lithology of the sample. It has been the writer's custom for years to withdraw a small piece of a sample (10 to 50 gm) from the acid before complete digestion. The etched surface (curved-surface sections, Ireland, 1950) shows fabric, grain size, boundaries of inclusions, and other features that give information on the history and environment of deposition of the original sample. The etched sections are especially valuable when studied in conjunction with thin-sections.

An insoluble residue of the carbonate material in a thin-section has value for certain types of study. The thin-section is immersed in dilute acid and in a few seconds all the carbonate is dissolved, leaving only the insoluble residue. This permits study of the noncarbonate content with regard to orientation, boundaries, distribution, and composition. A normal 30-μ thin-section can be used, but it is sometimes advantageous to use one 100 to 500-μ thick in order to have a thickness of each residue constituent adequate for study.

PROCEDURES

General

The following steps for making insoluble residues are recommended.

1. Weigh the sample to be dissolved if percentage of residue is desired.

2. Place sample in a receptacle (beaker, plastic bucket, or other acid-proof container).

3. Add properly diluted acid carefully to avoid frothing over the top of the container.

4. (a) If digestion is complete, wash well and decant several times to remove fine fraction. If weight or content of fine fraction are important, save the washings, allow clay and silt to settle, siphon off excess water. (b) If digestion is incomplete, pour off spent acid, decant or save fine fraction as desired, add new acid, and repeat the process until digestion is complete.

5. Wash the coarse fraction with distilled water and decant.

6. Dry at temperature below 100°C, weigh if desired, place in envelope or vial for filing and study.

7. If fine fraction is to be studied, remove excess water, dry, and weigh if desired or treat for clay analyses.

The sample to be dissolved can be as large as expedient or crushed into fragments 0.5 to 1.0 cm. For acquisition of fragile constituents in the sample large chunks of rock should be used, because fossils, replacement features, spongy material, clay aggregates, and other fragile contents could be broken or damaged if the sample is crushed. The writer has dissolved chunks weighing 25 to 30 kg submerged in a tank containing 100 or more liters of acid. If the carbonate rock is impure and the residue content is large, it may be necessary to decant the accumulated clay fraction to allow continued acid operation. Large undigested chunks should be lifted out and the accumulated residue and expended acid poured out, with later separation into coarse and fine fractions. New acid can then be added to the sample until digestion is complete. The process can be repeated as many times as necessary. Small bits of crushed sample or well cuttings cannot be lifted out and only a fine fraction should be decanted before additional applications of acid.

Types of Solvents

Insoluble residues may be extracted with several different solvents. The solvent used is determined largely by the composition of the sample and by the purpose of the study, which may be petrologic, paleontological, or environmental. For petrographic studies, hydrochloric acid is generally used because of the rapidity of solution and the lack of need to preserve phosphatic fossil material. Acetic or formic acid is used chiefly for studies in which phosphates need to be preserved. The use of Versene as a solvent for carbonates will be described later. Hydrofluoric acid dissolves silica, but research involving its use is limited.

Commercial grade hydrochloric acid (muriatic) is the most rapid and inexpensive solvent. Welch (1962) ran tests on unit cubes from the same limestone block using several acids in various concentration. He examined the fragile arenaceous Foraminifera recovered in order to determine the optimum dilution in reference to the rate of effervescence and damage to specimens. The results are also applicable to fragile structures which may be important for petrographic study. Acid as concentrated as 50% HCl in water produced no apparent damage to fragile residue consituents, but effervescence was so violent that the froth carried material over the top of the container. To prevent loss of material from frothing, the first application of acid should be about 10% concentration and fill about one-quarter of the receptacle. After the first flare of effervescence subsides, stronger acid can be added, but concentrations of less than 20% are recommended for all subsequent applications. A sample high in carbonate requires about 10 cc of 10% HCl for solution of each gram of sample.

Residues should not be left in spent HCl for more than a day or so, because compounds and impurities released or produced by solution of the carbonate may infiltrate porous fragments and leave undesirable deposits which are difficult to remove. It is important to flush out spent acid and included impurities before drying the residue. A final washing with distilled water flushes out mineral matter in the tap water and prevents damage to fragile constituents that may become attached to the wall and floor of the receptacle by precipitated mineral matter when a sample, wet with tap water, is dried. Alcohol can be applied to the sample after decanting the distilled water, and rapid volitization will provide dry samples within a few minutes without heat (Ireland, 1967a). If heat is used for drying, an oven temperature below 100°C must be used. Never use a bunsen burner or electric hot plate, because high temperature may oxidize iron compounds, roast sulfides, or cause alteration of some types of residues.

Procedures for the use of acetic acid are about the same as those for hydrochloric acid, but acetic acid reacts at a much slower rate. The recommended dilution of acetic acid is 20 to 25% concentration for the most effective ionization. Samples that have a high content of carbonate requires about 20 cc of 20% acetic acid to dissolve each gram of the sample. Acetic acid fumes are much less corrosive to the eyes or exposed skin than hydrochloric acid fumes, which is an advantage. It is used chiefly for samples in which phosphates need to be preserved.

The use of formic acid is not recommended, but some workers prefer it to acetic acid because it is more rapid than acetic acid. Formic acid does not attack phosphates and is generally more expensive than either acetic or hydrochloric acids.

Versene is a Dow Chemical Company trade name for EDTA (ethylene-diamene-tetra-acetic acid) which is an excellent solvent for carbonates, but is slow and expensive. Hill and Goebel (1963) have experimented with Versene and had excellent results. Commercial Versene is a very expensive powder ($8 to $10 per pound) which dissolves in water. When the solvent capacity of the Versene is spent, it can be rejuvenated by the addition of HCl. Although the original cost is high, the ultimate cost per unit dissolved can be reduced considerably if the chelate is recovered and rejuvenated. Ammonium, potassium, sodium, and lithium Versene are available. Ammonium Versene can be made cheaply in the laboratory by titration, but the ammonia fumes are undesirable. Potassium Versene is the most desirable of the chelates when applied to carbonates. The advantages are: that no foam of carbon dioxide bubbles is formed during solution, no fume hood is required during carbonate solution, if the Versene solution is heated the rate of solution of carbonate is about the same as 2% HCl, the chelate can be rejuvenated for subsequent use, and phosphates are not affected.

Hydrofluoric acid is rarely used for routine preparation of residues, but it does have its place when solution of silica is necessary. Special care needs to be taken of eyes and exposed parts of the body, because hydrofluoric is corrosive in a subtle way; although it causes no immediate pain or obvious damage, great pain and severe damage occur some hours after exposure.

Insoluble residues obtained by water solution have rarely been mentioned or considered. Water soluble samples of marine and nonmarine evaporites yield residues which may be important. Solution by water is simple and needs no discussion.

Treatment of Fine Fraction

If a sample contains a large volume of argillaceous matter, the spent acid and accumulated residue should be decanted and collected periodically as many times as is necessary until digestion is complete. The large volume of fine clay and silt tend to cover the undissolved remnant of the sample and prevent solution. If a large volume of residue accumulates and the acid is still strong, the fine fraction and acid may be decanted into a receptacle, allowed to settle for several minutes, and then the acid poured back for continued reaction. The decanted fine fraction may be saved for weighing and examination or all of it decanted down the drain, depending on the purpose for making the residue. Eventually, the coarse fraction is separated, washed, and dried for study as described previously. The fine fraction recovered from washed well cuttings or outcrop samples can be studied and interpreted as any other clay would be, by x-ray diffraction or electron microscopy, but, in dealing with well cuttings, the possibility of contamination by cavings from up the hole must be considered.

DESCRIPTION OF INSOLUBLE RESIDUES

The principal insoluble residues are detritals, nondetrital quartz, various types of chert, silt aggregates, rock fragments, microfossils, and siliceous replacements and casts of megafossils. Color, texture, transparency, luster, inclusions, dolomolds, and degree of crystallinity are the principal diagnostic properties of chert in insoluble residues. Other constituents that are sometimes diagnostic are pyrite, marcasite pseudomorphs, glauconite, micas, and a few other minerals.

Most of the terminology of residue types used by the writer are the same as, or are modifications of, those developed by McQueen (1932) for his work with the Missouri Geological Survey. Ireland (1946) arranged a conference attended by twenty experienced residue workers from central United States which resulted in publication of their agreement on standardized terminology and specific definitions for all types of residues, exclusive of taxonomy for fossils. The definitions in the *Glossary of Geology and Related*

Sciences (Howell, ed., 1957) are taken from that publication. Groskopf and McCracken (1949) and McCracken (1955) continued to use terms that were consistent with the earlier records of the Missouri Geological Survey, but added modifications to conform with the standardized terminology and in keeping with new developments in residue studies.

The descriptions and definitions established by the Midland Residue Conference (Ireland, 1946) are given in Table 1. Fig. 1 is a flow chart for analysis and detailed description of insoluble residues, using the standardized terminology.

TABLE 1   Definitions of insoluble residue terms

Abundant dolomolds or oomolds: See "dolomoldic."

Anhedral: No crystal faces developed.

Beekite: Botryoidal, subspherical, or discoid accretions of opaque silica replacing organic matter, generally white.

Chalcedonic: Transparent to translucent; smoky; milky; waxy to greasy; may be any color, generally buff or blue-gray; may be finely mottled.

Chalky: Uneven or rough fracture surface; commonly dull or earthy; soft to hard; may be finely porous; essentially uniform composition; resembles chalk or tripolite. (Formerly referred to as "dead" or "cotton chert." This includes dull, unglazed porcelaneous material which grades into glazed porcelaneous material of smooth chert.)

Chert: Cryptocrystalline varieties of quartz, regardless of color; composed mainly of petrographically microscopic fibers of chalcedony and/or quartz particles whose outlines range from easily resolvable to nonresolvable with binocular microscope at magnifications ordinarily used by geologists. Particles rarely exceed 0.5 mm in diameter.

Clay: Fine material of clay size.

Clustered: See "oolith."

Concentric: See "oolith."

Dolomold: Rhombohedral cavities in an insoluble residue. (Generally due to the solution of euhedral dolomite or calcite crystals.)

Dolomoldic: Containing dolomolds.

   Skeletal with dolomolds: Residues with rhombohedral openings in which the constituent material comprises less than 25 percent of the volume of the fragment. Openings vary from microscopic to megascopic.

   Abundant dolomolds: Residues with rhombohedral openings with the constituent material comprising from 25 to 75 percent of the volume of the fragment. Openings vary from microscopic to megascopic.

   Scattered dolomolds: Residues having rhombohedral openings in which constituent material comprises more than 75 percent of the volume of the fragment. Openings vary from microscopic to megascopic.

TABLE 1   Definitions of insoluble residue terms (Continued)

Dolomorphic: Used for describing residues where there has been replacement or alteration of dolomite or calcite by an insoluble mineral which assumes the crystal form of the soluble mineral, thus filling a dolomoldic cavity.

Drusy: Clusters or aggregates of crystals, generally incrustations.

Euhedral: Doubly terminated crystals; unattached.

Free: See "oolith."

Granular: Chert; compact, homogenous; composed of distinguishable relatively uniform-size grains, granules, or druses; uneven or rough fracture surface; dull to glimmering luster; hard to soft; may appear saccharoidal. (This type is frequently referred to as "crystalline.")

Granulated: Grains or granules partly cemented or loosely aggregated; saccharoidal; grades from angular to drusy; fine to coarse; particles rarely larger than 0.5 mm in diameter.

Lacy: Residues with irregular openings in which the constituent material comprises less than 25 percent of the volume of the fragment.

Massive: See "oolith." Used also to include fine or coarse granular anhydrite or gypsum.

Mottled: Residue fragments with two or more colors or different material interspersed and irregularly shaped with the boundaries between either sharp or gradational; often appears flocculated; grades into speckled residue.

Oolite: Composed of an aggregation of ooliths.

Oolith: Spheroidal bodies with nucleus or central mass enclosed by one or more surrounding layers of the same or different material; may be any color and of many kinds of material, generally less than 1.0 mm in diameter. Those over 2.0 mm are pisoliths.

    Concentric: Peripheral layers around a small, undetermined nucleus.

    Clustered: Attached ooliths without solid matrix.

    Drusy: Oolith covered with subhedral quartz; may be free or clustered.

    Free: Unattached oolith.

    Massive: Interior of granular, smooth, or chalk-textured material comprising nearly the entire mass of the spheroid.

    Radiate: Fibers radiating from small or large nucleus; may have several peripheral layers.

    Sand-centered: Nucleus, a quartz sand grain.

    Oomold: Spheroidal opening representing the former presence of ooliths.

Oomoldic: Containing oomolds.

    Skeletal with oomolds: Same definition as for "dolomoldic."

    Abundant oomolds: Same definition as for "dolomoldic."

    Scattered oomolds: Same definition as for "dolomoldic."

Ordinary: Smooth chert with even fracture surface; all colors, chiefly white, gray, or brown; may be mottled; approaches opaque; generally homogeneous, but may have slight evidence of granularity or crystallinity; grades into chalcedonic or granular chert.

TABLE 1   Definitions of insoluble residue terms (Continued)

Porcelaneous: Chert with smooth fracture surface; hard; opaque to subtranslucent; typically china-white resembling chinaware or glazed porcelain; grades to chalky.

Pseudoolithic: Rounded pellets with no peripheral layers or sharp distinction between pellets and matrix.

Pyrimolds: Cavities left after the removal of euhedral pyrite by weathering or otherwise.

Quartz: Clear, colorless quartz; not detrital.

Radiate: See "oolith."

Regenerated: Used in reference to quartz sand grains with secondary regrowth of crystal faces oriented with the original axis of the grain.

Rounded: Spheroidal or ellipsoidal sand grains, coarse to fine, may be polished, frosted, or etched.

Sand: Grains of sand size, chiefly quartz, but may be composed entirely or partly of other minerals.

Sand-centered: See "oolith."

Scattered: See "dolomoldic" and "oomoldic."

Silt: Grains of silt size, chiefly quartz, but may be composed entirely or partly of other minerals.

Skeletal: See "dolomoldic" and "oomoldic."

Smooth: Major type of chert with conchoidal to even fracture; surface devoid of roughness; may be botryoidal; homogeneous; no distinctive structure, crystallinity, or granularity.

Speckled: Disseminated fine spots of color or material different from that of the matrix and having relatively sharp boundaries.

Spicular: Containing inclusions of sponge spicules. Free spicules have been noted.

Subhedral: Crystal forms partly developed; may be loose, drusy, or granulated.

Subrounded: Polygonal grains or fragments but with well-rounded edges and corners.

Unmodified: Residue uniform with no modifying characteristics.

A condensed statement concerning some of the features of residues not included in the published definitions, and which are important, may be warranted here. Quartz may occur as detrital sand, or as authigenic, euhedral sometimes doubly terminated crystals, or as subhedral to anhedral crystals that develop in cavities, veins, or other openings. Quartz also occurs as siliceous replacement of fossils. Chert occurs as one of several types described as chalky, granular, or various kinds of smooth-textured chert which is subdivided into ordinary, porcelaneous, or chalcedonic. Ordinary and chalcedonic chert are the most common of the smooth-chert types and may grade rapidly into other types. A chert nodule may be essentially chalcedonic, or other smooth-textured microcrystalline type, but have an outer porous-leached crust or perimeter that is referred to as

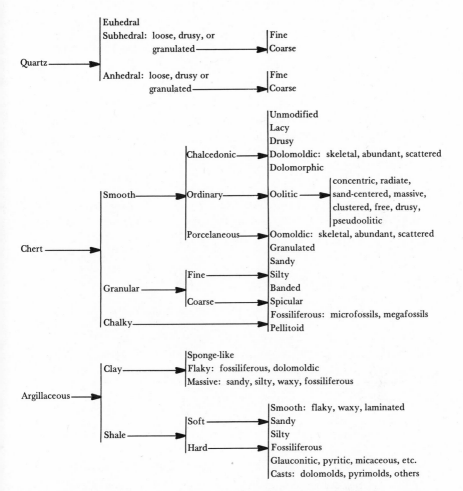

Fig. 1 Flow chart for analysis and detailed description of insoluble residues, using standardized terminology. Each term to the right of a line is applicable to terms having an arrow touching the line.

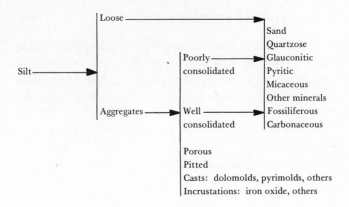

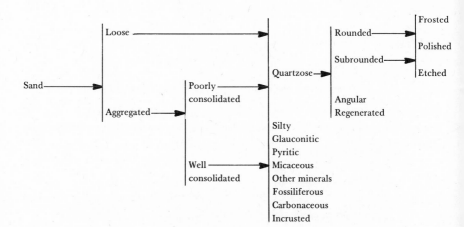

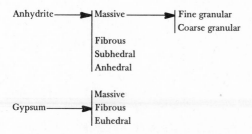

Accessories: Sulphur, pyrite, marcasite, magnetite, hematite, limonite, biotite, muscovite, chlorite, feldspar, barite, celestite, glauconite, other minerals, fossil material, pellets, beekite.

Fig. 1 (Continued)

"dead," "cotton," or tripolitic chert and classified as chalky chert. Dolomoldic chert has rhombohedral cavities left by solution of included calcite or, more often, dolomite rhombs. The dolomolds may be so abundant that they intersect and form what is called *skeletal dolomolds*, which form a lacy network or occur as loose, triangular fragments that represent interstitial silica between rhombs. Any chert type may be found alone or mixed with others in a sample. The presence of even a small quantity of a specific type of chert is sometimes more diagnostic than a large volume of some other type of chert. The color of chert is often a diagnostic feature. Multiple-colored or one-color siliceous oolites of blue, brown, buff, or white silica are significant markers in the Cambro-Ordovician carbonates of the midcontinent.

PLOTTING OF RESIDUE DATA

Written descriptions of residues require too much space, and symbols are desirable for plotting results on log strips or printed forms. The writer has devised a set of symbols combining color and overprints (Fig. 2), which is a shorthand for describing and plotting constituents according to percentage. Figure 3 shows a method (percentage-constituent) of recording data that shows total percentage of the constituents in one column, and the percentage of each constituent in relation to the others in another column. This method is considered the most satisfactory way to show diagnostic features for correlation. Figure 4 illustrates a method (percentage-percentage) in which each constituent is shown in proper relation to the entire original sample. It has some useful aspects, but it is more difficult to plot and does not clearly illustrate diagnostic characteristics. It requires too much space to be used on a standard log strip. In Fig. 4 it is obvious that a constituent which is 10% of a 10% total residue will not plot as very significant, that is, compare the 1010 to 1020-ft intervals in Figs. 3 and 4. Residues for many fairly pure limestones are generally less than 5% and, although the residues may be diagnostic, they are not obvious on a percentage-percentage plotted strip. The determination of percentage of each constituent is based on visual inspection, estimation, and good judgment.

Quartz (nondetrital)

  Euhedral

  Subhedral   Green

  Anhedral

Chert

  Ordinary   Red

  Chalcedonic   Red bars

  Porcelaneous   Pink

  Granular   Blue

  Chalky   Brown

Clastics

  Bentonite   Yellow border with green bars

  Clay or shale   Border same color as material

  Silt   Yellow border

  Sand   Yellow

  Anhydrite   Violet or purple

  Gypsum   Gray

Black Overprints on Yellow

  Coarse sand

  Medium sand

  Fine sand

  Very fine sand

  Silty

  Sand Rounded frosted

  Sand Rounded polished

  Sand Rounded etched

  Sand Subrounded frosted

  Sand Subrounded polished

  Sand Subrounded etched

  Sand Subangular

  Sand Angular

  Sand Regenerated

Black Overprints

  Aggregates   A

  Banded   ))

  Coquinoidal   C

  Dolomolds

    Skeletal

    Abundant

    Scattered

    Dolomorphic

  Drusy   Z

  Fibrous   Fi

  Fine   f

  Flaky

  Fossiliferous

  Glauconite   *

  Granular   gr

  Granulated   g

  Lacy

  Laminated

  Massive   M

  Mottled   m

  Ooliths

    Concentric   O

    Radiate   ⊕

    Sand-centered   ⊙

    Massive   ●

    Clustered   o⊗o

    Free   O

    Drusy   O

  Oomolds

  Porous   p

  Pseudoolith   ⦂

  Pseudomorphic   Ps

  Pyrimolds

Pyrite   +

Quartzose

  Euhedral

  Subhedral   v

  Annedral   X

  Sandy

  Silty   S

  Spicular   Y

  Unmodified   /

  Vein

Red Overprints

  Barite   Ba

  Chlorite   Ch

  Celestite   C

  Feldspar   F

  Gilsonite   G

  Hematite   H

  Limonite   L

  Magnetite   M

  Marcasite   Ma

  Mica   Mi

    Biotite   Bi

    Muscovite   Mu

  Sphalerite   Zn

  Sulfur   S

  Beekite   B

  Pellets   P

  Granite wash   XXXX

  Igneous   VVVV

  Detrital gravel   oooo

Fig. 2 Colors, overprints, and symbols for graphic description of insoluble residues.

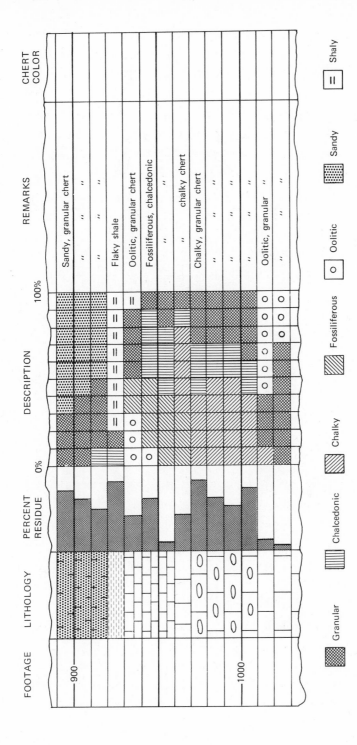

Fig. 3 Percentage constituent method for plotting descriptions of insoluble residues. Compare with Fig. 2. It is preferable to use color and overprints shown on Fig. 2 when plotting a section or well log.

495

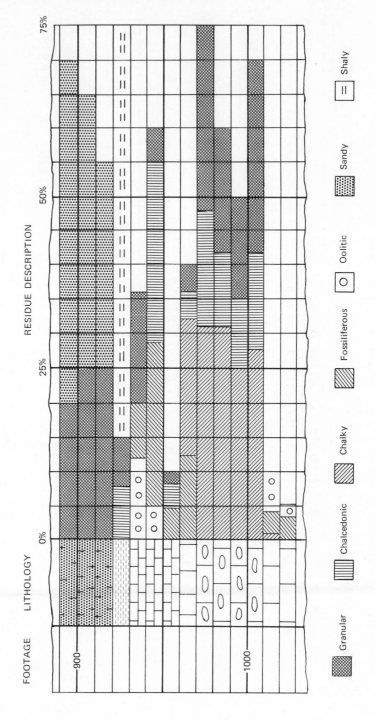

Fig. 4 Percentage-percentage method for plotting descriptions of insoluble residues. Plotting is based on the same data as used in Fig. 3.

REFERENCES

Groskopf, J., and E. McCracken, 1949, Insoluble residues of some Paleozoic formations in Missouri, their preparation, characteristics, and application, *Missouri Geol. Surv. Rept. of Inv. 10.*

Hill, W. E., and E. D., Goebel, 1963, Rate of solution of limestone using the chelating properties of Versene (EDTA) compounds, Bull. Kansas Geol. Surv., 165, Pt. 7, 15 pp.

Howell, J. V., ed., 1957, *Glossary of geology and related sciences*, Am. Geol. Inst., Washington, D.C., 325 pp.

Ireland, H. A., 1936, Use of insoluble residues for correlation in Oklahoma, *Bull. Am. Assoc. Petrol. Geologists,* 20, 1086–1121.

———, 1939, Devonian and Silurian Foraminifera from Oklahoma, *J. Paleon.,* 13, 190–202.

———, 1946, Terminology for insoluble residues, *Bull, Am. Assoc. Petrol. Geologists,* 31, 1479–1490.

———, 1950, Curved surface sections for microscopic study of calcareous rocks, *Bull. Am. Assoc. Petrol. Geologists,* 34, 1737–1739.

———, 1955, Precambrian surface in northeastern Oklahoma and parts of adjacent states, *Bull. Am. Assoc. Petrol. Geologists,* 39, 478–483.

———, 1958, Insoluble residues, Chapter 5, *in* Subsurface geology in petroleum exploration, *a symposium,* J. D. Haun and L. W. LeRoy, eds., Colorado School of Mines, 887 pp.

———, 1963, Insoluble residues for research, Proc. 21st Inter. Geol. Congr., Copenhagen, Part XXVII, pp. 233–241.

———, 1967a, Preparation techniques for microfossils and inorganic insoluble residues, *J. Paleon.,* 41, 523.

———, 1967b, Zonation and correlation of subsurface Hunton Group (Silurian-Devonian) in Kansas by Foraminifera and acid residues: Essays in Paleontology and Stratigraphy, R. C. Moore Commemorative Volume, Curt Teichert and E. L. Yochelson, eds., Univ. Kansas Press, Dept. of Geology, Spec. Publ. 2, 626 pp.

Kummel, B., and D. Raup, eds., 1965, *Handbook for paleontological techniques,* W. H. Freeman Co., 582 pp.

McCracken, E., 1955, Correlation of insoluble residue zones of Upper Arbuckle of Missouri and southeastern Kansas, *Bull. Am. Assoc. Petrol. Geologists,* **39**, 47—59.

McQueen, H. S., 1932, Insoluble residues as a guide to stratigraphic studies, *Missouri Bur. of Geology and Mines,* 56th Biennial Rept. (1931), Appendix I, 32 pp.

Moreman, W. L., 1930, Arenaceous Foraminifera from Ordovician and Silurian limestones of Oklahoma, *J. Paleon.,* 4, 52—59.

Welch, R., 1962, Extraction of arenaceous Foraminifera by solution of limestone, *Search,* Univ. of Kansas, pp. 5—8.

CHAPTER 21

GRAIN MOUNTS

DONALD J. P. SWIFT

Old Dominion University, Norfolk, Virginia

Sedimentologists engaged in petrographic analysis sooner or later run into the problem of grain mounting. The problem has existed since the advent of petrographic microscopy. Two early solutions have proved so satisfactory that they are still the commonest methods: temporary mounting in a succession of immersion oils and permanent mounting in thin-section. Both techniques interpose a medium of intermediate refractive index between the grains to be studied and air; the ensuing reduction in refraction and reflection permits the observer to see through the surface of the grain and determine the internal optical properties of the grain. These techniques are used in all phases of petrography and have been described in the literature. For the immersion method see Larsen and Berman (1934, pp. 11–13), Krumbein and Pettijohn (1938, pp. 379–390), Milner (1962, pp. 245–249), and Tickell (1965, pp. 72–95). For the manufacture of thin-sections see Reed and Mergner (1953), Krumbein and Pettijohn (1938, pp. 363–365), Milner (1962, pp. 83–88), or Tickell (1965, pp. 57–58). These techniques are common to all fields of geology; only aspects unique to the study of sediments are considered here.

AIR MOUNTS

Air mounts are used for visual examination of the surface texture of grains, for two-dimensional shape analysis, or as a preliminary step for immersion oil analysis or stain tests.

Cardboard slides for air mounts may be obtained commercially (Table 1). They consist of metal frames, cardboard bodies of 18, 28, 38, or 48 ply, and cover glasses. The bodies have one to four circular wells with the base painted black, or a rectangular-gridded well with 30 or 60 squares. The grains are mounted with a water-soluble gum (gum tragacanth or gum arabic) so that they can be easily moved. The gum is spread with a 00 water-color brush. The moistened tip of the brush is used to transfer or rotate specimens. Glue should never be used as a mounting medium, because it shrinks during drying and will crack fragile specimens. A drop of formalin, or the more pleasant-smelling oil of cinnamon, may be used to preserve the gum (Todd et al., 1965). A. C. Tester (1932) recommended cellophane as a well cover for permanent air mounts. It is only one-fifth as thick as a cover glass; hence it is better when used with a high-powered objective and stands up to much rougher treatment.

OIL AND RESIN MOUNTS

Oil and resin mounts are used for optical identification or two-dimensional shape analysis. Resin and related mounts are also used for stain tests or as a preliminary step in immersion oil analysis.

The fundamental optical parameter for mineral identification is the refractive index. Its determination by immersion of the grain in a sequence of index oils is a standard procedure, but a cumbersome one when a complex mineral suite must be identified in many samples. In this case grains are mounted in a specific oil, or in a resin of known index, so that each grain may be determined to have an index above, equal to, or below that of the medium. Supplementary optical techniques are then used to obtain a specific determination of refractive index. An immersion oil is used with a temporary mount. For light minerals, clove oil (n = 1.53) or cedar oil (n = 1.74) are commonly used. Useful liquids for heavy-mineral suites are mono-bromonaphthalene (N = 1.658) or methyl iodide (n = 1.74)

TABLE 1 Mounting Materials

Product	Nature and Purpose	Reference	Supplier	Response to Inquiry
Aroclor 4456	Artificial resin for heavy mineral mount	Keller, 1934	Swann Chemical Co.	Yes
Bakelite	Viscous mounting medium or impregnating material; $n = 1.63$ when hardened		Any hardware store	—
Bioplastic	Thermosetting plastic for 3D grain mounts	Ward's catalog	Ward's Natural Science Establishment, Rochester, New York	Yes
Borden's epoxy resin	Epoxy resin mounts for oil immersion studies	Langford, 1962	Local supplier	—
Caedax	Artificial resin for light mineral mounts; $n = 1.55$	Ward's catalog	Ward's Natural Science Establishment, Rochester, New York	Yes
Canada balsam	Resin for light mineral mounts; $n = 1.53$		Ward's Natural Science Establishment, Rochester, New York	Yes
Cardboard slides	Slides for air mounts	Krumbein and Pettijohn, 1938	R. P. Cargille, 26 Courtlandt St., New York	Yes
			Curtin and Co., Houston, Texas	
Castolite	Thermosetting plastic for 3D grain mounts	Hulbe, 1955	Ward's Natural Science Establishment, Rochester, New York	Discontinued
Cedar oil	Preservative for gum adhesives	Todd et al., 1965	Any chemical supply house	—
Clove oil	Immersion oil for temporary light mineral mount		Any chemical supply house	—
Flexible collodion	Gelatinous film for permanent mounts for oil immersion studies	Spencer, 1960	Local drugstore	—
Gum arabic	Water soluble adhesive for air mounts		Any chemical supply house	—
Gum tragacanth	Water soluble adhesive for air mounts			
Hyrax	Artificial resin for heavy mineral mounts; $n = 1.174$	Cameron, 1934	Einer and Amend New York	No
Kollolith	Artificial resin for light mineral mounts; $n = 1.535$	Russell, 1935, p. 104	Voigt and Hochgesand, Gottingen, Germany	No
Lakeside 70	Thermoplastic cement for thin sections	Tickell, 1965, p. 57	Hugh Courtright and Co., Chicago, Illinois	Yes
Lepage's epoxy resin	Epoxy resin mounts for oil immersion studies	Milner, 1963	Local supplier	—
Methyl iodide	Immersion oil for temporary heavy mineral mounts	Krumbein and Pettijohn, 1938	Any chemical supply house	—
Monobromo-naphthalene	Immersion oil for temporary heavy mineral mounts	Krumbein and Pettijohn, 1938	Any chemical supply house	—
Permount	Artificial resin for permanent light mineral mounts	Fisher catalog	Fisher Scientific Co.	Yes
Piperine	Artificial resin for permanent heavy mineral mounts	Martens, 1932	Eastman Kodak Co., Rochester, New York	Yes
Ward's clear epoxy resin	Epoxy resin mounts for oil immersion studies	Ward's catalog	Ward's Natural Science Establishment, Rochester, New York	Yes

(Krumbein and Pettijohn, 1938, p. 359). After examination, the grain may be washed with xylol, dried, and reserved for further study.

It is sometimes desirable to change the position of grains mounted in a temporary medium. For this purpose, Krumbein and Pettijohn (1938, p. 359) suggest using very viscous liquids such as Canada balsam (n = 1.54) or Bakelite varnish (n = 1.63). Mixing the grains with coarse glass fragments before mounting prevents the cover glass from holding the grains tight to the slide. Mounting with ground glass also tends to randomize the optic orientation of cleavable minerals such as calcite, which otherwise would present a preferred optical orientation. When it is necessary to turn the grain, the cover glass is touched with a pencil.

A variety of resins are available for permanent grain mounts. Canada balsam (n = 1.54) is the traditional material. Other low index resins are Kollolith (n = 1.535), Lakeside 70 (n = 1.54), Caedax (N = 1.55), and Permount (n = 1.53). Lakeside 70 is a transparent thermoplastic cement that some workers believe is superior (to Canada balsam) as an adhesive and cement for thin-section mounting (Tickell, 1965, p. 57). Permount and Caedax are advertised as being neutral (nonacid), thinner than Canada balsam, and less likely to form bubbles. van der Plas (1966, pp. 140–143), however, emphatically prefers Canada balsam for feldspar identification because of its suitability for distinguishing between alkali and plagioclase feldspars.

Canada balsam is used in the following manner (Krumbein and Pettijohn, 1938; Tickell, 1965).

1. The slide is cleaned by dipping into distilled water, alcohol, or a cleaning solution (concentrated sulfuric acid saturated with potassium dichromate), and dried with a lintless cloth.

2. The slide is placed on a hot plate. A piece of solid Canada balsam is placed on top of the slide and heated until it becomes fluid and spreads to the desired thickness and area. Care should be taken to heat slowly to prevent the formation of too many bubbles. Bubbles may be removed with a toothpick.

3. Grains are sprinkled on the slide uniformly and allowed to sink in.

4. After the resin has been heated enough so that a droplet cooled on the point of a probe is brittle enough to crush, it is moved off the hot plate.

5. One edge of a clean cover glass is placed on the edge of the puddle of resin. The cover glass is then rotated onto the resin and pressed gently down while the resin is cooling.

6. After the resin is cold, the part that has extruded from beneath the cover glass may be scraped off with a knife, and the slide cleaned with xylene.

For heavy minerals it is generally preferable to use a resin of high refractive index such as Piperine (n = 1.68) or Hyrax (n = 1.71). The strong dispersion of Piperine results in characteristic phenomena around the borders of the mineral grain which usually make it possible to estimate the refractive index without actually observing the direction of movement of the Becke line (Martens, 1932). For temporary mounts the Piperine has only to be melted. However, to avoid crystallization of the Piperine after a few days, it must be cooked for an hour at 180°C. This results in a darkening of the edges of the slide.

Hyrax has a higher surface tension than Canada balsam and tends to form bubbles (Cameron, 1934). These bubbles are minimized by cooking for an hour at 100°C. The temperature is then raised to 120°C while the cover glass is rotated onto the liquid.

Aroclor 4456 has been used by Keller (1934) and McMaster (1966). Its index of refraction is 1.66 and dispersion is not high. It melts at 70°C without bubbling and does not darken with heating.

For staining tests the resin is melted on a glass slide, then cooled. A drop of distilled water may be spread across the surface of the resin and grains sprinkled uniformly onto it. The slide is heated until the water evaporates and the grains begin to sink into the resin; then the mount is cooled. Some practice may be necessary to obtain the optimum mount between loose grains and those that have sunk in too far.

Several methods have been devised for permanent grain mounts and yet permit using successive immersion oils. Langford (1962) mounted grains on a film of epoxy resin so thin that only the lowest portion of the grain was immersed. This portion was visible as an "internal" Becke line. The rest of the grain's surface was then immersed in an index oil. Langford reports that Borden's and Lepage's epoxy resins are suitably clear resins that may be obtained locally. Ward's Natural Science Establishment also sells a clear epoxy. Langford used the Borden epoxy, which must be cured in an oven at 80°C for at least 2 hours. Langford recommends dipping a partially straightened paper clip into a reservoir of epoxy so that a small amount adheres to the wire. The epoxy is then smeared across the slide with heavy pressure so that only a thin film is left. Langford's tests show that Borden's epoxy resins, thus cured, are inert to immersion oils. The oil can be removed with xylene applied with an eyedropper. After application of the xylene, the slide is tipped and the solution is allowed to drain to a corner of the slide where it is wiped off. Blowing on the slide will rapidly evaporate xylene. The grain mount should be checked with a binocular microscope to insure that all xylene and immersion oil is removed. Langford has changed oils as many as 25 times without noticeable deterioration of the mount. Grains can be easily pried off with a needle if it becomes necessary to try other methods of analysis.

Spencer (1960) developed a similar technique while working with silt-sized heavy minerals. His mounting medium was flexible collodion, obtainable in any drugstore. His method is as follows.

1. Dip a glass rod into a bottle of USP flexible collodion and smear a uniform coating onto a microscope slide. Allow 20 seconds for the collodion to become tacky. If the collodion is applied too liberally, small bubbles form in the film; if too sparingly, then it dries in a few seconds.

2. Sprinkle grains to be mounted onto the slide. Allow to dry for about 30 minutes. Apply an appropriate immersion oil and a cover glass.

3. To change oils, wash both cover glass and oil from the mount using a wash bottle containing ethyl alcohol. Acetone is not suitable, for it will attack collodion. Dry at room temperature and apply a new oil and cover glass.

MOUNTS FOR THREE-DIMENSIONAL ANALYSIS

An intensive study of grain shape often entails measuring both the maximum and minimum projection areas on the same grain. Pye (1943) mounted grains in clove oil and then, by sliding the cover glass over the slide with a circular motion, rolled the grains, ". . . yielding a three-dimensional picture of their shape and roundness." Plumley (1948) cemented a slide to a lucite block (Fig. 1). He mounted the grains in a row along one edge of the glass slide so that

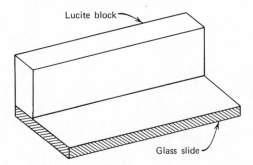

Lucite block

Glass slide

Fig. 1 Mounting device for determining three-dimensional grain properties. (Modified from Plumley, 1948, p. 562)

their planes of maximum projection area were approximately parallel to the surface of the slide. With this method it is necessary to ensure that the grains have every opportunity to achieve a stable position and expose their maximum projection area. After long and intermediate diameters are measured, the slide is turned on edge so that it rests on the lucite block and the short diameter can be measured.

Hulbe (1955) devised a technique in which grains were mounted in a rectangular plastic rod for easy viewing and storage. Equipment for this system has been modified and improved by Humphries (1966). The following procedure is taken from his report (pp. 241—245).

In brief, the technique of mounting sand grains is to split a sample of about 50 grains from a large sample using a microsplitter (*cf.* Humphries, 1961) or to collect initially samples of this size. These grains are placed on a bed of thermosetting plastic in a mould in such a way that their long and intermediate axes are horizontal and [then covered with layer of plastic]. When the plastic is hard the rod is transferred to a small clamp for polishing and is then ready for microscopic examination.

The mold [Fig. 2] ... consists of seven pieces of aluminium strip, 5/8 inch wide, 1/8 inch thick and 2 inches long with six pieces, 1/2 inch wide, 1/8 inch thick and 2 inches long, placed alternately between them. The strips are drilled and bolted together to form a block with six channels, each 1/8 inch square. The ends of the block are squared up by milling or filing. The end of the channels can be closed with a strip of scotch tape, though this is attacked by some plastics. It is better to close the channels with two lengths of angled aluminium strip bolted across the ends of the block (as shown). It should be noted that the use of copper or brass for the mould may inhibit the setting of some thermoplastic resins and has therefore been avoided.

The grains can be easily transferred to the mould by means of a [grain lifter, or] hypodermic needle mounted on a length of brass rod connected to a vacuum

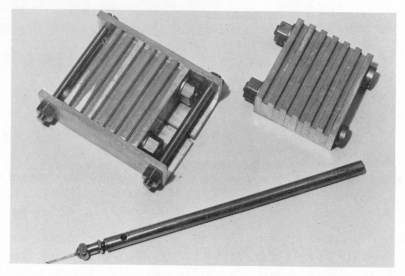

Fig. 2 Humphries' grain-mounting apparatus. *Left.* The complete mold with end pieces in place. *Right.* Mold showing laminated construction. *Below.* Grain lifter. (From Humphries, 1966)

line [Fig. 2]. The top of the needle is ground obliquely in manufacture but it is found better to grind it square on a carborundum stone. The needle is cemented onto the brass with hard wax (sealing wax or vacuum wax). A small hole in the side of the brass tube near to the needle allows fingertip control of the suction. A screw clip on the vacuum line gives additional control of the suction and once set rarely needs adjusting. The needle should be as small as possible otherwise the grains pass into the bore. Especially with very small grains the needle is preferable to a brush.

This clamp [polishing jig, Fig. 3] is made from three brass strips bolted and soldered to two narrow brass bars to form a plate with two longitudinal slots, 1 inch apart. Six thin (3/32 inch) brass strips are mounted across the plate and held in position by short screws (passing through and soldered to the bar) and pairs of locknuts. The locknuts are adjusted so that the bars can just slide in the slots. On the sliding bar at one end a nut is omitted from each screw and the remaining nut adjusted with a box spanner. The plastic rods are places [sic] between the bars, the assembly squeezed tight and the end bar clamped in place. This holds the rods sufficiently firmly for grinding and polishing.

The [microscope stage] jig . . . [Fig. 4] allows the rods to be turned through 90° while on the microscope stage in order to measure the three dimensions of the grain. The base of the jig is a 3 x 1 inch plate of Perspex (Lucite) on which two small blocks of Perspex are cemented. One of these blocks has a blind hole with a brass insert. This insert also has a blind hole (0.17 inch diameter) drilled in it which acts as the support for one end of the plastic rod. The other block is drilled in line with the blind hole and carries a short tube which also supports the rod. The tube can slide back and forth in its block to allow the insertion of the plastic rod and is retained in place with a spring loaded ball. The rod is held in place in the slot across the end of the table by a small set-screw in the side of the slot and also a spring loaded ball within the tube. The tube and the rod are turned by a brass quadrant which permits only $90°$ movement.

The mould is clamped tightly and the channels coated with a releasing agent; sufficient of the thermosetting plastic is then added to about half fill the channels. The plastic should be a clear colourless variety with a short gelling time. When the plastic has set, the channels are filled with water and the grains transferred to them by means of the grain lifter. Water is preferred to alcohol or acetone since it is not so volatile and is non-irritant. A small hand lens is useful in placing the grains in the center of the channel. The rods can at this stage be labelled by placing a small slip of thin card carrying details of the sample at the end of the channel. The water is driven off under an infra-red lamp and the channel filled with plastic. To economize in time and plastic a number of moulds can be prepared at the same time. When the plastic has been cured or has set firm (according to the type used) the rods can be removed from the mould by ... polishing. The rods are transferred to the polishing jig in the order and orientation in which they are removed from the mould and clamped in position. The surface of the rods is ground lightly on very fine carborundum paper and then polished on a rotating cloth lap using one of the polishes recommended for resins such as Perspex. Ordinary liquid metal polish can also be used. When the exposed sides of the six rods in the clamp are polished the clamp is slackened, the rods turned on to their sides and the grinding and polishing repeated. It is not necessary to polish the two remaining sides, though this is usually done and gives the rods a better appearance. Care should be taken while grinding that the rods are not ground to the level of the brass retaining strips, otherwise it is not possible to grind the reverse side of the rod. Once polished the rods are ready for examination.

Fig. 3 Polishing jig showing movable clamping strips and adjusting key. (From Humphries, 1966)

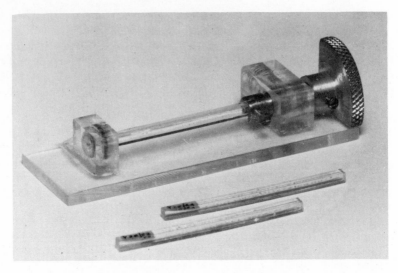

Fig. 4 Microscope stage jig with plastic rod in position. In the foreground grains
mounted in unpolished (upper) and polished (lower) plastic rods. (From
Humphries, 1966)

The placing of the grains in the mould is the slowest part of the operation, the
time taken depending chiefly on the size of the grains being handled. The finest
grains (e.g., 96μ) take the longest time. Usually the mounting of some 300 grains
in one mould takes between 15 and 30 min. The grinding and polishing can be
accomplished in about 10 min.

The . . . mould described will readily accommodate grains up to about 1 mm.
Larger·moulds can be made for larger grains; but it should be noted that the
microscope stage jig would need to be modified to accommodate the larger rods.

Hulbe (1955) used Castolite as the mounting medium, and
glycerine, recommended by the Castolite manufacturer, as the
releasing compound. He notes that a disadvantage of Castolite is its
refractive index, so close to that of quartz that definition of grain
edges is sometimes difficult. Hulbe's supplier (Ward's Natural Science
Establishment) no longer advertises Castolite, but it does stock a
similar "Bioplastic." Castolite is currently supplied by Buehler, Ltd.,
Chicago, Illinois.

REFERENCES

Cameron, E. N., 1934, Notes on the synthetic resin hyrax, *Am.
Mineralogist,* **19,** 375–383.

Hulbe, C. W. H., 1955, Mounting technique for grain size and shape measurements, *J. Sed. Pet.,* **25**, 302—303.

Humphries, D. W., 1961, A non-laminated miniature sample splitter, *J. Sed. Pet.,* **31**, 471—473.

Humphries, D. W., 1966, Mounting sand grains for three dimensional analysis, *Sedimentology,* **6**, 241—245.

Keller, W. D., 1934, A mounting medium of 1.66 index of refraction, *Am. Mineralogist,* **19**, 384—385.

Krumbein, W. C., and F. J. Pettijohn, 1938, Manual of sedimentary petrography, Appleton-Century-Crofts, 549 pp.

Langford, F. F., 1962, Epoxy resin for oil immersion and heavy mineral studies, *Am. Mineralogist,* **47**, 1479—1481.

Larson, F. S., and H. Berman, 1934, The microscope determination of the nonopaque minerals, 2nd ed., U.S. Geol. Surv. Bull. 848, 226 pp.

Martens, J. H. C., 1932, Piperine as an immersion medium in sedimentary petrography, *Am. Mineralogist,* **17**, 198—199.

McMaster, R. L., and L. E. Garrison, 1966, Mineralogy and origin of Southern New England shelf sediments, *J. Sed. Pet.,* **36**, 1131—1142.

Milner, H. B., 1962, Sedimentary petrography I, Methods in sedimentary petrography, Macmillan Co., 715 pp.

Plumley, W. J., 1948, Black Hills Terrace gravels: A study in sediment transport, *J. Geol.,* **56**, 526—577.

Pye, W. C., 1943, Rapid methods of making sedimentational analysis of arenaceous sediments, *J. Sed. Pet.,* **13**, 85—104.

Reed, F. S., and J. C. Mergner, 1953, Preparation of rock thin sections, *Am. Mineralogist,* **38**, 1184—1203.

Russell, R. D., 1935, Frequency percentage determinations of detrital quartz and feldspar, *J. Sed. Pet.,* **5**, 109—114.

Spencer, C. W., 1960, Method for mounting silt-size heavy minerals for identification by liquid immersion, *J. Sed. Pet.,* **30**, 498—500.

Tester, A. C., 1932, Cellophane as a slide cover, *J. Sed. Pet.,* **2**, 125.

Tickell, F. G., 1965, The techniques of sedimentary mineralogy, Elsevier Publishing Co., 220 pp.

Todd, R., D. Low, and J. F. Mella, 1965, Smaller foraminifers, *in* B. Kummel and D. Raup, eds., Handbook of Paleontological techniques, W. H. Freeman and Co., 582 pp.

Van der Plas, L. 1966, The identification of detrital feldspars, Elsevier Publishing Co., 305 pp.

CHAPTER 22

STAINING

GERALD M. FRIEDMAN

Rensselaer Polytechnic Institute, Troy, New York

Staining is one of the most useful techniques in the analysis of sedimentary rocks, particularly in carbonate analysis. It is indispensable in identifying the minerals that make up limestones and dolostones, and hence serves as a useful tool in fabric analysis. It can be used in Recent carbonate sediments and in partially or wholly consolidated Pleistocene rocks to identify the metastable carbonate minerals aragonite and magnesium calcite. It is equally useful in the differentiation of calcite, dolomite, gypsum, and anhydrite in Recent and ancient limestones, dolostones, and evaporites. Identification of carbonate minerals is effective in hand specimen, polished surface, and thin-section; in addition, stained peels of carbonate rocks can be prepared simply and effectively.

Staining techniques in the study of carbonate rocks date back to the classical paper of Lemberg (1887). A look at the references in this chapter shows the international scope and extensive literature that has developed in this field. Friedman (1959) reviewed the subject of staining, and Wolf et al. (1967) and Wolfe and Warne (1960) have added more references to this growing bibliography.

Additional papers on this subject have been published by Blazy and Cases (1963), Dickson (1965), Goni (1960), Goto (1961), Gundlach (1964), Schnitzer (1967), and Walger (1961). Carretero (1966) and Warne (1962) modified the Friedman (1959) staining schemes.

Many stains have been developed to promote identification of carbonate minerals; some are more effective and some "take" more rapidly than others. Staining schemes usually depend on the differences in rate of solution of carbonate minerals, for instance, between aragonite and calcite and between calcite and dolomite. Continuing exposure to a stain ultimately affects both minerals that one wishes to distinguish; hence the reaction has to be stopped before both are stained. Effective reaction time depends on the stain and on the mineralogy and fabric of the rock, and varies from one stain to another and often from one rock to another. Although staining techniques are easy, the beginner has to learn each technique and develop a feel for it. If he expects excellent results on the first try, he is often disappointed.

The stains and staining schemes presented in this chapter can be used in the field and in detailed laboratory examination (Friedman, 1959). Only the most effective staining techniques are presented.

Staining schemes given in this chapter are for the common minerals of carbonate sediments, limestones, dolostones, and evaporites. Warne (1962) expanded the author's original (1959) staining scheme to witherite, rhodochrosite, smithsonite, cerussite, and siderite. Because these minerals are not common in sedimentary rocks, they are not discussed here.

In sandstones, limestones, and dolostones in which feldspar occurs, staining technique serves to distinguish feldspar and quartz. This is particularly important in quantitative modal analysis. This method was developed by Gabriel and Cox (1929) and improved by Bailey and Stevens (1960). An extensive literature has grown around this subject and has been reviewed by van der Plas (1966).

ETCHING

Hand samples, cores, or drill cuttings should be etched with dilute hydrochloric acid and washed in running water before staining. The acid solution is made up of eight to ten parts by volume of concentrated hydrochloric acid (commercial grade can be used) diluted with water to 100 parts (Lamar, 1950, p. 2; Ives, 1955, p. 8). Exposure to acid should vary depending on fabric and mineralogy. An etching period of 2 to 3 minutes is usually effective; it is best, however, to examine the sample with a hand lens or under a binocular microscope to make sure that relief develops. A well-etched sample appears in three dimensions under a binocular microscope.

STAINING OF CARBONATES

Organic Stains Specific for Calcite

Alizarine Red S (Friedman, 1959)

Alizarine red S is the most effective stain for calcite. It is prepared by dissolving 0.1 gm in 100 ml 0.2% hydrochloric acid (add 2 ml of commercial grade concentrated hydrochloric acid to 998 ml of water). Calcite is stained deep red within 2 to 3 minutes, whereas dolomite is not stained, except on excessive exposure. However, staining time should be considered flexible, depending on the texture of the rock studied. For some studies a half minute exposure is ample; it results in a pale red coloration of the calcite.

This stain is suited particularly for thin-sections. Staining time varies slightly, but is usually no longer than a minute. The operator should stain his thin-section until sufficient contrast between calcite, which is stained red, and other minerals, such as dolomite, which remain unstained, is developed.

Harris' Hematoxylin (Friedman, 1959)

Harris' hematoxylin can be purchased commercially. The staining solution is prepared by mixing 50 ml of commercial grade Harris' hematoxylin with 3 ml of 10% hydrochloric acid (10 ml commercial grade concentrated hydrochloric acid added to 90 ml of water). The solution becomes more effective with age. A fresh solution may require 9 to 10 minutes to stain a sample. After the solution has been used repeatedly, staining time is reduced to 3 or 4 minutes or even less. The stain imparts an even purple coat over the surface of calcite, whereas dolomite is unaffected.

Inorganic Stains Specific for Calcite

Ferric Chloride (Lemberg, 1887, 1888, 1892; Keller and Moore, 1937; Rodgers, 1940)

Ferric chloride is the original Lemberg stain. The sample is immersed in a 2.5% ferric chloride solution (97.5 ml of water to 2.5 gm of ferric chloride) for a few seconds. Calcite is stained brown, whereas dolomite is not affected. After staining, the sample is washed thoroughly in water and immersed in commercial grade ammonium sulfide solution, $(NH_4)_2S$, for a few seconds. A black color is imparted to calcite, whereas dolomite remains unaffected. The black disappears on drying and the brown of ferric chloride reappears.

Copper Nitrate (Hinden, 1903; Mahler, 1906; Spangenburg, 1913; Rodgers, 1940)

The specimen is immersed in a molar solution of $Cu(NO_3)_2$ (add 188 gm of $Cu(NO_3)_2$, or 225 gm of $Cu(NO_3)_2 \cdot 3H_2O$, or 332 gm of $Cu(NO_3)_2 \cdot 6H_2O$ to 1000 gm of water) for 2.5 to 6 hours, depending on the desired intensity of coloration. After the speciman is removed from the nitrate solution, it is immersed in concentrated commercial grade NH_4OH for a few seconds and then washed and dried. The stain imparts a green or bluish-green color to calcite but does not affect dolomite.

Organic Stains Specific for Dolomite (Friedman, 1959)

Alizarine red S, alizarine cyanine green, and titan yellow are the recommended organic stains, although 16 other organic stains have been used with equal success (for the names of these other stains see Friedman, 1959, Table 3). Two-tenths gram of dye is dissolved in 25 ml methanol, if necessary by heating. Methanol lost by evaporation should be replenished. Fifteen milliliters of 30% NaOH solution (add 70 ml of water to 30 gm of sodium hydroxide) is added to the solution and brought to a boil. The sample is immersed in this boiling solution for about 5 minutes (occasionally it may take even more time). Dolomite is stained purple in alizarine red, deep green in alizarine cyanine green, and deep orange-red in titan yellow alkaline solution. Inadequate staining imparts a yellow to yellow-orange color with titan yellow.

Amaranth Stain for Dolomite
(Laniz, Stevens, and Norman, 1964)

This stain is primarily for distinguishing plagioclase feldspar from other minerals. However, it also stains dolomite and does not affect calcite; hence it can be used for differentiating dolomite from calcite. Dolomite is stained deep red after etching, whereas calcite is not stained. The procedure for the preparation of this stain is given in the section on "Staining of Feldspar."

Inorgani Stain Specific for Calcite and
Dolomite Containing Ferrous Iron (Heeger,
1913; Friedman, 1959; Evamy, 1963;
Katz and Friedman, 1965)

This is a routine analytical test for iron. A staining solution is prepared by dissolving 5 gm of potassium ferricyanide in distilled water containing 2 ml of concentrated hydrochloric acid, followed by dilution to 1 liter with distilled water. A black color will be imparted to the specimen, the deepness of color being proportional to the Fe^{2+} concentration. In the literature this test is described as specific for dolomite containing ferrous iron (see Friedman, 1959, p. 95). However, since this is a test for iron, both ferroan calcite and ferroan dolomite are stained, as pointed out by Evamy (1963). This stain is suitable for use on thin-sections.

Combined Organic and Inorganic Stains
Specific for Calcite, Ferroan Calcite,
and Ferroan Dolomite (Evamy, 1963)

A solution consisting of alizarine red S and potassium ferricyanide will stain calcite as well as ferroan calcite and ferroan dolomite. The reactions of alizarine red S and potassium ferricyanide in the combined reagent are the same as those in the individual stain solutions. According to Evamy (1963, p. 165) this solution consists of 0.2% hydrochloric acid, 0.2% alizarine red S, and 0.5 to 1.0% potassium ferricyanide. Katz and Friedman (1965) recommend that the solution be made up as follows: dissolve 1 gm of alizarine red S with 5 gm of potassium ferricyanide in distilled water containing 2 ml concentrated hydrochloric acid and bring the solution to 1 liter with distilled water. The following colors are obtained: iron-free calcite, red; iron-poor calcite, mauve; iron-rich calcite, purple; iron-free dolomite (dolomite *sensu stricto*), not stained; ferroan dolomite, light blue; ankerite, dark blue. Evamy (1963) distinguishes between ferroan dolomite and ankerite as follows:

$$\text{ferroan dolomite } \frac{Fe^{2+}}{Mg^{2+}} < 1 \text{ and ankerite } \frac{Fe^{2+}}{Mg^{2+}} > 1.$$

The combined stain can be used successfully in thin-sections.

Stains Specific for Aragonite (Feigl, 1958)

Of several stains specific for aragonite (see Friedman, 1959, pp. 95-96) the most sensitive was developed by Feigl (1937, pp. 328-329). In the writer's 1959 paper, he named this reagent "Feigl's solution." The preparation of this solution is as follows:

1 gm of solid (commercial grade) Ag_2SO_4 is added to a solution of 11.8 gm $MnSO_4 \cdot 7H_2O$ in 100 ml of water and boiled. After cooling, the suspension is filtered and one or two drops of diluted sodium hydroxide solution is added. The precipitate is filtered off after 1 or 2 hours (Feigl, 1958, p. 470). It is important that only distilled water be used; tap water leaves a white precipitate of silver chloride (Katz and Friedman, 1965).

Stains Specific for Magnesian-Calcite (Friedman, 1959)

Magnesian-calcite stains with alizarine red S in acid solution like any other calcite, that is, the stain imparts a deep red color to the mineral. It also stains purple with Harris' hematoxylin like any calcite. The preparation of these two stains has been explained under the heading of "Organic Stains Specific for Calcite." However, in alkaline solution, as indicated under the heading of "Organic Stains Specific for Dolomite," magnesian-calcite reacts in the same manner as dolomite. Alizarine red S in alkaline solution imparts a purple color and titan yellow a yellow color. The intensity of staining reflects the amount of magnesium present in the mineral.

Stains Specific for Gypsum (Friedman, 1959)

One-tenth to 0.2 gm of alizarine red S or of one of several other organic stains (see Friedman, 1959, p. 92, Table 4) is dissolved in 25 ml of methanol. Fifty milliliters of 5% sodium hydroxide (5 gm of sodium hydroxide added to 95 ml of water) are added. The specimen is immersed in the cold solution. Staining imparts a deep color to gypsum within a few minutes and a very faint tint of the same color to dolomite. Alizarine red will stain gypsum purple. Heating the solution increases the intensity of the stain. Anhydrite and calcite are not stained.

Stains Specific for Anhydrite (Friedman, 1959)

No effective stains were developed for anhydrite. However, schemes have been worked out whereby the presence of anhydrite can be determined by the process of elimination, as indicated in Figs. 1 and 2.

Staining Schemes for Carbonates (Friedman, 1959)

Staining schemes have been proposed for routine carbonate staining analysis using a combination of alizarine red S and Feigl's solution (Fig. 1), and titan yellow, Harris' hematoxylin, and Feigl's solution (Fig. 2). Figures 1 and 2 have been taken from Friedman (1959). More recent work has shown that alizarine cyanine green is preferable to titan yellow in the second staining scheme (Fig. 2). The two staining procedures can be used to differentiate between dolomite, calcite, aragonite, magnesium-calcite, gypsum, and anhydrite. Warne (1962) extended one of the writer's schemes (Fig. 1) to include witherite, rhodochrosite, smithsonite, strontianite, cerussite, and siderite.

Staining Carbonates in Thin-Section

Of the stains developed, the following are most useful for work in thin-section: potassium ferricyanide for calcite and dolomite containing ferrous iron, alizarine red S for calcite, and the potassium ferricyanide-alizarine red combination stain for calcite and for calcite and dolomite containing ferrous iron. Procedures for preparing these stains have been given under the appropriate headings.

STAINED PEELS (Katz and Friedman, 1965; Davies and Till, 1968; Ali, 1968)

The advantage of peels combined with staining was effectively demonstrated by Katz and Friedman (1965). Germann (1965) developed independently a similar technique, but his success was confined to ferric chloride, Harris' hematoxylin, and a modified Lemberg stain. The preparation of peels and their usefulness in the

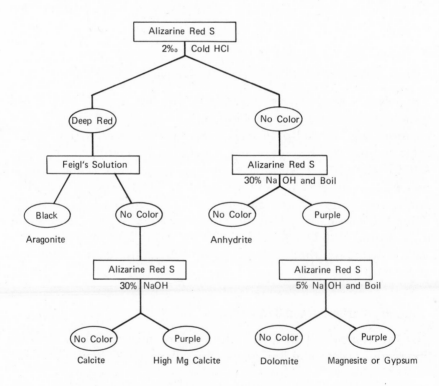

Fig. 1 Recommended staining procedure I; alizarine red S and Feigl's solution. [a]Or faint stain. (After Friedman, 1959)

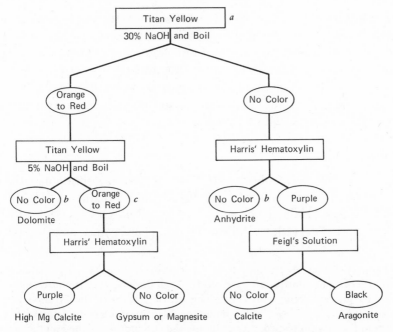

Fig. 2 Recommended staining procedure II; titan yellow, Harris' hematoxylin, and Feigl's solution. Alizarine cyanine green may be substituted for titan yellow. [a] Dolomite has a wide range of colors, any of the 20 excellent stains from Table 3 (Friedman, 1959) may be substituted for titan yellow to provide color contrast between dye and dolomite tested. Alizarine cyanine green is particularly recommended as an excellent stain. [b] Or faint stain (light orange). [c] High magnesian calcite used in this study was very fine grained. The behavior of coarse-grained high-magnesian calcite was not studied. (After Friedman, 1959)

study of carbonate rocks have been discussed in Chapter 10. In stained peels, the actual layer of stain is transferred from a polished rock surface to a peel.

Rock Preparation

Rock slabs are sawed and one surface is ground with carborundum powder. A final polish is given with 800 grade powder and the polished surface is thoroughly washed and etched with dilute hydrochloric acid solution (10 ml of commercial grade hydrochloric acid diluted with water to 100 ml). Etching is the most important single step in preparing good quality peels. Etching time varies depending on the grain and crystal size of the rock and on the relative amount of calcite and dolomite present. Dolomitic rocks take longer than calcitic ones and fine-grained calcilutites take longer than do coarse-grained calcarenites. For most purposes etching time varies from a minimum of 20 seconds to a maximum of about 1 minute; usually some experimentation is necessary. Etching is stopped by washing with tap water.

Peels

Commercially available acetate sheets are most commonly used in the preparation of stained peels (Katz and Friedman, 1965; Ali, 1968). Frank (1965) recommends ordinary plexiglass, 1/8-in. thick. Davies and Till (1968) follow Honjo (1963) and make peels as follows:

The peel sheets are made of a 7 percent solution of ethyl cellulose in trichloroethylene. Slowly, 10.01 g of ethyl cellulose powder is poured into 100 ml of trichloroethylene with continuous stirring. Initially the solution is very thick, but after standing for two hours the viscosity is markedly reduced and all air bubbles have escaped. A little of the trichloroethylene-ethyl cellulose solution is poured onto a glass plate. An even thickness is obtained by first binding the edge of the plate with cellotape and spreading the liquid by rolling a glass rod across the plate. The peel sheet dries in about 5 minutes and is easily removed after cutting around the cellotape with a scalpel. The thickness of the peel sheet can be varied by changing the number of layers of cellotape binding the edge of the glass plate. Thin peel sheets ($<30\mu$) tend to tear on removal from calcarenites and can only be used on nonporous micritic rocks. However, peel sheets 100μ thick duplicate the most minute detail in the rock and can be easily removed from all types of carbonates.

Transfer of Stain to Peel

The preparation of the staining solutions has been discussed in the section on "Staining of Carbonates." The following stains are recommended: alizarine red S for calcite, alizarine cyanine green for dolomite, potassium ferricyanide for ferroan calcite and ferroan dolomite, combined potassium ferricyanide and alizarine red S for calcite, ferroan calcite and ferroan dolomite, and Feigl's solution for aragonite.

For all stains, except the dolomite stain, the technique is as follows. The etched rock slab is immersed in the staining solution, polished side up. Staining takes about 4 minutes, but some experimentation is recommended. The solution is poured off and the slab is washed with water. For Feigl's solution, only distilled water should be used. For staining dolomite, the etched rock slab is immersed in a boiling solution for 5 to 10 minutes.

Next, the slab is ready for immersion in a solvent. The solvent for acetate sheets is acetone, for plexiglass, ethylene dichloride, and for the peel sheets of Davies and Till, ethyl acetate. If acetate sheets are used, the slabs are immersed in acetone for not more than 2 or 3 seconds and taken out. A piece of acetate sheet is placed immediately on the stained and etched surface. The sheet is pressed firmly and evenly with the fingers. No air bubbles should remain under the film if the operation is done quickly and sufficient acetone is applied. The rock with the attached acetate sheet is left to dry for about 40 minutes and then removed carefully from the rock.

Ali (1968) recommends flooding of stained surface with acetone and then progressive rolling of the acetate sheet onto the stained surface.

The preparation of peels using plexiglass is almost identical with that of acetate sheets except that the solvent is ethylene dichloride. For the Davies and Till peel sheets, ethyl acetate is used as a solvent to soften the peel sheet when it is placed on the rock surface. The stained and etched rock surface is flooded with a minimum of ethyl acetate from a teat pipette. The sheet is pressed firmly on the rock surface, dries in about 3 minutes, and is removed.

The advantage of this method is the short drying time.

Mounting

Stained peels are mounted between glass slides for permanent storage. Glass slides of 35-mm size are commonly used; the peels, in their glass mountings, are fastened with aluminum ties and stored in boxes designed for 35 mm slides. Plexiglass peels do not require mounting.

Examination

Stained peels can be projected on a screen such as 35-mm slides and thus be examined. They can also be studied under binocular and petrographic microscopes. The advantage of plexiglass is that it can be studied at the highest possible magnification because the objective lens of the microscope can approach and even touch the plastic surface. The thickness of the glass-mounted peels prevents examination with higher powers.

STAINING OF FELDSPARS

Gabriel and Cox (1929) proposed a staining method to differentiate quartz from alkali feldspar. The feldspar is etched by hydrofluoric acid vapor and treated with a concentrated solution of sodium cobaltinitrite. Potassium feldspar is coated by yellow potassium cobaltinitrite, whereas quartz remains unaffected. This method is simple in theory but difficult to apply as indicated by several articles giving details of procedure (Keith, 1939; Chayes, 1952; Rosenblum, 1956; Hayes and Klugman, 1959).

Bailey and Stevens (1960) proposed a method for staining plagioclase feldspar with barium chloride and potassium rhodizonate after etching. Plagioclase takes on a red coloration. This staining technique was combined with yellow staining of potassium feldspar with sodium cobaltinitrite.

Laniz, Stevens, and Norman (1964) advocate a method for sequentially staining plagioclase red with F.D. and C. Red No. 2 (amaranth) and K-feldspar with cobaltinitrite.

Reeder and McAllister (1957) proposed the staining of the aluminum ion in feldspar with hemateine after etching. Doeglas et al. (1965) and van der Plas (1966) combined successfully the techniques of Gabriel and Cox (1929) with those of Reeder and McAllister (1957) based on experiments by Favejee (van der Plas, 1966, p. 50).

Caution on Hydrofluoric Acid

Staining feldspars involves etching with hydrofluoric acid vapor. Extreme care must be taken when this reagent is used. The acid reacts rapidly with tissue, but pain and other overt signs of deep burns may not be noticed for several hours. To avoid painful burns, gloves are advisable and hood ventilation is imperative.

Staining with Barium Chloride, Potassium Rhodizonate, and Sodium Cobaltinitrite (Bailey and Stevens, 1960)

For rock slabs, the steps taken in this technique are as follows.

1. Saw rock slabs. If the rock is porous, soak the specimen in molten paraffin for about 15 minutes or impregnate with Lakeside 70 before polishing. The flat surface is polished with No. 400 grit and dried.

2. Pour concentrated hydrofluoric acid (52% HF) into an etching vessel to within about 5 mm of the top. (*Note:* Hydrofluoric acid must be used in a well-ventilated hood.)

3. Put the rock slab, polished surface down, across the top of the etching vessel. Leave for 3 minutes.

4. Cover the etching vessel and specimen with an inverted plastic cover to prevent drafts.

5. Remove the slab from the etching vessel, dip it in water, and quickly dip twice in and out of the barium chloride solution (5%).

6. Rinse the slab in water and immerse the polished surface for 1 minute in sodium cobaltinitrite solution (saturated).

7. Remove excess cobaltinitrite in tap water. K-feldspar is stained bright yellow. If the feldspar is not well stained, remove the etch by rubbing the surface under water, dry, and etch again for a longer period; then repeat the process again, starting with step 5.

8. Rinse the slab briefly in distilled water and cover the polished surface with rhodizonate reagent (0.05 gm of rhodizonic acid potassium salt dissolved 20 ml of distilled water; the reagent is unstable; make fresh in a small dropping bottle). Plagioclase feldspar takes on a red stain.

9. Remove excess stain in tap water.

For thin sections the steps are the following.

1. Etch the uncovered thin section for 10 seconds over hydrofluoric acid at room temperature.

2. Immerse the section in saturated sodium cobaltinitrite solution for 15 seconds. The K-feldspar is stained light yellow.

3. Rinse the section in tap water to remove cobaltinitrite.

4. Dip the section quickly in and out of barium chloride solution (5%).

5. Rinse the slide quickly with distilled water.

6. Cover the thin-section with rhodizonate reagent from a dropping bottle (see previous paragraph for preparation of this reagent). Plagioclase feldspar will be stained pink.

7. Wash the slide in tap water, dry, and mount with cover glass.

Staining with F.D. and C. Red No. 2 (Amaranth) (Laniz, Stevens, and Norman, 1964)

This method is combined with a barium chloride and sodium cobaltinitrite treatment. In this technique a red coloration is obtained on plagioclase by absorbing barium ion on the etched plagioclase and then dipping the specimen in the amaranth dye. Various washings in this technique (see below) lead to a purple-red coloration of the plagioclase that contrasts sharply with the yellow color of the K-feldspar (stained yellow by cobaltinitrite); quartz remains unstained. Pure albite does not stain, but can be stained by first dipping the sample in calcium chloride solution.

For rock slabs the steps in this technique are as follows.

1. Saw rock slabs and polish them on a lap with No. 400 to 800 grit. If the rock is porous, impregnate with Lakeside 70 before polishing.

2. Etch the polished surface for 10 to 15 seconds in concentrated (52%) hydrofluoric acid.

3. Dip the slab in water.

4. Immerse the slab in a saturated solution of sodium cobaltinitrite for 1 minute.

5. Wash the slab in tap water to remove excess cobaltinitrite.

6. Dry the slab under a heat lamp.

7. Immerse the slab for 15 seconds in 5% barium chloride solution.

8. Dip the slab once quickly in water and dry gently with compressed air.

9. Immerse the slab for 16 seconds in F.D. and C. Red No. 2 solution (1 ounce of 92% pure coal-tar dye in 2 liters of water).

10. Dip the slab once quickly in water.

11. Remove excess dye from the surface of the slab with a gentle stream of compressed air.

12. When white grains, crystals, or patches suggestive of albite remain after the above treatment, repeat steps 1 to 3, dip in calcium chloride, dry and then proceed as in steps 4 to 11.

For thin-sections, the steps are as follows.

1. Etch the uncovered rock section for 15 seconds in hydrofluoric acid vapor.

2. Immerse the section in cobaltinitrite solution for 15 seconds.

3. Wash quickly in tap water.

4. Immerse the section for a few seconds in barium chloride solution.

5. Dip the section once in distilled water.

6. Immerse the section for one minute in the F.D. and C. Red No. 2 solution.

7. Dip the section once in water.

8. Remove excess dye from the surface of the section with a gentle stream of compressed air.

For sand grains, the steps are as follows.

1. Mount the grains in melted Lakeside 70 containing lampblack to make it opaque.

2. Cool.

3. Grind a smooth surface to expose the sand grains.

4. Etch and stain the same as directed for rock slabs.

Staining with Cobaltinitrite and Hemateine (van der Plas, 1966)

Cobaltinitrite Staining

For rock slabs and thin-sections, the steps in this technique are as follows.

1. Polish rock slab or thin-section. Remove Canada balsam, resin, or grease with organic solvent or ultrasonic cleaner.

2. Put rock slab or section on a small, flat lead plate in the etching vessel. If it is a thin-section, the glass surface must be covered carefully with a grease resistant to hydrofluoric acid vapor.

3. Etch the surface for 1 minute in hydrofluoric acid vapor at 90°C.

4. If it is a rock slab, it is heated in an electric furnace for about 5 minutes at 400°C; if a thin-section, heating is unnecessary, in fact, inadvisable.

5. Pour cobaltinitrite solution on slab or section (1 gm of sodium cobaltinitrite in 4 ml of distilled water) and leave for 2 minutes.

6. Wash in distilled water and dry. Alkali feldspar shows a yellow stain.

For sand grains, the steps are as follows.

1. Etch the grains in hydrofluoric acid vapor for 1 minute at 90°C. The grains must be in the vapor above the liquid hydrofluoric acid (35%). The procedure should be carried out in a well-ventilated hood.

2. After etching, the sample is heated in an electric furnace for about 5 minutes at 400°C.

3. Pour cobaltinitrite on the grains (1 gm of sodium cobaltinitrite in 4 ml of distilled water) and leave for about 1 minute.

4. Wash the sample in distilled water. Alkali feldspar grains show a yellow stain.

Hemateine Staining

For slabs or thin-sections, the steps are as follows.

1. Prepare a hemateine solution (50 mg of hemateine in 100 ml of 95% ethanol) and a buffer solution (20 gm sodium acetate in 100 ml of distilled water to which are added 6 ml of glacial acetic acid. This solution is diluted to 200 ml and buffered at pH 4.8 with an acidity of 0.5 N).

2. Mix hemateine and buffer solutions in the proportions 2:1 prior to use.

3. Etch the sample in 1:10 HCl; after etching, pour the mixed hemateine and buffer solutions (of step 2) on the slab, or uncovered thin-section, and leave for 5 minutes.

4. Rinse with 95% ethanol and with acetone. Feldspars show a bluish stain.

For sand grains, the steps are the following.

1. Etch the grains and heat them in an electric furnace for about 5 minutes at 400°C.

2. Add about 10 drops of hemateine solution and 5 drops of buffer solution to the sample (for preparation of these two reagents see previous paragraph). Swirl the container for 2 or 3 minutes to mix the solution. Leave the sample in the solution for about 5 minutes.

3. Remove the solution by washing it away with 95% ethanol. Siphon off any supernatant liquid; wash the sample twice with acetone. The feldspar shows a purple-bluish stain.

Combined Staining

Feldspar grains can be stained simultaneously for potassium and aluminum ions. In this technique, cobaltinitrite staining must precede hemateine staining. If the grains are not well stained, they can be cleaned with dilute hydrochloric acid, washed in distilled water, dried in acetone, and etched and stained a second time.

ACKNOWLEDGMENTS

The author extends his thanks to S. Ali, R.E. Carver, M.G. Gross, P. Rose, and S. Sommer for reviewing this manuscript.

REFERENCES

Ali, S. A., 1968, Acetate peels for the study of carbonate rocks, *Pakistan J. Scientific and Industrial Res.,* 11, 213–214.

Bailey, E. H., and R. E. Stevens, 1960, Selective staining of K-feldspar and plagioclase on rock slabs and thin sections, *Am. Mineralogist,* 45, 1020–1026.

Blazy, P., and J. Cases, 1963, Coloration selective de carbonates et traitement des minerais, *Soc. Francaise de Minér. êt de Cristallographie Bull.,* 86, 200–201.

Carretero, P. A., 1966, Differentiation of the main sedimentary minerals by selective staining techniques, *Annales Edafologia Agrobiologia (Madrid),* 25, 11–12, 689–696, Instituto de Edafologia y Biologia Vegetal, Madrid.

Chayes, Felix, 1952, Notes on the staining of potash feldspar with sodium cobaltinitrite in thin section, *Am. Mineralogist,* 37, 337–340.

Davies, P. J., and R. Till, 1968, Stained dry cellulose peels of ancient and Recent impregnated carbonate sediments, *J. Sed. Pet.,* 38, 234–237.

Dickson, J. A. D., 1965, A modified staining technique for carbonates in thin section, *Nature,* 205, 587.

Doeglas, D. J., J. Ch. L. Favejee, D. J. G. Nota, and L. van der Plas, 1965, On the identification of feldspars in soils, *Mededelingen van de Landbouwhogeschool te Wageningen,* 65(9), 14 pp.

Evamy, B. D., 1963, The application of a chemical staining technique to a study of dedolomitization, *Sedimentology,* 2, 164–170.

Feigl, Fritz, 1937, Qualitative analysis by spot tests, Nordemann Publishing Company, 400 pp.

———, 1958, Spot tests in inorganic analysis, 5th ed., Elsevier Publishing Co., 600 pp.

Frank, Ruben M., 1965, An improved carbonate peel technique for high powered studies, *J. Sed. Pet.,* 35, 499–500.

Friedman, G. M., 1959, Identification of carbonate minerals by staining methods, *J. Sed. Pet.,* 29, 87–97.

Gabriel, A., and E. P. Cox, 1929, A staining method for the quantitative determination of certain rock minerals, *Am. Mineralogist,* 14, 290–292.

Germann, Klaus, 1965, Die technik des folienabzuges und ihre ergänzung durch anfärbemethoden, *Neues Jahrb. Geologie Paläontologie, Abh.,* 121, 293–306.

Goni, J. C., 1960, Nouvelle méthode de différenciation entre calcite et dolomite, *Soc. Française de Minér. et de Cristallographie Bull.,* 83, 254–256.

Goto, M., 1961, Some mineralogical-chemical problems concerning calcite and aragonite, with special reference to the genesis of aragonite, *J. Faculty Sci., Hokkaido Univ.,* 1, 571–640.

Gundlach, Heinrich, 1964, New field test to distinguish limestone-dolomite, *Neues Jahrbuch Geologie Paläontologie, Monatshefte,* 10, 626–628.

Hayes, J. R., and M. A. Klugman, 1959, Feldspar staining methods, *J. Sed. Pet.*, 29, 227–232.

Heeger, J. E., 1913, Ueber die mikrochemische Untersuchung fein verteilter Carbonate im Gesteinsschliff, *Centralblatt Min. Geologie u. Palaontologie*, 44–51.

Hinden, F., 1903, Neue Reaktionen zur Unterscheidung von Calcit und Dolomit, *Verh. Naturforschend Ges. Basel*, 15, No. 2, 201.

Honjo, S., 1963, New serial micropeel technique, Kansas Geol. Surv. Bull. 165, pt. 6.

Ives, William, Jr., 1955, Evaluation of acid etching of limestone, Kansas Geol. Surv. Bull. 114, pt. 1.

Katz, Amitai, and G. M. Friedman, 1965, The preparation of stained acetate peels for the study of carbonate rocks, *J. Sed. Pet.*, 35, 248–249.

Keith, M. L., 1939, Selective staining to facilitate Rosiwal analysis, *Am. Mineralogist*, 24, 561–565.

Keller, W. D., and G. E. Moore, 1937, Staining drill cuttings for calcite-dolomite differentiation, *Bull. Am. Assoc. Petrol. Geol.*, 21, 949–951.

Lamar, J. E., 1950, Acid-etching in the study of limestones and dolomites, Illinois State Geol. Surv. Circ. 156, 47 pp.

Laniz, R. V., R. E. Stevens, and M. B. Norman, 1964, Staining of plagioclase and other minerals with F. D. and C. Red No. 2, U.S. Geol. Surv. P.P. 501-B, B152–B153.

Lemburg, J., 1887, Zur microchemischen Untersuchung von Calcit, Dolomit und Predazzit, *Zeits. Deutschen Geologischen Ges.*, 39, 489–492.

_____, 1888, Zur microchemischen Untersuchung von Calcit, Dolomit und Predazzit, *Zeits. Deutschen Geologischen Ges.*, 40, 357–359.

_____, 1892, Zur microchemischen Untersuchung einiger Minerale, *Zeits. Deutschen Geologischen Ges.*, 44, 224–242.

Mahler, O., 1906, Ueber das Chemische Verhalten von Dolomit und Kalkspat, dissertation, Freiburg.

Reeder, S. W., and A. L. McAllister, 1957, A staining method for the quantitative determination of feldspars in rocks and sands from soils, *Canadian J. Soil Sci.*, 37, 57–59.

Rodgers, J., 1940, Distinction between calcite and dolomite on polished surfaces, *Am. J. Sci.,* **238,** 788–798.

Rosenblum, Samuel, 1956, Improved techniques for staining potash feldspars, *Am. Mineralogist,* **41,** 662–664.

Schnitzer, W. A., 1967, Bromophenol blue for distinguishing between limestone and dolomite, *Zement-Kalk-Gips,* 20(1), 31–32.

Spangenburg, K., 1913, Die küenstliche Darstellung des Dolomites, *Zeits. Kristallographie,* **52,** 529–567.

Van der Plas, Leendert, 1966, The identification of detrital feldspars, Elsevier Publishing Co., 305 pp.

Walger, Eckart, 1961, Zur mikroskopischen Bestimmung der Gesteinsbildenden Karbonate im Dünnschliff, *Neues Jahrb. Min., Monatshefte,* 182–187.

Warne, Slade, 1962, A quick field or laboratory staining scheme for the differentiation of the major carbonate minerals, *J. Sed. Pet.,* **32,** 29–38.

Wolf, K. H., A. J. Easton, and Slade Warne, 1967, Techniques of examining and analyzing carbonate skeletons, minerals, and rocks, *in* G. V. Chilingar, H. J. Bissell, and R. W. Fairbridge, eds., Carbonate rocks, developments in sedimentology 9B, Elsevier Publishing Co., 413 pp.

Wolf, K. H., and Slade Warne, 1960, Remarks on the application of Friedman's staining methods, *J. Sed. Pet.,* **30,** 496–497.

CHAPTER 23

X-RAY DIFFRACTION MOUNTS

RONALD J. GIBBS

Northwestern University, Evanston, Illinois

In mineralogical studies utilizing x-ray diffraction, the type and technique of mounting used are extremely critical and should therefore be thoroughly evaluated for suitability to the particular study with which it is to be used. There is no "all-purpose" mounting technique that can be used for all the wide variety of problems encountered. Each problem requires a suitable mounting method, and each type of mount is especially well suited for a particular purpose. This discussion will be limited to those mounting methods commonly used with x-ray diffractometers. For information regarding the various mounting methods for use with cameras, the reader is referred to studies by Buerger (1962) and Nuffield (1966).

In investigating the type of mounting to be used in any x-ray diffraction analysis, it should be remembered that, just as the sample used in preparing the mount must be representative of the material studied, the portion of the mounted material "seen" by the x-ray beam must be representative of the entire mount. For example, 90% of the pattern at the position for analysis of montmorillonite at the 17 Å peak is derived from the upper 4.8μ thickness, and for kaolinite at 7 Å from the upper 11.9μ thickness (Gibbs, 1965).

The long dimension of the exposed area of a mount varies with the 2Θ angle setting of the diffractometer as seen in Fig. 1. The mount dimensions recommended in this chapter are the generally adequate average of 20 to 25 mm length for 2Θ goniometer settings below 60°, with divergent, scattering, and receiving slits of 1°, 1°, and 0.006 in. (0.15 mm), respectively. For detailed investigation of glycolated montmorillonite at the 17 Å peak, mounts of average length can be used with narrower slits, or mounts 55 to 60 mm long can be used with the divergent, scattering, and receiving slits given above. Study of a peak at 30° 2Θ requires a mount only 11 mm long at standard slit settings and requires much less sample material. If peak intensities in a sample are to be compared, the x-ray mount should be as long as the lowest 2Θ angle exposure area to be investigated, and all slits should be of equal width across the entire 2Θ range.

The upper surface of the sample mount must be positioned exactly on the focusing circle of the diffractometer. For the Norelco diffractometer the focusing circle is represented by the plane of the lower surface of the mount holder. If the upper surface of a sample is above this plane by as little as a few thousandths of an inch, it is not generally acceptable and should be corrected by placing spacers (microscope cover slides, sheet metal, or metal shim stock) between the holder and the mount.

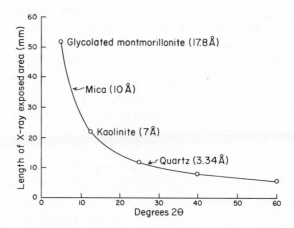

Fig. 1   Relation of length of x-ray-exposed area of mounts to degrees 2θ using $1°$ slits.

ORIENTED-PARTICLE MOUNTS

Basal reflections of clay minerals are especially diagnostic of the mineral species and variety. Therefore their platy nature can be used to advantage by preparing mounts in which the majority of particles is oriented with the basal surface of each particle parallel to the sample holder. Most techniques for orientation of clay minerals parallel to the mount require sedimentation of the material on the mount. However, in the sedimentation process larger particles settle faster than smaller particles, producing an upper surface that is not representative of the entire sample, with resulting quantitative errors as great as 250% (Gibbs, 1965; 1968).

Among the numerous oriented-particle mounting techniques proposed for x-ray diffraction analysis are several acceptable with regard to freedom from error due to segregation and with regard to precision and accuracy. The following are mounting techniques found acceptable.

1. The smear-on-glass slide technique (Gibbs, 1965).
2. The suction-on-ceramic tile technique.

Smear-on-Glass Slide Technique

To a portion (about 1/3 cc) of thoroughly mixed sample material add enough water to make a paste, but not a fluid. Place the paste on the edge of a small (25 x 45mm) glass slide using a spatula (Fig. 2a). With a single stroke of a special spatula, spread the paste across the slide in a thin, even layer (Figs. 2b and 2c). A special spatula can be constructed by mounting a 15 x 40 mm rectangular piece of plastic, 0.2 to 0.4-mm thick, in a slotted plastic handle so that the extended portion can be adjusted according to the consistency of the clay paste and the flexibility of the plastic. After the clay paste has been smeared in the manner illustrated, allow the mount to dry at room temperature and humidity.

Suction-on-Ceramic Tile Technique

Tile for use with this technique can be obtained from Coor's Porcelain Co., 600 Ninth Street, Golden, Colorado, by ordering porous plates of various sizes which must be cut to size and lapped to 25 x 45 x 5 mm or obtained from a local ceramic shop by purchasing

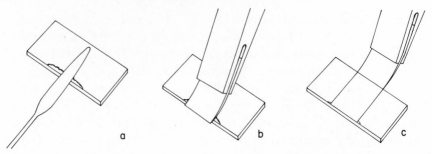

Fig. 2   Procedure for making smear mounts for x-ray diffractometers.

"green" tile blanks, cutting them to rough size (easily done with knife and sandpaper), firing them in a high-temperature oven or kiln, and finishing them to the desired size on a lap. During the mounting procedure (Fig. 3) the tile should be clamped in a holder made of noncorrosive rigid material (stainless steel, aluminum, or plexiglass) which has two neoprene gaskets that seal the edges of the top and bottom faces of the tile. One to 2 cc of clay suspension should be placed on the dry tile, the liquid portion drawn through the tile from below by means of a vacuum, assisted by the capillary action of water soaking into the tile, which leaves a solid coat of clay on the tile.

The user of the suction-on-ceramic tile technique must be alert to three problems. Some tiles have a permeability so low that the length of time required to suck the water through the tile allows settling of the coarser particles, and therefore the occurrence of size segregation. Note that the suspension is only a few millimeters deep. Since 2μ particles fall 1 mm in 4.8 minutes, 5μ particles fall 1 mm in about 30 seconds. Therefore, if sample material $< 2\mu$ in diameter falls for more than 5 minutes, segregation can occur. The rate of fall for 5μ-diameter particles illustrates the unsuitability of the technique for use with material greater than 2μ in diameter.

A second problem is the possible "clogging-up" of the tile, which also lengthens the processing time excessively. A third danger lies in the possibility of superimposition of a diffraction pattern for minerals in the tile on the diffraction pattern of the sample material being analyzed; this is due to the sample material on the mount being too thin and portions of the tile being exposed to the x-ray beam.

There are several advantages of the suction-on-ceramic tile technique. The mounts can be heat treated without damage, the exchange cations can be changed by passing various solutions through the tile, and the tile can act as a storage chamber for excess glycol, thus preventing deglycolation of clays.

Other Oriented-Particle Methods

Oriented-particle methods for x-ray diffraction analysis shown by Gibbs' (1965) data as not being generally acceptable for mounting sample material, because of segregation error and with regard to precision and accuracy, are as follows:

1. Pipette or dropper on glass-slide technique.
2. Beaker on glass-slide technique.
3. Centrifuge on glass-slide technique.
4. Centrifuge through ceramic-tile technique.

Variations of these four techniques are equally unacceptable.

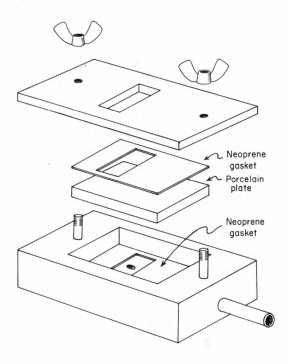

Neoprene gasket

Porcelain plate

Neoprene gasket

Fig. 3 Mounting device for suction-on-ceramic tile technique.

RANDOMLY ORIENTED PARTICLE MOUNTS

For many x-ray diffraction analyses, random particle orientation of sample material is advantageous. For example, in studying nonbasal reflections, use of randomly oriented mounts is essential. Two of the many techniques for random orientation are given below. The powder-press technique has proven to be more reliable with respect to precision, and the side-loading technique produces better randomness of particle orientation.

Powder-Press Technique

In the powder-press technique, thoroughly mixed, dry, powdered sample material is loaded into an aluminum holder from the back, with the surface to be exposed to the x-ray beam placed face downward on the mounting plate (Fig. 4). The powder is then compressed, using a piston as illustrated. An aluminum powder holder that has proven satisfactory for general work on the Norelco diffractometer has a well size of 10 x 20 mm, overall dimensions of 25 x 35 mm, and is 5 mm thick.

For randomly oriented specimens, the surface to be exposed is placed on filter paper and the sample is lightly packed with low pressure (10 psi). For partially oriented specimens, the surface to be exposed is placed on polished metal and the sample is packed with high pressure (180 psi) in an arbor press. The compressed cake of powder has remarkable strength and adhesion to the walls of the mount. It therefore does not require physical support from below when removed from the mounting plate to be placed on the diffractometer. If sufficient sample material to produce this adhering cake is not available, the mount can be strengthened by adding powdered coffee as a filler and backing before packing. This additive cannot interfere with the sample diffraction pattern because the x-ray pattern represents only the uppermost surface of the mount and because the powdered coffee is x-ray amorphous. Powdered coffee is also convenient because, if necessary, it can be washed out of the sample with water. With the powdered coffee backing technique, excellent patterns can be obtained from as little as 50 mg of sample material.

It is difficult to attain complete random orientation of particles using the powder-press technique even when filter paper is used as

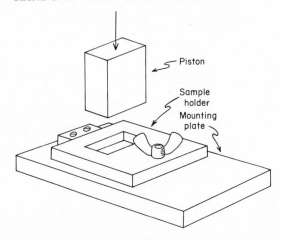

Fig. 4 Mounting device for powder-press technique.

the mounting plate. However, random orientation can be appreciably increased by treating the sample to coat the platy mineral particles, prior to mounting, thus producing a sphere of x-ray amorphous material around each particle. In one technique, thoroughly tested and described by Brindley and Kurtossy (1961), one part of Lakeside 70 cement is ground with five parts of sample material for 7 minutes. Dioxane is added to the mixture (on a hot water bath) to dissolve the Lakeside 70. The mixture is stirred continuously with a spatula until it cools to a granular texture. When completely cool and hard, the mixture is lightly ground to a fine powder (in about 5 minutes). It is suggested that coarse filter paper be used on the mounting plate when using the powder-press technique to mount this powder mixture.

Side-Loading Powder Technique

Better randomness of particle orientation is produced by the side-loading powder technique (described by Niskanen, 1964; and Bystrom-Asklund, 1967) than by the nonpretreated powder-press technique. In the side-loading powder technique, thoroughly mixed powder is loaded into a holder having one temporary side of ground glass, held in place with tape or by a clamp (Fig. 5). As the mount is being filled it should be tapped gently to help pack the powder. When the mount has been filled with sample material, it is set down with the glass side up and the tape is carefully cut or the clamp

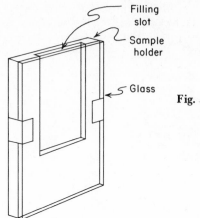

Filling slot

Sample holder

Glass

Fig. 5 Mounting device for side-loading powder technique.

carefully released. The glass side is then lifted directly upward without the least sliding motion. Because this type of prepared mount is delicate, care must be used to prevent cracks from forming in the loose powder. The method can be used to prepare mounts for the Norelco diffractometer, but cannot be used for mounting samples for the General Electric diffractometer because its mounts must be held on edge in a vertical position.

Rock-Slab and Thin-Section Mounting Techniques

In studying fine-grained rocks there are instances, such as in particle orientation studies, when it is advantageous to x-ray the rock directly. If a rock slab cut to about 5-mm thickness has sufficient strength, it can be used directly as a mount with the x-ray diffractometer. If the rock slab does not have sufficient strength alone, it can be mounted on a glass slide, using any standard thin-section mounting technique, and then it can be ground away to about 1-mm thickness.

CONCLUSION

The techniques described here do not include all the proposed techniques. Rather, the intention has been to include a variety of techniques adequate for the wide assortment of problems encountered in x-ray diffraction studies. The techniques included have been tested for accuracy and, with proper preparation of mounts, the x-ray patterns obtained will be representative of the total sample material. The precision (repeatability) of these techniques can also be excellent when they are used carefully.

REFERENCES

Brindley, G. W., and S. S. Kurtossy, 1961, Quantitative determination of kaolinite by x-ray diffraction, *Am. Mineralogist,* **46,** 1205–1215.

Buerger, M. J., 1962, X-ray crystallography; an introduction to the investigation of crystals by their diffraction of monochromatic x-radiation, John Wiley and Sons, 531 pp.

Bystrom-Asklund, A. M., 1967, Sample cups and a technique for sideward packing of x-ray diffractometer specimens, *Am. Mineralogist,* **51,** 1233–1237.

Gibbs, R. J., 1965, Error due to segregation in quantitative clay mineral x-ray diffraction mounting techniques, *Am. Mineralogist,* **50,** 741–751.

_____ , 1968, Clay mineral mounting techniques for x-ray diffraction analysis, *J. Sed. Pet.,* **38,** 242–244.

Niskanen, E., 1964, Reduction of orientation effects in the quantitative x-ray diffraction analysis of kaolin minerals, *Am. Mineralogist,* **49,** 705–714.

Nuffield, E. W., 1966, X-ray diffraction methods, John Wiley and Sons, 409 pp.

CHAPTER 24

INTERPRETATION OF X-RAY DIFFRACTION DATA

GEORGE M. GRIFFIN

University of South Florida, Tampa, Florida

Interpretation of modern x-ray diffractograms requires several steps during which the nameless electronic peaks of the diffractogram are converted into significant geologic data.[1] The following steps are discussed. Obviously, some of these procedures are optional, whereas others are essential.

Step 1. Measurement of molecular plane repeat distances (*d*-spacings).

Step 2. Identification of mineral species.

Step 3. Quantitative or semiquantitative interpretation of mineral abundance.

Step 4. Measurement of average crystallite size of selected minerals.

---

[1] It is assumed throughout this chapter that the samples have been collected in a meaningful manner, that adequate preparatory procedures have been followed, and that instrumental settings have been optimized for the particular x-ray equipment used.

STEP 1. MEASUREMENT OF MOLECULAR PLANE REPEAT DISTANCES (d-SPACINGS)

The x-ray diffractogram consists of an x-y plot of diffraction *angle* versus intensity of diffracted radiation (Fig. 1). Analysis begins with an accurate measurement of these quantities.

Generally, the recorder will register an unlabeled tick mark on the x-axis of the diagram every degree or half-degree. The identity of at least one of these degree marks must be known; to accomplish this the operator must label the first tick mark on the diagram when a run is started. Later, during the interpretive stage, the exact position of each peak can be measured by reference to the known starting mark. Interpolation between marks can be made with a suitable engineer's scale. (The scale depends on the chart and goniometer speeds. For a presentation of 4°/minute, an engineer's scale divided into 40ths of an inch is convenient.) The values should be listed immediately *below* the appropriate peaks.

Next, the angles just measured (which are in terms of "degrees 2θ") must be converted into molecular plane repeat distances (d-spacings) in Angstrom (Å) units. This conversion is based on the Bragg equation ($n\lambda = 2d\sin\theta$), which relates diffraction angle (θ) to molecular plane repeat distances (d) for any multiple (n) of any x-ray wavelength (λ). However, in actual practice the Bragg calculation is rarely made; prepared tables of 2θ angle versus d-spacing are available for all the common x-ray wavelengths (e.g., Table 1 for copper K_α radiation). Simply enter the table with the measured 2θ angle and read directly[2] the d-spacing in Å. Record these values immediately *over* the appropriate peaks on the diffractograms.

Considerable time can be saved in the measurement and conversion process by using a transparent overlay (preferable of flexible plastic or a film positive) on which are inscribed the most important d-spacings of common sedimentary minerals. The scale will, of course, depend on the available radiation source and recorder, and the mineral x-ray peaks included in the transparency will vary with the problem. In practice, the overlay is positioned over the diffractogram and degree markings on the overlay are aligned with the equivalent degree marks on the pattern.

[2]*Note:* Some tables list values in terms of "degrees θ" instead of "2θ"; for these, halve the 2θ angle before entering the table.

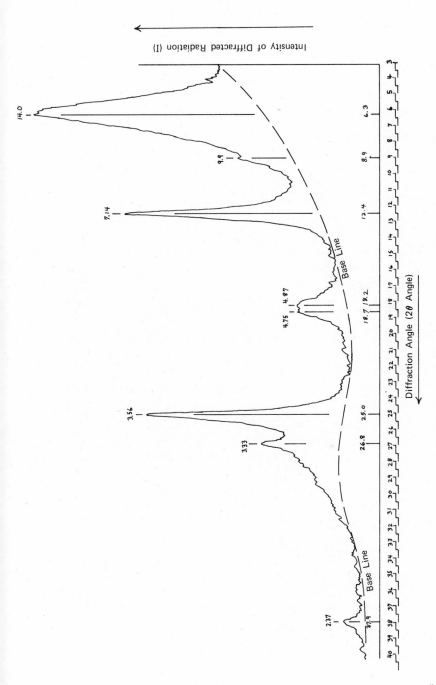

Intensity of Diffracted Radiation (I)

Diffraction Angle (2θ Angle)

Fig. 1 Typical x-ray diffractogram of a sediment clay fraction

TABLE 1 d-Spacings as a function of 2θ angle for copper, K_α radiation ($\lambda = 1.5418$ Å) (after Switzer et al., 1948)

2θ	.0	.1	.2	.3	.4	.5	.6	.7	.8	.9
2	44.171	42.068	40.156	38.410	36.810	35.338	33.979	32.721	31.552	30.464
3	29.449	28.499	27.609	26.773	25.985	25.243	24.542	23.879	23.251	22.655
4	22.089	21.550	21.037	20.548	20.082	19.636	19.209	18.800	18.409	18.034
5	17.673	17.327	16.994	16.673	16.365	16.068	15.781	15.504	15.237	14.979
6	14.730	14.488	14.255	14.029	13.810	13.598	13.392	13.192	12.998	12.810
7	12.628	12.450	12.277	12.109	11.946	11.787	11.632	11.481	11.334	11.191
8	11.051	10.915	10.782	10.652	10.526	10.402	10.281	10.163	10.048	9.9355
9	9.8254	9.7176	9.6122	9.5091	9.4082	9.3093	9.2126	9.1178	9.0250	8.9341
10	8.8450	8.7576	8.6720	8.5880	8.5057	8.4249	8.3456	8.2678	8.1915	8.1166
11	8.0430	7.9708	7.8998	7.8302	7.7617	7.6944	7.6283	7.5634	7.4995	7.4367
12	7.3750	7.3142	7.2545	7.1957	7.1379	7.0810	7.0251	6.9699	6.9157	6.8624
13	6.8098	6.7580	6.7071	6.6569	6.6074	6.5587	6.5107	6.4634	6.4168	6.3708
14	6.3256	6.2809	6.2369	6.1935	6.1507	6.1085	6.0669	6.0259	5.9854	5.9454
15	5.9060	5.8671	5.8288	5.7909	5.7535	5.7166	5.6802	5.6442	5.6088	5.5737
16	5.5391	5.5049	5.4711	5.4378	5.4049	5.3723	5.3402	5.3084	5.2771	5.2461
17	5.2154	5.1852	5.1552	5.1257	5.0964	5.0675	5.0390	5.0107	4.9828	4.9552
18	4.9279	4.9009	4.8742	4.8478	4.8216	4.7958	4.7702	4.7450	4.7199	4.6952
19	4.6707	4.6465	4.6225	4.5988	4.5753	4.5521	4.5291	4.5063	4.4838	4.4615
20	4.4394	4.4175	4.3959	4.3744	4.3532	4.3322	4.3114	4.2908	4.2704	4.2502
21	4.2302	4.2104	4.1907	4.1713	4.1520	4.1329	4.1140	4.0953	4.0767	4.0583
22	4.0401	4.0220	4.0042	3.9864	3.9689	3.9515	3.9342	3.9171	3.9001	3.8833
23	3.8667	3.8502	3.8338	3.8176	3 8015	3.7855	3.7697	3.7540	3.7385	3.7231
24	3.7078	3.6926	3.6776	3.6627	3.6479	3.6332	3.6187	3.6043	3.5900	3.5758
25	3.5617	3.5477	3.5339	3.5201	3.5065	3.4930	3.4796	3.4662	3.4530	3.4399
26	3.4269	3.4140	3.4012	3.3885	3.3759	3.3634	3.3510	3.3386	3.3264	3.3143
27	3.3022	3.2903	3.2784	3.2666	3.2549	3.2433	3.2318	3.2203	3.2090	3.1977
28	3.1865	3.1754	3.1644	3.1534	3.1426	3.1318	3.1210	3.1104	3.0998	3.0893
29	3.0789	3.0685	3.0582	3.0480	3.0379	3.0278	3.0178	3.0079	2.9980	2.9882
30	2.9785	2.9688	2.9592	2.9497	2.9402	2.9308	2.9214	2.9122	2.9029	2.8938
31	2.8847	2.8756	2.8666	2.8577	2.8488	2.8400	2.8312	2.8225	2.8139	2.8053
32	2.7968	2.7883	2.7798	2.7715	2.7631	2.7549	2.7466	2.7385	2.7303	2.7223
33	2.7143	2.7063	2.6984	2.6905	2.6827	2.6749	2.6671	2.6595	2.6518	2.6442
34	2.6367	2.6292	2.6217	2.6143	2.6069	2.5996	2.5923	2.5851	2.5779	2.5707
35	2.5636	2.5565	2.5495	2.5425	2.5355	2.5286	2.5218	2.5149	2.5081	2.5014
36	2.4947	2.4880	2.4813	2.4747	2.4682	2.4616	2.4551	2.4487	2.4422	2.4358
37	2.4295	2.4232	2.4169	2.4106	2.4044	2.3982	2.3921	2.3860	2.3799	2.3738
38	2.3678	2.3618	2.3559	2.3500	2.3441	2.3382	2.3324	2.3266	2.3208	2.3151
39	2.3094	2.3037	2.2981	2.2924	2.2869	2.2813	2.2758	2.2703	2.2648	2.2593
40	2.2539	2.2485	2.2432	2.2378	2.2325	2.2273	2.2220	2.2168	2.2116	2.2064
41	2.2012	2.1961	2.1910	2.1859	2.1809	2.1759	2.1709	2.1659	2.1609	2.1560
42	2.1511	2.1462	2.1414	2.1365	2.1317	2.1270	2.1222	2.1175	2.1127	2.1080
43	2.1034	2.0987	2.0941	2.0895	2.0849	2.0804	2.0758	2.0713	2.0668	2.0623
44	2.0579	2.0534	2.0490	2.0446	2.0402	2.0359	2.0316	2.0273	2.0230	2.0187
45	2.0144	2.0102	2.0060	2.0018	1.9976	1.9935	1.9893	1.9852	1.9811	1.9770
46	1.9729	1.9689	1.9649	1.9609	1.9569	1.9529	1.9489	1.9450	1.9411	1.9372
47	1.9333	1.9294	1.9255	1.9217	1.9179	1.9141	1.9103	1.9065	1.9028	1.8990
48	1.8953	1.8916	1.8879	1.8842	1.8806	1.8769	1.8733	1.8697	1.8661	1.8625
49	1.8589	1.8554	1.8519	1.8483	1.8448	1.8413	1.8378	1.8344	1.8309	1.8275
50	1.8241	1.8207	1.8173	1.8139	1.8105	1.8072	1.8039	1.8005	1.7972	1.7939

544

TABLE 1 d-Spacings as a function of 2θ angle for copper, K_α radiation (λ = 1.5418 Å) (after Switzer et al., 1948) (Continued)

51	1.7906	1.7874	1.7841	1.7809	1.7776	1.7744	1.7712	1.7680	1.7648	1.7617
52	1.7585	1.7554	1.7523	1.7491	1.7460	1.7430	1.7399	1.7368	1.7338	1.7307
53	1.7277	1.7247	1.7217	1.7187	1.7157	1.7127	1.7098	1.7068	1.7039	1.7009
54	1.6980	1.6951	1.6922	1.6894	1.6865	1.6836	1.6808	1.6779	1.6751	1.6723
55	1.6695	1.6667	1.6639	1.6612	1.6584	1.6556	1.6529	1.6502	1.6474	1.6447
56	1.6420	1.6393	1.6367	1.6340	1.6313	1.6287	1.6260	1.6234	1.6208	1.6182
57	1.6156	1.6130	1.6104	1.6078	1.6053	1.6027	1.6002	1.5976	1.5951	1.5926
58	1.5901	1.5876	1.5851	1.5826	1.5801	1.5777	1.5752	1.5728	1.5703	1.5679
59	1.5655	1.5631	1.5607	1.5583	1.5559	1.5535	1.5512	1.5488	1.5465	1.5441
60	1.5418	1.5395	1.5371	1.5348	1.5325	1.5302	1.5279	1.5257	1.5234	1.5211
61	1.5189	1.5166	1.5144	1.5122	1.5099	1.5077	1.5055	1.5033	1.5011	1.4989
62	1.4968	1.4946	1.4924	1.4903	1.4881	1.4860	1.4839	1.4817	1.4796	1.4775
63	1.4754	1.4733	1.4712	1.4691	1.4670	1.4650	1.4629	1.4609	1.4588	1.4568
64	1.4547	1.4527	1.4507	1.4487	1.4467	1.4447	1.4427	1.4407	1.4387	1.4367
65	1.4347	1.4328	1.4308	1.4289	1.4269	1.4250	1.4231	1.4211	1.4192	1.4173
66	1.4154	1.4135	1.4116	1.4097	1.4079	1.4060	1.4041	1.4023	1.4004	1.3985
67	1.3967	1.3949	1.3930	1.3912	1.3894	1.3876	1.3858	1.3840	1.3822	1.3804
68	1.3786	1.3768	1.3750	1.3733	1.3715	1.3697	1.3680	1.3662	1.3645	1.3628
69	1.3610	1.3593	1.3576	1.3559	1.3542	1.3524	1.3507	1.3491	1.3474	1.3457
70	1.3440	1.3423	1.3407	1.3390	1.3373	1.3357	1.3340	1.3324	1.3308	1.3291

Exact positioning of the overlay is made best by using quartz or some other common natural component as an internal standard. For most detrital silicate fractions, the 4.26 Å peak of quartz is a convenient reference point, and the 4.26 Å line of the overlay should be moved carefully into coincidence with this peak. Often the diffractometer will prove to be out of calibration by several tenths of a degree. This miscalibration is not objectionable so long as the overlay method and a natural internal standard are used, but it can be a significant source of error if only direct measurement from the starting mark is used.

The apex of each peak should be labeled with its d-spacing in Angstrom units. The location of all maxima of bimodal peaks should also be indicated as well as the range of broad peaks without well-defined maxima. It is usually helpful to sketch in the approximate shape of each peak that is suspected of contributing to polymodal peaks; in doing this consider both peaks symmetrical.

STEP 2. IDENTIFICATION OF MINERAL SPECIES

Any mineral species can, in theory, occur in sediments or sedimentary rocks; but, in fact, relatively few are at all common in the sediment size-fractions and sedimentary-rock types to which x-ray diffraction is most usefully applied. In general, the dominant minerals which will occur can be predicted fairly well if the general rock type and size fraction are known (Table 2). This prediction is a definite aid in interpreting diffractograms.

If the geographic location and age are also known, an experienced worker can further limit or expand the number of expected principal minerals. Surprise occurrences of unusual minerals are more common in some areas than others. For example, sediments from the Tertiary of California often contain unusual minerals; whereas samples of equivalent age from coastal Texas and Louisiana rarely do.

For interpretation, tables of d-spacings for minerals must be available. The most complete set of these tables is published by the American Society for Testing Materials and is known as the *A.S.T.M. Powder Data File*. In the file all known crystalline materials are arranged systematically according to d-spacings of their three most intense peaks. An index is issued yearly, together with revisions and additions to the basic file. The proper method of using the A.S.T.M. file is described in the preface to the yearly indexes.

The A.S.T.M. file is comprehensive, including data for all known natural and artificial organic and inorganic crystalline materials. However, it does not, for many materials, indicate degree of isomorphism or normal ranges of d-spacings. In practice, the A.S.T.M. file contains too many unusual materials to use on a routine basis, and most workers in sediments find it more convenient to refer to smaller groups of cards or lists extracted from the A.S.T.M. file and other primary sources. The writer has found the lists included in Brindley (1951); Brown (1961); Deer et al. (1963); and Grim (1968) to be very useful. Data from several of these sources are included in Table 3—a list of d-spacings for 27 common minerals in sediments and sedimentary rocks.

In Step 1, d-spacings were listed above appropriate peaks on the diffractograms. Each of these peaks must now be assigned systematically to a specific crystalline material which can be either organic or inorganic. In this assignment it is best to proceed from the common to the uncommon. That is, first assign peaks to the most common mineral groups (such as quartz in a shale analysis or calcite in a limestone analysis), then assign the remaining peaks to the less common materials.

**TABLE 2   Minerals often encountered in the size fractions
and rock types commonly subjected to x-ray
diffraction and analysis**

A. Clastic Rocks

   1. Coarse Fraction ($>$ 0.002 mm)

     Quartz
     Potassium feldspars
     Plagioclase feldspars
     Muscovite
     Zeolites

   2. Fine Fraction ($<$ 0.002 mm)

     Quartz
     Montmorillonite (Smectite) group
     Kaolinite (Kandite) group
     Illite group
     Chlorite group
     Vermiculite group
     Potassium feldspars
     Plagioclase feldspars
     Goethite
     Gibbsite
     Gypsum
     Pyrite
     Calcite
     Siderite
     Opal (as disordered cristobalite)
     Zeolites

B. Carbonate Rocks

   1. Bulk (ground)

     Calcite (including high-Mg type)
     Dolomite
     Aragonite
     Apatite
     Quartz

   2. Insoluble Residue (after HCl)

     Quartz
     Muscovite
     Chlorite group
     Montmorillonite (Smectite) group
     Kaolinite (Kandite) group
     Illite group
     Opal (as disordered cristobalite)

C. Carbonaceous Organic Rocks, Pore Fillings, etc.

   1. Humic Coal Insoluble Residues (after HF)

     Graphite (disordered and expanded; crystallites with relatively large dimensions parallel to the *ab* plane)

   2. Asphaltic Insoluble Residues (after HF)

     Graphite (disordered and expanded; crystallites with relatively large dimensions perpendicular to the *ab* plane)

D. Evaporites

   1. Bulk (ground)

     Halite
     Anhydrite
     Gypsum

   2. Insoluble Residue (after H_2O)

     Gypsum
     Anhydrite
     Clay minerals of all types

TABLE 3 d-spacings, intensities, and hkl of common sedimentary minerals.

1			2			3			4			5			6			7		
Quartz, Low			Orthoclase			Plagioclase, Albite (low)			Analcite			Chlorite, Monoclinic			Muscovite, 2M$_1$			Biotite		
d(Å)	I	hkl	d(Å)	I	hkl	d(Å)	I	hkl	d(Å)	I	hkl	d(Å)	I	hkl	d(Å)	I	hkl	d(Å)	I	hkl
4.26	35	100	6.66	10	110	6.39	20	001	6.87	<10	200	14.2	80	001	9.99	S	002	10.1	100	001
3.34	100	101	6.52	20	020	5.94	2	11̄1	5.61	80	211	7.12	100	002	4.98	M	004	4.59	20	110 / 020
2.46	12	110	5.87	20	11̄1	5.59	2	1̄11	4.86	40	220	4.75	80	003	4.47	VS	110 / 1̄11			
2.28	12	102	4.24	60	2̄01	4.03	16	20̄1	3.67	20	321	3.56	100	004	4.29	W	111	3.37	100	003
2.24	6	111	3.94	30	111	3.86	8	11̄1	3.43	100	400	2.85	40	005	4.11	W	022	3.16	20	112
2.13	9	200	3.87	10	200	3.78	25	111	2.93	80	332	2.58	30	131 / 2̄02	3.95	VW	112	2.92	20	11̄3
1.980	6	201	3.79	100	130	3.68	20	130	2.80	20	422	2.55	50	132 / 2̄01	3.87	M	11̄3	2.66	80	201̄ / 130
1.817	17	112	3.62	20	13̄1	3.66	16	13̄1 / 1̄30	2.69	50	431 / 510	2.44	40	132 / 2̄03	3.72	M	023	2.52	40	004 / 113
1.801	<1	003	3.56	20	22̄1				2.51	50	521				3.55	VW	113			
1.672	7	202	3.46	60	11̄2	3.51	10	11̄2	2.43	30	440	2.38	20	133 / 2̄02	3.48	M	114̄	2.45	80	201
1.659	3	103	3.33	100	220	3.48	2	22̄1	2.23	40	611 / 532				3.32	VS	024 / 006	2.28	20	040 / 132
1.608	<1	210	3.28	70	20̄2	3.80	8	1̄12				2.27	30	133 / 2̄04	3.20	MS	114			
1.541	15	211	3.26	50	040	3.20	100	002	2.17	<10	620				3.1	VW	115̄	2.18	80	
1.453	3	113	3.22	90	002	3.151	10	2̄20	2.12	<10	541	2.04	20	007	2.98	S	025	2.00	80	
			3.00	60	131	2.96	10	13̄1	2.02	10	631	2.01	40	135 / 2̄04	2.86	M	115	1.91	20	
			2.93	20	22̄2	2.93	16	02̄2	1.94	<10	543 / 550 / 710				2.78	M	116̄	1.75	20	
			2.91	40	041	2.87	8	131				1.891	20	135 / 2̄06	2.59	W	131̄ / 200	1.67	80	
			See note under column 3.			2.84	2	132	1.90	50	640	1.833	20	136 / 2̄05	2.56	VS	202̄ / 131	1.54	80	
						2.79	2	022	1.87	40	633 / 721 / 552	1.732	10	136 / 2̄07	2.49	W	008	1.47	20	
						2.64	6	1̄32	1.83	<10	642	1.672	10	137 / 2̄06	2.46	W	202 / 133̄	1.43	20	
						2.56	8	24̄1	1.74	60	732 / 651	1.577	20	137 / 2̄08	2.39 / 2.38	M	204̄ / 133	1.36	60	
						2.54	2	31̄2	1.72	30	800	1.541	60	060 / 3̄31	2.25	W	204 / 135̄	1.33	40	
						2.51	2	11̄2	1.69	40	741 / 811 / 554	1.507	20	062 / 331	2.19	W	22̄3	1.31	40	
						2.50	6	22̄1	1.66	10	820 / 644	1.429	10	0,0,10	2.14 / 2.13	M	206̄ / 135			
						2.46	6	221	1.62	20	822 / 660				2.05	VW	044			
						2.44	4	24̄1	1.60	30	831 / 743 / 750				1.99	S	00 10			
						2.43	2	15̄1	1.498	20	842				1.95	W	206 / 137̄			
						2.41	2	240	1.480	20	761 / 921 / 655				1.83	VW	?			
						2.39	4	310							1.76	W	138			
						2.32	4	33̄1							1.65	W	2,0,1̄0			
						2.28	2	1̄13							1.64	M	139			
						2.19	4	042							1.504	S	060 / 33̄1			
						2.13	8	060												
						2.12	6	151												
						2.08	2	24̄1												
						2.04	2	241												
						2.00	2	202												
						1.980	4	061												
						1.927	2	42̄1												
						1.889	8	222												

Plus 6 lines to 1.785

Note: The ~4.03 line distinguishes plagioclase from K-feldspars, but specific identification of plagioclase species is difficult.

Note: Basal peak intensities vary with the Al/Fe^{+++} ratio.

Note: Basal peak intensities vary with the Mg/Fe^{++} ratio.

TABLE 3 d-spacings, intensities, and hkl of common sedimentary minerals.
(Continued)

8			9			10			11			12			13			14		
Vermiculite			Montmorillonite, Wyo. Bent.			Glauconite, Well Ordered			Talc			Kaolinite			Hematite			Goethite		
d(Å)	I	hkl	d(Å)	I	hkl	d(Å)	I	hkl	d(Å)	I	hkl	d(Å)	I	hkl	d(Å)	I	hkl	d(Å)	I	hkl
14.4	VVS	002	~15	100	00l / 001	10.1	100	001	9.3	48	002	7.16	100⁺	001	3.67	35	102	4.98	15	020
7.20	VW	004	~5	80	003	4.98	10	002	4.58	64	020 / 111	4.46	40	020	2.69	100	104	4.18	100	110
4.79	VW	006	~3	100	005 (hk)	4.53	80	020	4.11	5	113	4.36	50	1$\bar{1}$0	2.51	75	110	3.38	10	120
4.60	S	02l / 11l	4.61	100	11 / 02	4.35	20	11$\bar{1}$	3.13	40	006	4.18	50	1$\bar{1}$1	2.20	25	113	2.69	30	130
3.59	M	008				4.12	10	021	2.62	32	130	4.13	30	1$\bar{1}$1	2.07	3	202	2.58	8	021
2.87	M	00 10 / 130	2.56	80	13 / 20	3.63	40	11$\bar{2}$	2.49	100	132 / $\bar{2}$04	3.85	40	02$\bar{1}$	1.838	30	204	2.52	3	101
266	MW	200 / 202	2.22	30	22 / 04	3.33	60	003 / 022	2.22	14	134 / $\bar{2}$06	3.74	20	021	1.692	45	116	2.49	15	040
		132 / $\bar{2}$04	1.692	60	31 / 15 / 24	3.09	40	112	2.10	8	$\bar{1}$36 / $\bar{2}$04	3.57	100⁺	002	1.635	2	121	2.45	25	111
260	M	134 / $\bar{2}$02	1.492	100	33 / 06	2.89	5	113	1.95	3	136 / $\bar{2}$08	3.37	40	111	1.597	15	108	2.25	10	121
2.55	MW	00 12	1.289	60	26 / 40	2.67	10	023	1.87	3	0,0,10	3.14	30	11$\bar{2}$	1.484	20	214	2.19	20	140
2.39	MS	136 / 204				2.59	100	130 / 13$\bar{1}$ / 200	1.72	11	$\bar{3}$11 / $\bar{3}$13	3.10	30	11$\bar{2}$	1.452	25	300	2.01	2	131
2.27	VVW	136 / 208				2.40	60	13$\bar{2}$ / 201	1.68	5	138 / 2,0,1$\bar{0}$	2.75	30	022	1.346	6	208	1.920	6	041
2.21	VW	138 / 206				2.26	20	040 / 22$\bar{1}$	1.60	2	208	2.56	60	$\bar{1}$30 / $\bar{2}$01 / 130	1.310	10	1,0,10 / 119	1.799	7	211
2.08	W	138 / 20 1$\bar{0}$				2.21	10	220 / 041	1.56	2	0,0,12	2.53	40	13$\bar{1}$ / 112 / $\bar{1}$31	1.257	8	220	1.770	2	141
2.05	VW	00 14				2.15	20	13$\bar{3}$ / 202	1.53	64	060 / $\bar{3}$32	2.49	80	$\bar{2}$00 / 112				1.721	20	221
2.01	VW	208				1.994	20B	005	1.46	2		2.38	60	003				1.694	10	240
1.835	VVW	1,3,1$\bar{2}$ / 2,0,1$\bar{0}$				1.817	5	224	1.40	5		2.39	90	$\bar{2}$0$\bar{2}$ / $\bar{1}$31				1.661	4	060
1.748	W	2,0,1$\bar{4}$				1.715	10	31$\bar{1}$ / 24$\bar{1}$ / 240	1.32	11		2.29	80	$\bar{1}$31 / 131				1.606	6	231
1.677	MW	1,3,1$\bar{4}$ / 2,0,1$\bar{2}$				1.66	30B	31$\bar{2}$ / 310 / 241	1.30	13		2.25	20	132 / 040				1.564	15	160 / 151
1.574	VVW	1,3,1$\bar{4}$ / 2,0,1$\bar{6}$				1.511	60	060 / 33$\bar{1}$	1.27	3		2.19	30	2$\bar{2}$0				1.509	10	250
		060 / 1,3,1$\bar{6}$ / 2,0,1$\bar{4}$				1.495	10	330				2.13	30	02$\bar{3}$				1.467	4	320
1.537	MS	330 / 33$\bar{2}$ / 33$\bar{4}$				1.307	30	260 / 400 / 170				2.06	20	2$\bar{2}$2				1.453	10	061
1.508	VVW	332 / 33$\bar{6}$				1.258	10	350 / 420				1.989	60	$\bar{2}$03 / $\bar{1}$32						
												1.939	40	132						
												1.896	30	13$\bar{3}$						
												1.869	20	042						
												1.839	40	$\bar{1}$33 / $\bar{2}$02 / 22$\bar{3}$						
												1.809	20	114						
												1.781	40	004						
												1.707	20	222						
												1.685	20	24$\bar{1}$						
												1.662	70	$\bar{2}$04 / 133						
												1.619	60	133						
												1.584	40	13$\bar{4}$						
												1.542	50B	134						
												1.489	80	060 / $\bar{3}$31 / 33$\bar{1}$						

Notes (Column 9): 00l Positions vary with moisture and interlayer cation population. Most peaks are broad, suggesting mixed hydration stages. Ethylene glycol solvation must be used for certain identification. The method of Brunton (1955) is suggested (1 hr over glycol heated to 60°C in a closed chamber).

Note (Column 10): 4.98 Å Peak is obtained only on an oriented sample; 10 is a typical intensity value.

15			16			17			18			19			20			21		
Gibbsite			Phillipsite			Heulandite			Cristobalite, Low			Aragonite			Calcite			Dolomite		
d(Å)	I	hkl	d(Å)	I	hkl	d(Å)	I	hkl	d(Å)	I	hkl	d(Å)	I	hkl	d(Å)	I	hkl	d(Å)	I	hkl
4.85	100		7.64	100		8.90	100		4.04	100	101	4.21	2	110	3.86	12	102	4.03	3	101
4.37	40		6.91	100		7.94	20		3.14	12	111	3.40	100	111	3.04	100	104	3.69	5	102
4.31	20		6.34	20		6.80	10		2.85	14	102	3.27	52	021	2.85	3	006	2.89	100	104
3.35	6		5.24	50		6.63	10		2.49	18	200	2.87	4	002	2.50	14	110	2.67	10	006
3.31	10		4.91	50		5.92	10		2.47	6	112	2.73	9	121	2.29	18	113	2.54	8	105
3.18	7		4.56	20		5.58	10		2.34	<1	201	2.70	46	012	2.10	18	202	2.41	10	110
3.10	4		4.25	70		5.24	10		2.12	4	211	2.48	33	200	1.927	5	204	2.19	30	113
2.45	15		4.07	70		5.09	10		2.02	3	202	2.41	14	031	1.913	17	108	2.07	5	201
2.42	4		3.54	50		4.89	10		1.932	4	113	2.37	38	112	1.875	17	116	2.02	15	202
2.38	25		3.18	100		4.69	20		1.874	4	212	2.34	31	130	1.626	4	211	1.848	5	204
2.29	4		2.94	50		4.45	20		1.756	1	220	2.33	6	022	1.604	8	212	1.804	20	108
2.24	6		2.71	70		4.36	10		1.736	1	004	2.19	11	211	1.587	2	1,0,10	1.786 }	30 {	116
2.17	8		2.52	50		3.97	20		1.692	3	203	2.11	23	220	1.525	5	214	1.781 }		009
2.08	1		2.40	50D		3.89	30		1.642	1	104	1.977	65	221	1.518	4	208	1.567	8	211
2.04	15		2.16	20		3.83	10		1.612	5	301	1.882	32	041	1.510	3	119	1.545	10	212
2.02	1		2.07	20		3.71	10		1.604	2	213	1.877	25	202	1.473	2	215	1.496	2	1,0,10
1.991	8		1.97	50		3.56	10		1.574	1 {	310 / 222 }	1.814	23 {	132 / 230 }				1.465	5	214
1.916	6		1.91	20		3.47	10		1.535	2	311	1.759	4	141				1.445	4	208
1.801	10		1.84	20		3.40	20		1.495	3	302	1.742	25	113				1.431	10	119
1.750	9		1.78	50		3.12	10		1.432	2	312	1.728	15	231				1.413	4	215
1.685	7		1.72	50		3.03	10													
1.655	2		1.67	20D		2.97	40		*Note:* Opal											
1.590	2		1.61	20		2.80	10		produces a											
1.574	1		1.55	20		2.72	10		disordered											
1.555	1		1.49	20		2.67	10		cristobalite											
1.533	1		1.38	50		2.48	10		pattern;											
1.485	1		1.34	50		2.43	10		otherwise											
1.477	1		1.28	50D		2.35	10		rare in											
1.457	8					2.28	20		sediments.											
1.440	4		*Note:*																	
1.411	5		D = doublet																	
1.402	4																			

Note: Additional zeolites will be found in Deer et al. (1963) **4**, pp. 408–417.

Zeolites are usually characterized by sharp, intense peaks in the low angle region.

TABLE 3 d-spacings, intensities, and *hkl* of common sedimentary minerals.
(Continued)

22			23			24			25			26			27		
Siderite			Apatite, Hydroxy-			Gypsum			Anhydrite			Halite			Pyrite		
d(Å)	I	*hkl*	d(Å)	I	*hkl*	d(Å)	I	*hkl*	d(Å)	I	*hkl*	d(Å)	I	*hkl*	d(Å)	I	*hkl*
3.59	25	102	8.17	11	100	7.56	100	020	3.87	6	111	3.26	13	111	3.13	36	111
2.79	100	104	5.26	5	101	4.27	51	12$\bar{1}$	3.50	100	{020	2.82	100	200	2.71	84	200
2.56	2	006	4.72	3	110	3.79	21	{031			{002	1.994	55	220	2.42	66	210
2.34	15	110	4.07	9	200			{040	3.12	3	200	1.701	2	311	2.21	52	211
2.13	20	113	3.88	9	111	3.16	3	11$\bar{2}$	2.85	33	210	1.628	15	222	1.92	40	220
1.962	15	201	3.51	1	201	3.06	57	14$\bar{1}$	2.80	4	121	1.410	6	400	1.633	100	311
1.794	10	204	3.44	40	002	2.87	27	002	2.47	8	022	1.294	1	331	1.564	14	222
1.736	20	108	3.17	11	102	2.79	5	21$\bar{1}$	2.33	22	{202	1.261	11	420	1.503	20	230
1.730	20	116	3.08	17	210	2.68	28	{022			{220				1.445	24	321
1.526	5	121	2.81	100	211			{051	2.21	20	212				1.243	12	331
1.504	10	212	2.78	60	112	2.59	4	{150	2.18	8	103				1.211	14	420
1.438	3	1,0,10	2.72	60	300			{20$\bar{2}$	2.09	9	113				1.182	7	421
			2.63	25	202	2.53	<1	060	1.993	6	301				1.155	6	332
			2.53	5	301	2.50	6	200	1.938	4	222				1.106	6	422
			2.30	7	212	2.45	4	22$\bar{2}$	1.869	15	230				1.043	27	511
			2.26	20	310	2.40	4	14$\bar{1}$	1.852	4	123						
			2.23	1	221	2.22	6	15$\bar{2}$	1.749	11	004						
			2.15	9	311	2.14	1	24$\bar{2}$	1.748	10	040						
			2.13	3	302	2.08	10	12$\bar{3}$	1.648	14	232						
			2.07	7	113	2.07	8	{112	1.594	3	1̇33						
			2.04	1	400			{25$\bar{1}$	1.564	5	{024						
			2.00	5	203	1.990	4	170			{042						
			1.943	30	222	1.953	2	211	1.525	4	{204						
			1.890	15	312			{080			{240						
			1.871	5	320	1.898	16	{062	1.515	1	{313						
			1.841	40	213			{013			{331						
			1.806	20	321	1.879	10	14$\bar{3}$	1.490	5	214						
			1.780	11	410	1.864	4	31$\bar{2}$	1.424	3	{402						
			1.754	15	{402	1.843	1	231			{420						
					{303	1.812	10	26$\bar{2}$	1.418	1	323						
			1.722	20	{004				1.398	3	242						
					{411												
			1.684	3	104												
			1.644	9	{322												
					{223												
			1.611	7	313												
			1.587	3	{501												
					{204												

Sources of Data:
Brown (1961), Minerals No. 1, 2, 5, 6, 8, 9, 12, 13, 14, 15, 18, 19, 20, 21, 22, 23, 24, 25, 27.
Deer, Howie, and Zussman (1963) 4, Minerals No. 4, 16, 17.
A.S.T.M. file, Minerals No. 3, 7, 10, 11, 26.
Some *d*-spacings have been rounded off and, where possible, intensities restated on a 0 to 100 scale.

Using the d-spacing tables, label each assignable peak with a code letter (Q = quartz, etc.). In complex mixtures it is helpful to shade the peaks lightly with a different color pattern for each mineral. Compare the list of relative peak intensities in the table with the peak heights on the diffractogram. If any peak appears relatively more intense than the d-spacing table indicates, or a peak exhibits an asymmetrical or polymodal character, two or more minerals may be contributing to it.

Enhanced intensities can also be caused by either accidental or intentional preferential orientation of mineral grains in the specimen mount. If preferential orientation occurs, a whole series of peaks based on the same lattice planes will be enhanced: for example, the $(00l)$ series in micas and clay minerals.

It is well to make a categorical list, on the diffractogram, of the major, minor, and questionable minerals encountered. If important to the study, means should be devised for identifying the questionable species more positively (e.g., heating, ethylene glycol solvation, magnetic separation, acid leaching, DTA, chemical analysis, and electron microscopy).

Following assignment of peaks, there may be a residuum of completely unknown and unassigned peaks. A list of d-spacings and relative intensities for unidentified peaks should be made, and the most likely laboratory contaminants and possible accessory minerals should be checked. One should consider the possibility that extraneous crystalline materials may have been manufactured inadvertently during sample preparation (e.g., calcium acetate during preparation of insoluble residues of limestone with acetic acid, or calcium fluoride compounds produced by solution of silicates for concentration of organic compounds).

STEP 3. QUANTITATIVE OR SEMIQUANTITATIVE INTERPRETATION OF MINERAL ABUNDANCE

The geologic uses of x-ray diffraction data are so varied that it is impossible to present a single quantitative analysis system that can be applied to all problems. Specific procedures must depend on the material examined and the type of geologic results desired. Several useful quantitative and semiquantitative procedures are outlined below and the suggested uses are indicated.

A. Typical polymineralic clay fraction extracted from silicate clastics and containing an assemblage of common clay minerals (kaolinite, chlorite, illite, and expandable montmorillonite, either as a distinct mineral or as part of a mixed-layer complex).

1. Semiquantitative analysis by peak-height ratios

For most stratigraphic problems involving clay minerals the principal interest is in fluctuations in the relative proportions of two or more end-member clay mineral populations. For example, in Tertiary and Quaternary sediments of the Gulf Coast region, relative proportions of end-member suites dominated, respectively, by montmorillonite and kaolinite are sensitive to source area fluctuations and thus to paleogeographic changes in distribution factors (Griffin, 1962, 1964). Changes in relative proportions of this type are easily measured by peak-height ratios. In the instances cited the 15 Å/7 Å peak-height ratio would be used; these peaks represent the (001) peaks of montmorillonite and kaolinite.

Peaks are measured relative to a base line that must be carefully drawn by hand before measurement. A typical base line for the silicate clay fraction is *not* a straight line. Typically, it rises sharply toward the low angle region of the pattern and also forms a very broad swell beneath the 3.35 Å (26.6° 2θ) quartz peak (Fig. 1). In general, the base line is drawn as a smooth curve along the base of the peaks. Because its exact shape is somewhat variable, the base line is a nonsystematic source of error in quantitative work based on peak heights or peak areas, and careful attention should be given to its placement.

Once the baseline has been drawn, the vertical distance from the base line to the apex of the critical peaks must be measured. This is done most accurately by constructing vertical lines from each critical apex to the base line. A simple division then yields the peak-height ratio. This value should be considered only as a *function* of the relative quantities present. If *absolute* quantities are required, more elaborate procedures are necessary.

It is theoretically more accurate to compare peak areas (or, better still, actual counts per second) rather than peak heights. However, these parameters are considerably more difficult to measure, and the geologic usefulness of the data is usually not increased in proportion to the required additional effort.

Perhaps the easiest method for making peak area measurements is to overlay the diffractogram on a grid-pattern background and to count whole and partial squares. Alternatively, the peaks can be traced onto paper of uniform thickness, cut out with scissors, and weighed on an analytical balance. The ratio of weights will approximate closely the ratio of peak areas. This method is quite accurate if sufficient care is taken in making the tracings and cuts and in drawing the base line.

As a second alternative, a polar planimeter can be used to trace the peak outlines and convert them to peak areas. This is a relatively rapid method, but probably is less precise than the weighing method described above.

In any event, peak area measurements are still only functions of the absolute values. Complex correction factors, often somewhat questionable themselves, must be applied to convert to absolute values.

2. Quantitative Analysis

No universally accepted system has been developed for the routine quantitative analysis of clay-mineral mixtures. Chemical and structural variability of the possible end members make such an ideal system unattainable.

For special situations it may be possible to use the precise and accurate methods of Gibbs (1967), which involves end members extracted from the samples themselves. However, for much routine work the necessary extractions will prove too time consuming. The system proposed below was developed in essence by the writer's former colleagues in the Shell Development Company laboratory (D. B. Shaw, R. G. Stevenson, C. E. Weaver, and W. F. Bradley). It has been used routinely for the analysis of thousands of samples, and if used exclusively it provides values that are internally consistent. However, the values are not strictly comparable to values derived by the use of other systems. The following assumptions inherent in its use are well to keep in mind: (a) that the reported clay minerals comprise 100% of the sample, whereas in many cases, there are other minerals present as well as amorphous material; (b) that the refracting ability of the clay minerals, which is generally dependent on the composition, polytype, and degree of crystallinity, is consistent; and (c) that there is a 1:1 linear relationship between the ratio of the 3.58 Å kaolinite peak to the 3.54 Å chlorite peak. If these assumptions are acceptable, proceed as in the following.

Diffractograms of the sample after ethylene glycol (EG) treatment and immediately after 180°C heating for 1 hour must be available in addition to the diffractogram of the untreated sample. In the formulas that follow, the symbols K, C, I, and M refer to kaolinite, chlorite, illite, and expandable-layer clay minerals, respectively. The symbol $h7Å$ (180°) refers to the height of the 7 Å peak on the 180°C pattern; other peak heights are noted by similar subscripts.

(a) On the 180°C pattern, measure the heights of the 7 and 10 Å peaks. Calculate the total percent kaolinite and/or chlorite as follows:

$$\% \text{ K} + \text{C} = \frac{\dfrac{h7Å(180°)}{2.5}}{\dfrac{h7Å(180°)}{2.5} + h10Å(180°)} \times 100$$

(b) On the 180°C pattern measure the heights of the 3.54 and 3.59 Å peaks. These heights are used to apportion the 7 Å peak into percent kaolinite and percent chlorite as follows:

$$\% \text{ K} = \frac{h3.59Å(180°)}{h3.59Å(180°) + h3.54Å(180°)} \times \% \text{ K} + \text{C}$$

$$\% \text{ C} = (\% \text{ K} + \text{C}) - \% \text{ K}$$

(c) Calculate the total percent illite and/or montmorillonite as follows, using peak heights from the 180°C pattern:

$$\% \text{ I} + \text{M} = \frac{h10Å(180°)}{\dfrac{h7Å(180°)}{2.5} + h10Å(180°)} \times 100$$

(d) Measure the 7 and 10 Å peak heights on the ethylene glycol pattern. Use these measurements and the previous ones on the 180°C pattern to calculate percentages of illite and montmorillonite as follows:

$$\% \, I = \frac{{}^h 10\text{Å (EG)} \left(\dfrac{{}^h 7\text{Å } (180°)}{{}^h 7\text{Å (EG)}} \right)}{{}^h 10\text{Å} (180°)} \times \% \, I + M$$

$$\% \, M = (\% \, I + M) - \% \, I$$

Sample Calculation

This is a sample calculation for the clay fraction of a typical shale with a clay mineral suite composed of kaolinite, chlorite, illite, and an illite/montmorillonite mixed-layer complex. Measurement indicates the following peak heights in millimeters:

$${}^h 7\text{Å } (180°) \quad = 120 \qquad\qquad {}^h 3.59\text{Å } (180°) = \quad 81$$

$${}^h 10\text{Å } (180°) \quad = \quad 65 \qquad\qquad {}^h 7\text{Å (EG)} \quad = 115$$

$${}^h 3.54\text{Å } (180°) = \quad 31 \qquad\qquad {}^h 10\text{Å (EG)} \quad = \quad 35$$

$$\% \, K + C = \frac{\dfrac{12.0}{2.5}}{\dfrac{120}{2.5} + 65} \times 100 = 42.5\%$$

$$\% \, K = \frac{81}{81 + 31} \times 42.5 = 30.7\%$$

$$\% \, C = 42.5 - 30.7 = 11.8\%$$

$$\% \, I + M = \frac{65}{\dfrac{120}{2.5} + 65} \times 100 = 57.5\%$$

$$\% \, I = \frac{35 \left(\dfrac{120}{115} \right)}{65} \times 57.5 = 32.3\%$$

$$\% \, M = 57.5 - 32.3 = 25.2\%$$

Therefore:

kaolinite	=	30.7%
chlorite	=	11.8%
illite	=	32.3%
expandable (Mont.)	=	25.2%

$$\overline{}$$
$$100.0\%$$

B. Percentage aragonite and percentage calcite in rocks composed only of these two minerals.

1. Prepare a cavity mount of the finely crushed, sieved (320 mesh), and split sample. It is important that grinding not be severe or undesirable mineralogical transformation may occur, such as the alteration of calcite to aragonite.[3]

2. Mount in diffractometer and scan from 26.3 to 30.5° 2θ. Select a sensitivity that will retain the entire peaks on the chart.

3. Sketch in the background curve; in this spectral region it will be a nearly straight line connecting valleys between peaks.

4. Measure the maximum heights above background of the following peaks:

	hkl	$d(\text{Å})$	2θ (Cu $K\alpha$)
calcite (normal)	(104)	3.03	29.5
calcite (magnesian)	(104)	3.03 to 2.98[4]	29.48 to 29.98[4]
aragonite	(111)	3.40	26.2

5. Determine the ratio:

$$\frac{\text{height aragonite peak}}{\text{height calcite peak}}$$

[3]Burns and Bredig (1956) reported that after grinding calcite for 30 minutes at room temperature in a mechanical mortar, an x-ray pattern indicated a small aragonite peak; after 38.3 hours, aragonite had become the principal ingredient. Thus, although some grinding is obviously necessary, it should be kept to a minimum and should be standardized for large groups of samples.

[4]Varies with magnesium content (Chave, 1952); see Fig. 4.

6. Refer to Fig. 2 and read percent aragonite; calcite is obtained by difference. The character of the working curve will vary slightly depending on the purity and crystallinity of the particular end members used in its construction. Therefore, in critical applications, such as with highly magnesian calcites, it may be necessary to construct a working curve using end member minerals similar in perfection to those actually in the unknown samples.

C. Percentage dolomite and percentage calcite in rocks composed only of these two minerals.

1. Prepare a cavity mount of the finely crushed, sieved (320 mesh), and split sample. Grinding should be minimal to avoid mineral alterations such as the transformation of calcite to aragonite.[3]

2. Mount in diffractometer and scan from 28.5 to 32.0° 2θ using a sensitivity that will retain the entire peaks on the chart.

3. Sketch in the background curve; in this spectral region it will be a nearly straight line connecting the valleys between peaks.

4. Measure the maximum heights of the calcite and dolomite peaks at the following approximate locations:

	hkl	$d(\text{Å})$	2θ (Cu$K\alpha$)
calcite (normal)	(104)	3.03	29.48
calcite (magnesian)	(104)	3.03 to 2.98[4]	29.48 to 29.98[4]
dolomite	(104)	2.88	31.05

5. Determine the ratio:

$$\frac{\text{height dolomite peak}}{\text{height calcite peak}}$$

6. Refer to Fig. 3 and read percent dolomite; calcite is obtained by difference. The character of the working curve will vary slightly depending on the purity and crystallinity of the particular end members used in its construction. Therefore in critical applications it may be necessary to construct a working curve using end-member minerals similar in crystallinity and atomic ratios to those actually in the unknown samples.

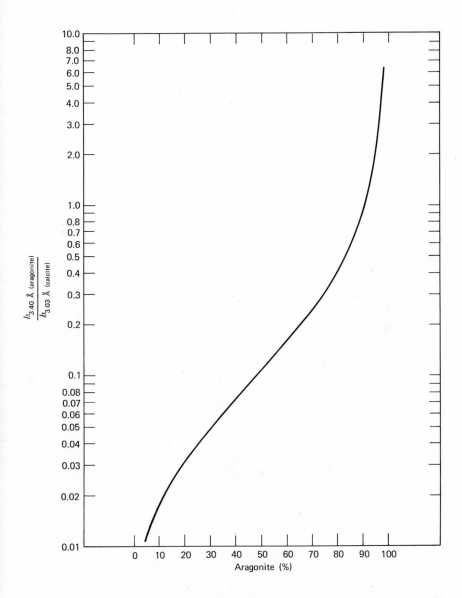

Fig. 2 Working curve for determination of percent aragonite in a sample composed entirely of aragonite and calcite

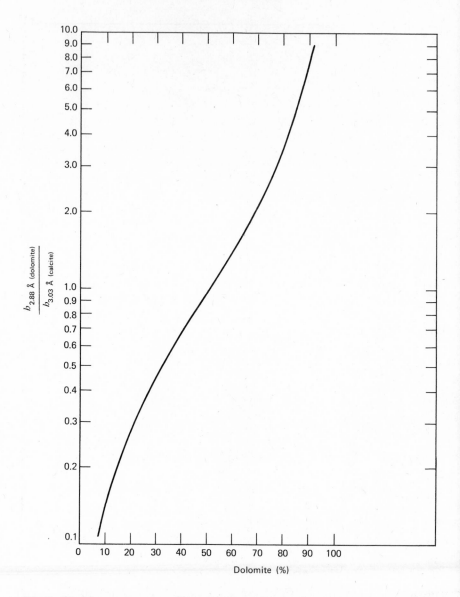

Fig. 3 Working curve for determination of percent dolomite in a sample composed entirely of dolomite and calcite

560

E. Percent $MgCO_3$ incorporated into calcite $CaCO_3$ structures (i.e., percent $MgCO_3$ in high-magnesian calcite).

The following procedure was in part suggested to the writer by K. S. Deffeyes of Princeton University and is based on the fundamental work of K. E. Chave (1952). When used in conjunction with Fig. 4 (from Chave, 1952), it provides a value for the percentage of $MgCO_3$ incorporated diadochically into the calcite-type $CaCO_3$ structure.

1. Select a representative sample of the unknown; mix this well with an equal weight of standard reagent grade CaF_2 and grind the mixture thoroughly but not excessively.[3]

2. Extract a representative split of the ground mixture and pack into a cavity mount.

3. Place the mount in the diffractometer and scan the spectral range from 27.5 to 30.5° 2θ (CuK_α). The optimum instrumental constants will vary with the equipment available. The writer has used the following settings with the Norelco diffractometer equipped with a geiger counter:

radiation: $CuK\alpha$
scan speed: 0.25° 2θ/minute   = record: 0.25°/in.
paper speed: 1.0   in./minute
sensitivity: 2-0.6-4
slits (tube to counter):   1° — 0.006 ——— 0.003

In any event, the settings should allow the critical peaks to remain on scale and provide for a high degree of peak resolution.

4. Locate the center of the (111) CaF_2 standard peak at 28.3° 2θ (CuK_α) and draw a vertical line at this point. Adjust the horizontal scale of the pattern to conform with the standard 28.3° peak; this can be done by sliding an appropriate engineer's scale along the base.

5. Construct perpendicular lines through the calcite peaks located in the 29.5 to 30.5° 2θ range. Determine their exact angular location by direct measurement from the 28.3° CaF_2 standard peak. Write these values immediately below each peak.

6. Convert the angular peak measurement to d-spacings and write the values above the peaks. These values will be between 3.04 and 2.98 Å.

7. Enter Fig. 4 (from Chave, 1952) with the d-spacing of each peak and read the weight percent $MgCO_3$ included in the calcite structure. There will often be more than one calcite peak in the 29.5 to 30.5° range, each representing a different proportion of $MgCO_3$. If the peaks overlap considerably, it is well to sketch in the entire outline of each peak before measurement of angular location. To do this, assume that the "hidden" peak flank will be a mirror image of the exposed flank and that arithmetic additive affects occur in the region of overlap.

8. If desired, weight percent $MgCO_3$, obtained from Fig. 4, can be converted into number of atoms as follows:

Mg atoms per 100 Ca atoms = 1.185 $\times$ weight % $MgCO_3$.

D. Percentages of dolomite, calcite, and quartz in carbonate rocks (After Raish, 1964, slightly modified).

1. Grind sample thoroughly but not excessively.[3] Sieve (320 mesh) and extract a representative 4.0 gm sample of the unknown.

2. Mix 0.4 gm of thoroughly ground and sieved (320 mesh) pure fluorite (CaF_2) standard with the unknown and mix thoroughly.

3. Prepare a cavity mount and insert into the diffractometer.

4. Intensity of diffracted radiation must now be measured for one characteristic peak from each mineral and from the standard. The method of measurement will depend on the equipment available; either direct counting of diffraction pulses or peak height measurements can be used (a *or* b below).

(a) Counting of diffraction pulses, if possible, will yield the most precise results. To do this the control panel should be set for a fixed 64-second counting period, and scans *started* at the angles noted below, the goniometer being adjusted to move *toward* higher angles. Record the total count for each peak.

Mineral (hkl)	Starting point for 64-second scan (°2θ)
Quartz (101)	26.43
Fluorite (111)	28.10
Calcite (104)	29.22
Dolomite (104)	30.80

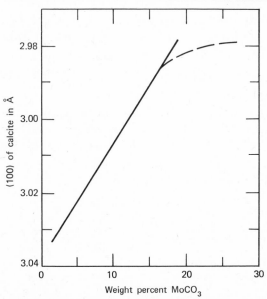

Fig. 4 Weight percent $MgCO_3$ incorporated diadochtically in a calcite type $CaCO_3$ structure as a function of peak dislocation. (After Chave, 1952)

Background ratiation must be subtracted from each peak intensity measurement. To determine the background count, two measurements should be made in a spectral location near the peaks being examined but entirely free from the peak radiation. The most appropriate location will depend on the samples involved. Ideally, 64-second counts can be started at 25 and 32° 2θ and the average count used. The background count for the two runs must then be averaged and subtracted from each of the four peak-counts.

(b) Ratios of arithmetic peak heights should give results only slightly less precise than counting, and peak-height ratios are adequate for most uses. Also, the peak-height method is more rapid than counting. Simply construct a straight base line connecting the average position of the intra-peak valleys. Measure the perpendicular distance from the base line to the apex of each peak and record the values below the peaks.

5. Compute the following intensity ratios:

$$I_{dolomite}/I_{fluorite}$$

$$I_{quartz}/I_{fluorite}$$

$$I_{calcite}/I_{fluorite}$$

6. Refer to the working curves of Fig. 5 and read directly the percentages of dolomite, quartz, and calcite. If these three components do not total 100%, the presence of one or more additional components can be inferred; in many carbonate rocks this component will represent the clay fraction.

On Fig. 5 the solid lines represent the least squares best fit and the dashed lines indicate the 95% confidence level for the curves. Actually these confidence limits were developed by Raish (1964) on the basis of five counting runs per sample; if fewer runs are made, the precision will be less than is indicated.

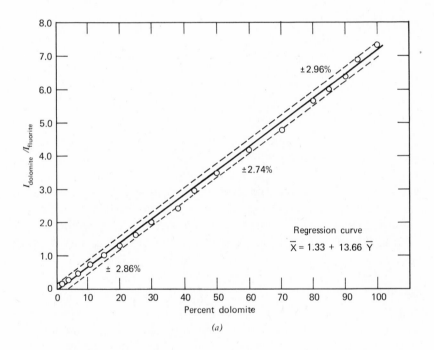

Fig. 5 Percentages of (a) dolomite, (b) quartz, and (c) calcite as a function of peak intensity ratios. (After Raish, 1964)

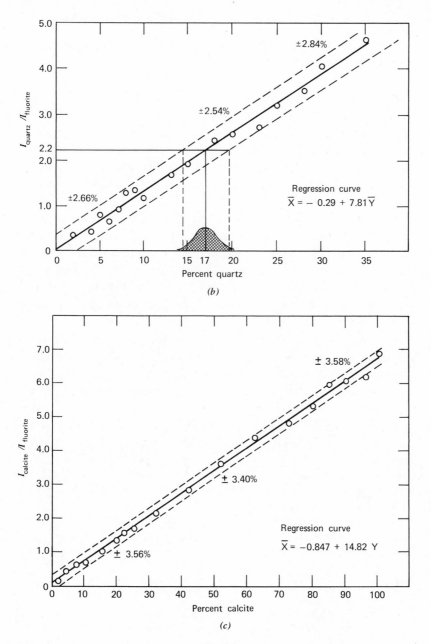

Fig. 5 (Continued)

STEP 4. MEASUREMENT OF AVERAGE CRYSTALLITE SIZE OF SELECTED MINERALS

In a general way the width of diffraction peaks from powder samples indicates the average size of the tiny crystallites in the analyzed specimen. Large crystallites produce narrow peaks; small crystallites produce broad, ill-defined peaks. If several assumptions concerning the condition of the specimen can be accepted[5] and if suitable correction factors are applied,[6] peak breadth measurements can yield a quantitative measure of the average dimensions of the crystallites producing the peaks. These dimensions have been used in several instances by sedimentary petrologists to interpret the state of development of the crystalline organic components of coals and asphalts, for example, Miller (1949); Siever (1952); Young (1954); French (1964); and Griffin (1967). Their application to mineral analysis has been in the author's opinion somewhat neglected and deserves more attention.

Rau (1963) has developed a series of correction curves which shorten the process of correcting peak widths for instrumental factors; he has simplified further the process by use of a curve that allows observed peak widths to be converted directly into an approximate measure of average crystallite size. The writer has recomputed and extended Rau's curve for use with broad peaks such as are often produced by the fine-grained components of sediments (Fig. 6).[7]

Procedure

1. Identify the minerals in the sample using the techniques suggested in Steps 1 to 3. The same patterns can be used for crystallite size measurements.

---

[5] Assumptions include that (a) the material is not in a strained condition and (b) that it does not have a mosaic structure.

[6] These factors generally correct for (a) the angular separation of the K_{α_1} and K_{α_2} components of the radiation and (b) instrumentally induced peak broadening.

[7] The curve is based on the Scherrer equation (see Rau, 1963) using a value of 0.9 as a shape constant and a standard instrumental correction factor of $-0.08°$ to convert observed peak widths at half-height (B_0) into corrected peak widths (β). In critical applications it is suggested that the basic literature be consulted (e.g., Klug and Alexander, 1954) and special curves drawn, but for routine geologic uses the curve of Fig. 6 will probably yield data of suitable quality.

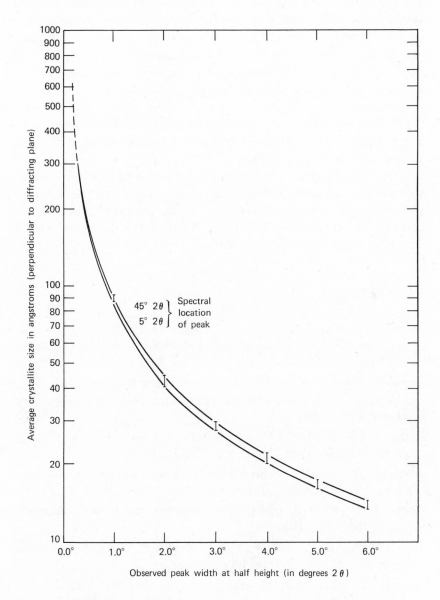

Fig. 6 Average crystallite size as a function of peak width at half height

2. Select one or more peaks to be measured. The peaks should be symmetrical and unimodal with no interference from neighboring peaks. The dimension measured will be perpendicular to the planes producing the peak; therefore for c-axis measurement a basal $(00l)$ peak should be selected. Measurements of the a and b dimensions, if desired, will require that planes perpendicular to those axes be measured.

3. Construct a base line and a perpendicular to the peak apex. At a point midway between the base-line and apex measure the peak-width in terms of degrees 2θ.

4. Refer to Fig. 6. Enter with the observed peak width and read the average crystallite dimension perpendicular to the refracting plane measured. Separate curves are shown for peaks near 5 and $45°$ 2θ in the diffraction spectrum; other curve locations can be approximated between these curves.

REFERENCES

Brindley, G. W., ed., 1951, *X-ray identification and crystal structures of clay minerals,* a symposium, The Mineralogical Soc., London, 345 pp.

Brown, George, ed., 1961, *The x-ray identification and crystal structures of clay minerals,* a symposium, The Mineralogical Soc., London, 544 pp.

Brunton, George, 1955, Vapor-pressure glycolation of oriented clay minerals, *Am. Mineralogist,* **40,** 124–126.

Burns, J. H., and M. A. Bredig, 1956, Transformation of calcite to aragonite by grinding, *J. Chemical Phys.,* **25,** 1281.

Chave, K. E., 1952, A solid-solution between calcite and dolomite, *J. Geol.,* **60,** 190–192.

Deer, W. A., R. A. Howie, and J. Zussman, 1963, *Rock-forming minerals,* Vol. 1–5, John Wiley and Sons.

French, B. M., 1964, Graphitization of organic material in a progressively metamorphosed precambrian iron formation, *Science,* **146,** 917–918.

Gibbs, R. J., 1967, Quantitative x-ray diffraction analysis using clay mineral standards extracted from the samples to be analyzed, *Clay Minerals,* **7,** 79–90.

Griffin, G. M., 1962, Regional clay mineral facies—products of weathering intensity and current distribution in the northeastern Gulf of Mexico, *Bull. Geol. Soc. America,* **73**, 737—768.

———, 1964, Development of clay mineral zones during deltaic migration, *Bull. Am. Assoc. Pet. Geologists,* **48**, 57—69.

———, 1967, X-ray diffraction techniques applicable to studies of diagenesis and low rank metamorphism in humic sediments, *J. Sed. Pet.,* **37**, 1006—1011.

Grim, R. E., 1968, Clay mineralogy, 2nd ed., McGraw-Hill Book Co., 596 pp.

Klug, H. P., and L. E. Alexander, 1954, X-ray diffraction procedures, John Wiley and Sons, 716 pp.

Miller, H. P., 1949, The problems of coal geochemistry, *Econ. Geol.,* **44**, 649—662.

Raish, H. D., 1964, Quantitative mineralogical analysis of carbonate rocks, *Texas J. Sci.,* **16**, 172—180.

Rau, R. C., 1963, Measurement of crystallite size by means of x-ray diffraction line-broadening, *Norelco Reporter,* **10**, 114—118.

Siever, Raymond, 1952, "X-ray diffraction study of rank increase in coal," pp. 341—362, *in* On the origin and constitution of coal, 2nd Conf. Proc., Nova Scotia Dept. of Mines, June 18-20, 1952, Nova Scotia Res. Foundation.

Switzer, George, J. M. Axelrod, M. L. Lindberg, and E. S. Larsen, 3[d], 1948, Table of d-Spacings for angle 2θ, CuK_{α}, CuK_{α_1}, CuK_{α_2}, FeK_{α}, FeK_{α_1}, FeK_{α_2}, *United States Geol. Surv. Circ. 29.*

Young, R. S., 1954, Preliminary x-ray investigation of solid hydrocarbons, *Bull. Am. Assoc. Pet. Geologists,* **38**, 2017—2020.

SECTION
VI

CHEMICAL ANALYSIS

CHAPTER 25

CARBON DETERMINATION

M. GRANT GROSS

State University of New York at Stony Brook, Stony Brook,
New York

ANALYSIS OF CARBONACEOUS ORGANIC MATTER IN SEDIMENTS AND SEDIMENTARY ROCKS

Most sediments and sedimentary rocks contain some remains of the
organisms that were deposited with the sediment. When skeletal
remains are well preserved, they may be studied as fossils and thus
provide information about the age and depositional environment of
the sediment. In many sediments and rocks, organic remains are no
longer recognizable but occur as widely disseminated decomposition
products of tissues or skeletal units. Still the abundance of tissue or
decomposition products can provide information about the depositional environment, such as dissolved oxygen concentration, produc-
tivity of the overlying water, and at least indirectly, the strength of
bottom currents.

Unlike fossil remains, which are usually separated from the
sediment for study, complete separation by physical means of the
carbonaceous organic matter from the sediment is usually impossible.
Consequently, most of our information about the abundance and
composition of these organic disseminated remains depends upon

chemical analysis. In general, applicable analytical techniques detect elements or compounds that were incorporated in the tissues or skeleton of the organism while it was alive.

This chapter discusses some analytical techniques that have been used to estimate the abundance of organic matter in sediments and sedimentary rocks and briefly evaluates the significance of the data obtained from each technique. A list of necessary equipment and reagents and a detailed discussion of each procedure are given at the end of the chapter.

Source and Composition of Organic Matter

Organic matter in sediments or sedimentary rocks can be considered to be derived from either soft tissues or hard tissues (skeletal remains). Tissues and their decomposition products are primarily carbonaceous compounds, whereas skeletal remains may be composed of carbonates, silica, or phosphate compounds. Let us consider the elemental composition of the tissues of various organisms (Table 1). As a first approximation we can regard the elemental composition of marine phytoplankton as typical of the composition of marine organisms.

TABLE 1   Relative composition of tissue

| | Soft Tissues | | | |
	Carbon	Nitrogen	Phosphorus	Hard Tissues
Phytoplankton	100	16	1.7	Opal (diatoms) $CaCO_3$ (coccolithophores)
Zooplankton	100	16	2.4	Chitin Phosphate (apatite)
Bacteria	100	18	5.5	
Mollusks	100	21	1.5	$CaCO_3$
Fish	100	23	3.8	Phosphate (apatite)
Mammals	100	18	8.9	Phosphate (apatite)
Angiosperms	100	6	0.5	

Data recalculated after Bowen (1966, pp. 68−73).

Carbon, the most abundant element, constitutes about half (45 to 55%) of the tissue on an ash-free basis (Strickland, 1960, pp. 21—22). Thus organic carbon analysis provides the most sensitive and reliable test for the abundance of biogenous material in the sediment. Nitrogen and phosphorus are much less abundant than carbon and exhibit much greater variation. Analytical techniques that determine the abundance of nitrogen and phosphorus can be expected to be less sensitive than carbon determination methods, because of the low concentration of these elements, and less useful, because of their variable concentration in the organisms. Of these two elements, only nitrogen is used to estimate the abundance of organic matter. Sulfur also occurs as a constituent of certain organic compounds, such as amino acids (Degens, 1965, p. 214), but its abundance in sediments is not a useful indicator of the abundance of organic matter.

Although we speak of organic matter in terms of carbon, nitrogen, phosphorus, and other elements, it actually consists of a complex mixture of high molecular weight compounds including lipids, carbohydrates, proteins, pigments, and lignins (Mortenson and Himes, 1964). A complete analysis of the organic matter in sediments or sedimentary rocks would ideally include determining the abundance of all these compounds. Such analyses have been made for specific compounds (Black et al., 1965, pp. 1397—1451; Degens, 1965, pp. 202—312), but the analytical procedures require highly specialized techniques and equipment and are rarely included in petrologic studies of sediments.

SAMPLING AND INITIAL SAMPLE HANDLING

Sampling frequently presents major problems in the study of organic matter in sediments and sedimentary rocks. The distribution of organic matter is affected by physical processes that control sediment grain size and sorting (Trask, 1939). In addition, complications arise from biological factors that control the production and destruction of organic matter during and after deposition. Consequently, careful attention must be given to sampling procedures to ensure that a sample is representative of the sediment being studied (Krumbein and Pettijohn, 1938, pp. 11—42; Trask and Patnode, 1942, pp. 12—15; Black et al., 1965, pp. 1—81; Griffiths, 1967, pp. 10—30).

After a representative field sample has been carefully subsampled (Krumbein and Pettijohn, 1938, pp. 43–46; Jackson, 1958, p. 34; Black et al., 1965, pp. 54–72) and air-dried at room temperature, the sample should be crushed, if necessary, and ground until fine enough to pass through a 200-mesh sieve. Normally, grinding is one of the most time-consuming parts of the analytical procedure, and it is desirable to prepare enough material at this stage to avoid having to grind more samples later. Usually 20 to 50 gm will permit a replicate sample analysis. After sieving, samples are normally stored dry in labeled vials. Samples should be well mixed and subsamples dried at 110°C for at least 4 hours just prior to analysis. Variable water content resulting from humidity change in the laboratory is often a major source of error in quantitative analyses of sediments and sedimentary rocks.

Laboratory Equipment

The necessary laboratory equipment is simple and not overly expensive (see Table 2). The most important part is the balance; it must be sensitive to 0.1% of the sample weight, preferably weigh to 1 mg and have a large tare range.

If the number of samples is large, or if the type of analysis warrants, it may be desirable to obtain one of the commercially available single element analyzers. This equipment is designed for analysis of carbon in steel and permits combustion at higher temperatures than can be achieved with normal laboratory equipment. Moreover, such analyzers are usually combined with gas analysis systems, which permit much faster, and often more precise, analyses than can be easily obtained with standard laboratory equipment.

Method of Approach

After choosing an analytical technique appropriate to the problem (Table 3), it is necessary to plan the actual analytical procedure as outlined at the end of the chapter. In an involved procedure, it is advisable to consult the reference cited before undertaking the analysis.

In general, duplicate samples should be analyzed and the average reported as the result. If the results of two analyses of the same sample are outside previously set limits of agreement, additional analyses should be made and either an average reported or, if the difficulty can be identified and corrected, the average of the acceptable analyses reported.

It is not only essential to perform the analyses carefully but also to evaluate and report both the precision (reproducibility) and the accuracy (measure of error of the analysis). This means that appropriate standard samples must be analyzed in the same way as the samples being studied. One or more standard samples should be included in the analyses made each day to permit immediate detection and correction of difficulties in the analytical procedure. With data from many analyses, it is possible to evaluate analytical precision. The analysis of standards will permit evaluating the accuracy of the method as well as precision. Typical results for modern Potomac River sediments are given in Table 4. Standards for the various analytical procedures are suggested in Table 5.

TABLE 2 Laboratory equipment needed for analysis of organic matter in sediments and sedimentary rocks

Basic Equipment
    Drying oven ($110°C$ or higher), infrared heaters may also be used
    Desiccators
    Balance, sensitive to 0.1% of sample weight
    Hot plate
    Mortar and pestle (sample crushers may be needed for lithified materials)
    Sieves
    Shakers
    Muffle furnace ($1000°C$)
    Crucibles, usually porcelain
    Laboratory glassware (beakers, flasks, graduated cylinders, pipets, burets)

Desirable Equipment, Relatively Low Cost
    pH meter with electrodes and buffer solutions
    Automatic grinding equipment
    Vacuum oven and vacuum pump or aspirator

Desirable Equipment, Moderate to High Cost
    High-temperature combustion furnaces ($T > 1200°C$)
    Automatic carbon analyzers
    Gasometric analyzers
    CO_2 detector (nondispersive infrared or volumetric analyzers)

TABLE 3 Summary of analytical techniques

Quantity Estimated	Quantity Measured	Possible Sources of Error	Remarks
Total organic matter	Weight loss of sample heated to 550°C for 4 hours.	Partial decomposition of certain minerals, for example, carbonates, clays.	Simple. Rapid. Minimal equipment required. Poor reproducibility. Not useful in clay-rich sediments.
	Weight loss after oxidization with H_2O_2.	O_2 uptake by Fe^{2+} or Mn^{2+} and precipitation of new phases may cause weight gain.	Simple. Rapid. Minimal equipment. Reproducibility may be fair to poor. Detects only easily oxidized compounds.
Total carbon	CO_2 derived from high-temperature combustion (T>1500°C) CO_2 detected by infrared CO_2 analyzer or gasometry.		Good reproducibility. Equipment moderately expensive (more than $2000).
Organic carbon	CO_2 derived through wet combustion or dry combustion at 550°C.	Partial decomposition of carbonates may occur if temperature exceeds 600°C.	Variable recovery with fair reproducibility.
Carbonate-carbon	CO_2 derived from acidification of sample. CO_2 detected by infrared CO_2 analyzer or gasometry.		Easy analysis, good reproducibility.
Carbonate content	Weight loss of sample after acid treatment.	Partial decomposition of other minerals, for example, sulfides.	Simple. Rapid. Reproducibility moderately good. Minimal equipment. Not useful in sulfide-bearing sediments.
	Differential weight loss after heating at 550°C and 1000°C.		Good for carbonate rocks and calcareous materials.

TABLE 4   Results of analyses for organic carbon and total organic matter in samples of Potomac River sediments.
Values (± one standard deviation) are given in weight percent and the number of analyses is indicated in parentheses.

Potomac River Sample	Organic Carbon by Dry Combustion		Total Organic Matter by Weight Loss	
	Method 1[a]	Method 2[b]	H_2O_2 Digestion	Ignition (700°)
1	2.2 ± 0.1 (6)	2.23 ± 0.03 (2)	2.9 ± 0.3 (3)	6.5
2	1.9 ± 0.2 (6)	1.86 ± 0.04 (2)	2.3 ± 0.1 (3)	6.4
3	1.6 ± 0.2 (8)	1.62 ± 0.01 (2)	2.0 ± 0.3 (3)	6.3
4	2.4 ± 0.1 (3)	2.50 ± 0.07 (2)	2.4 ± 0.1 (3)	7.2

[a] Sample combusted at 900°C in dry, CO_2-free oxygen. CO_2 released was determined by absorption. Equipment used was Colemen C-H analyzer Model 33.

[b] Sample combusted at temperatures exceeding 1500°C in dry, CO_2-free oxygen. CO_2 released was determined by difference in thermal conductivity between pure carrier gas and carrier gas containing CO_2. Equipment used was Leco induction furnace Model 523-000 and Leco low-carbon analyzer Model 589-600.

TABLE 5   Standard materials suitable for calibration purposes

Total Carbon
   Standard steel samples (1.2 to 0.02%)
      (National Bureau of Standards, Washington, D.C.)

Organic Carbon or Total Organic Matter
   Sucrose[a]   (42.1% C)
   Glucose[a]   (40.0% C)
   Galactose[a]   (40.0% C)

Carbonate Carbon
   Calcium carbonate[a]   (12.0% C)

[a] Reagent grade, preferably diluted to approximate percentage expected in the samples by mixing with an inert, carbonate-free refractory material (i.e., silica or alumina).

ESTIMATION OF TOTAL ORGANIC MATTER

In general, these qualitative techniques involve destruction of the organic material in the sample by oxidation and the estimation of its original abundance by comparing initial and final weights. The abundance of organic matter can be estimated by destroying the organic matter through oxidation at 550°C (ignition loss) or through the use of H_2O_2.

Although high-temperature ignition may alter or destroy many minerals, especially clays, low-temperature oxidation by H_2O_2 does not affect most minerals. Nevertheless, oxidation techniques suffer from relatively poor reproducibility because the oxidation of organic matter is usually incomplete. Only the more readily oxidized compounds are destroyed. The technique has the advantage that is is readily combined with the weight-loss technique for determining carbonate; thus both sets of data can be obtained from the same sample.

Ignition loss techniques tend to be less accurate because of many interfering reactions, for example, partial decomposition of carbonates or partial dehydration of minerals, such as clays that contain H_2O or OH. This technique is useful for groups of samples with uniform mineral composition and less than 10% $CaCO_3$.

Carbon Analyses

Although the carbon content should provide the most sensitive index of the amount of organic matter in a sediment or sedimentary rock, the situation is complicated by the fact that carbon occurs in several forms: (a) carbon-containing compounds, little altered from their initial composition in living organisms; (b) highly altered and fairly resistant decomposition products of the original tissues, such as coal or graphite; and (c) carbonate minerals, such as $CaCO_3$, or rarely, as soluble salts containing HCO_3.

The most satisfactory method for analysis of the total carbon content of sediments or sedimentary rocks appears to be combustion of a sample at temperatures exceeding 1500°C in an atmosphere of dry, CO_2-free oxygen. Such temperatures can be reached in high-frequency induction furnaces in which the sample is mixed with iron and heated (Fig. 1) or in special high-temperature resistance furnaces. The combustion gases commonly pass through a dust trap, a sulfur trap, and a catalyst furnace that converts CO to CO_2. This equipment gives about 95% recovery of total carbon with a coefficient of variation of 3% or less. Depending on the amount of CO_2 evolved from the sample, combustion gases may be passed through a gasometric analyzer, through absorption trains for gravimetric analyses, or detected by nondispersive infrared CO_2 analyzers.

Organic carbon may be estimated using dry combustion techniques that differ from total-carbon analyses in that the sample is combusted at lower temperatures (e.g., 550°C) in an atmosphere of dried, CO_2-free oxygen. The combustion gases are then passed over a catalyst to complete oxidation of CO to CO_2 and through traps to

Fig. 1 High-frequency induction furnace for combustion of sediment samples at temperatures exceeding 1500°C shown on left. Traps for dust and sulfur compounds and a catalyst furnace for oxidation of CO to CO_2 are mounted on the side of the furnace. Unit on right determines the amount of CO_2 by measuring the difference in thermal conductivity between pure carrier gas and carrier gas containing CO_2. (Photo courtesy of Laboratory Equipment Corporation, St. Joseph, Michigan)

remove halogens and sulfur compounds. Water and CO_2 are absorbed in preweighed absorption tubes, which are weighed to determine the amount of hydrogen and carbon oxidized during the analyses. A simplified flow diagram of this type of instrument is shown in Fig. 2 and a commercially available instrument in Fig. 3.

Wet-oxidation techniques have been used for many years in studies of carbon in soils (Allison, 1935; Walkley, 1947; Jackson 1958, pp. 205–222) and in sediments (Trask and Patnode, 1942, p. 22), Basically these techniques involve using a strong oxidizing agent such as $K_2Cr_2O_7$ or MnO_4 in an acid, usually H_2SO_4 to oxidize organic matter. Modifications include various combinations of oxidizing agents (Van Slyke, 1954). The degree of heating can also be varied; and as more heat is applied more resistant carbon-containing compounds are decomposed (Jackson, 1958, pp. 205–222).

Other modifications of the techniques for organic carbon determination include (a) back-titration of excess oxidizing agent; (b) precipitation in alkali of CO_2 released and gravimetric determination of the amount; (c) measurement of the volume of CO_2 evolved in a gasometric buret (Hillebrand et al., 1953).

The carbonate content of a sediment or sedimentary rock can be simply and rapidly determined, with fair precision, by determining the weight loss after treating the sample with dilute acid to drive off CO_2 from the carbonates. Heating for 1 to 5 minutes is necessary to break down the less soluble carbonates. This technique can be combined with determination of organic matter by weight loss on oxidation by H_2O_2 to obtain two sets of data from the same sample, which is often an advantage when the quantity of sample is limited.

Carbonate can also be determined by taking the difference between total carbon content and the amount of organic carbon as determined by other methods, including ignition at $550°C$ or wet combustion.

To avoid some of the errors introduced by using weight loss after acidification or ignition as a measure of the amount of carbonate, one can collect and measure the volume of gas released during acidification. Through the use of the simple apparatus illustrated in Fig. 4, it is possible to obtain a rapid determination of the amount of carbonate in sediment if $CaCO_3$ is abundant. However, the technique does not work well in sediment where dolomite is the dominant carbonate.

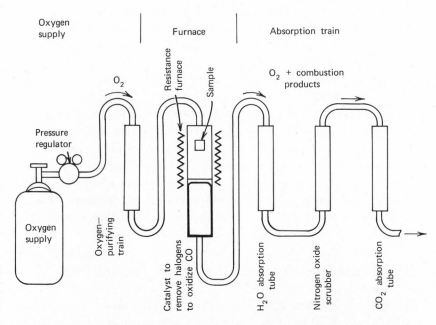

Fig. 2 Simplified flow diagram of a resistance furnace used to combust a sediment sample in dry CO_2-free oxygen with an absorption train to determine the abundance of CO_2.

With smaller amounts of carbonate in the sample, it may be necessary to use more refined techniques. There are several possible modifications. For example, a stream of CO_2-free gas (or CO_2-free air) may be used to sweep the gas out of the reaction vessel. Also a more sensitive means of measuring the amount of CO_2 evolved may be employed such as (a) absorption trains (Hillebrand et al., 1953, p. 769) in which the CO_2 is absorbed in an absorption bulb and the amount of CO_2 determined by the weight difference before and after analysis; (b) volumetric gas analyzers in which the gas evolved from the reaction fills a gas buret, then passes through a strong alkali solution which absorbs the CO_2, and the amount of CO_2 is indicated by the change in gas volume (Simons et al., 1955); (c) passing the gas through a strong alkali solution, such as $Ba(OH)_2$, thereby forming a heavy, insoluble precipitate which is weighed to give a measure of the amount of CO_2 released (Hillebrand et al., 1953, p. 776); and (d) the use of nondispersive infrared analyzers which detect very small amounts of CO_2 in the carrier gas used to sweep the reaction (Menzel and Vaccaro, 1964). Sensitivity, simplicity, and freedom from errors make the infrared analyzer an appealing technique, especially when small samples are involved.

An alternative technique for determining the carbonate content of sediments involves reacting a weighed sample with a known excess of standardized acid. When the reaction is complete, the amount of excess acid is determined by back-titration with a standardized solution of sodium hydroxide (Herrin et al., 1958). This simple technique is highly reproducible and reasonably accurate. It avoids loss of constituents other than CO_2 during the acidification and heating of a sediment. There is, however, some possibility of error because of clays or sulfides in the sample reacting with the acid.

SUMMARY OF PREFERRED AND USABLE TECHNIQUES

The following section provides information about equipment, reagents, general procedures, and literature references for analyses of carbon content of sediment and sedimentary rocks. The various techniques differ substantially in simplicity of the analysis, precision and reproducibility of the analyses, freedom from errors, and type and cost of the analytical equipment required. Based on experience, gained primarily from analysis of modern marine and river sediments, Table 6 was prepared to indicate those techniques which seemed to provide the most reliable data.

In general, those analytical procedures that selectively oxidize or break down carbon compounds of interest and measure directly the CO_2 released have been found to be the most satisfactory. For total carbon determination, it is essential that all carbon compounds are broken down or oxidized so that the higher the temperature the more reproducible the results. It is noted that the infrared CO_2 analyzer is an exceedingly useful instrument for this purpose in that it is simple to operate, relatively free of errors, and permits analysis of very small samples.

Although the preferred analytical techniques indicated generally require moderately expensive apparatus (more than $2000), quite good results can be obtained using simpler equipment when there is little chance of interference from other sources of error. For

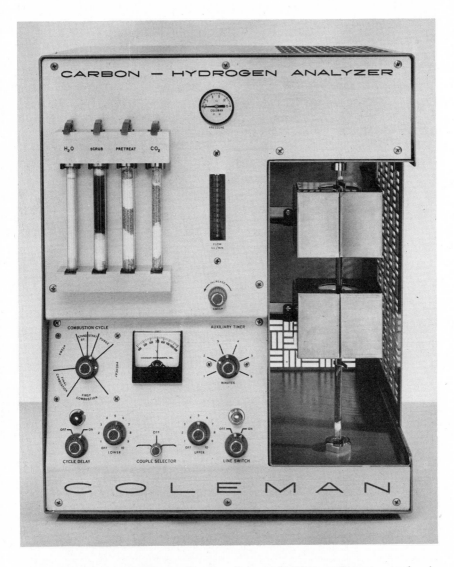

Fig. 3 Automated resistance furnace for combusting sediment samples in CO_2-free oxygen at temperatures less than $1100°C$. The furnace is visible in the right side of the unit and the absorption tubes for determining the abundance of CO_2 and H_2O released during the combustion are mounted on the front near the left side of the unit. (Photo courtesy of Coleman Instruments Division of the Perkin-Elmer Corporation, Maywood, Illinois)

example, the loss on ignition at 550°C gives good results when used with sand-sized sediments, but very poor (often unusable) results when used on clay-rich sediment or sedimentary rock because of the breakdown and release of water from the clay minerals.

ORGANIC MATTER BY IGNITION LOSS

Apparatus

Furnace, capable of operating continuously at 600°C, porcelain crucibles, desiccator.

Reagents

None

Procedure

1. Transfer a weighed sample (approximately 1 gm) of ground sediment that has been dried at 110°C for 8 hours or more and place in crucible.

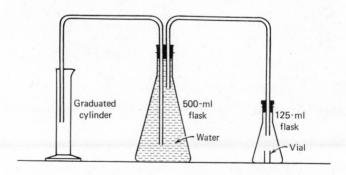

Fig. 4 Apparatus for gasometric determination of $CaCO_3$ by measuring the volume of CO_2 released following acidification of the sample. (From Jackson, 1958, by permission of Prentice-Hall, Englewood Cliffs, New Jersey)

2. Place crucible in electric muffle furnace and heat at 550°C for 1 to 2 hours.

3. Cool partially in air and then transfer to desiccator for cooling to room temperature.

4. Remove from desiccator and reweigh; calculate weight loss as percentage of initial weight of sample.

5. Report as ignition loss (550°C).

References

American Public Health Association, 1965, p. 425; Schopf and Manheim, 1967, pp. 1203–1204.

TABLE 6 Preferred and usable analytical techniques

	Total Carbon
Preferred	Dry combustion at T > 900°C in CO_2-free atmosphere. CO_2 released is determined by infrared analyzer or by gasometric analysis, either volumetric (usually absorption in hydroxide) or gravimetric techniques. Commercially available apparatus for determining carbon in metals can be used for these analyses.
Usable	Weight loss on ignition (usable for sands, not for clayey rocks or sediments)
	Carbonate Carbon
Preferred	Sample reacted with acid and heated to release CO_2 which is determined as above.
Usable	Weight loss on acid leach (clay-free sediments)
	Titration of excess acid after reaction with sample
	Differential weight loss after ignition at 550 and 1000°C
	Organic Carbon
Preferred	Dry combustion at 550°C in CO_2-free atmosphere, CO_2 determined as above.
Usable	Calculating difference between results for total carbon and carbonate carbon
	H_2O_2 digestion of sample
	Weight loss after ignition at 550°C (sands only)
	Wet combustion techniques, using strong oxidizing agents. CO_2 released may be determined as above or excess oxidizing agent can be determined by back titration

TOTAL ORGANIC MATTER BY H_2O_2 OXIDATION

Apparatus

Hot plate (or infrared heaters), tall-form 250-ml Pyrex beakers with cover glasses.

Reagents

H_2O_2, 30%, reagent grade and residue free (should not leave a weighable residue after evaporation); distilled water.

Procedure

1. Transfer approximately 2 gm (for sediment containing less than 2.5% organic carbon) of ground sediment dried at 110°C into a preweighed tall-form 250-ml beaker; cover with a watch glass or petri dish. Smaller samples may be used if organic carbon exceeds 2.5%.

2. Add 30 ml of 10% H_2O_2 (1 part 30% H_2O_2 mixed with 2 parts distilled water) replace cover, and begin heating either with infrared heaters or on hot plate. Watch carefully when reaction begins to avoid boiling over.

3. Continue heating until reaction has ceased.

4. Rinse down sides of beaker and watch glass with additional 5 ml H_2O_2 (30%), replace cover, and heat again.

5. Continue adding H_2O_2 until there is no weight loss or the character of the reaction changes from a violent, frothy boiling to a steady production of small bubbles with no foam. Evaporate to dryness on a hot plate, the sample often changes color to a rusty red at this point.

6. Cool beaker in desiccator and reweigh.

7. Calculate results as follows:

$$\% \text{ organic matter} \quad (H_2O_2) = \frac{100 \times \text{weight loss}}{\text{initial sample weight after drying at } 110°C}$$

Accuracy may be improved by applying (if necessary) a correction for any residue remaining after evaporating a known volume of the reagent grade H_2O_2 at 110°C.

Reference

Jackson, 1958, pp. 222–225.

ORGANIC CARBON BY WET COMBUSTION

Apparatus

Erlenmeyer flasks, 500-ml pipettes, 50-ml burets, Meker or Fisher burners, thermometer (0° to 200°C).

Reagents

Potassium dichromate (1 N). Dissolve 49.04 gm reagent-grade $K_2Cr_2O_7$ in distilled water and dilute to 1 liter.

Sulfuric acid-silver sulfate solution. Add 25 gm reagent-grade Ag_2SO_4 per liter of concentrated H_2SO_4.

Ferrous sulfate-sulfuric acid solution. Dissolve 278.0 gm of reagent-grade $FeSO_4 \cdot 7H_2O$ in water, add 15 ml concentrated H_2SO_4, and dilute to 1 liter. Standardize by titration against the 1 N $K_2Cr_2O_7$ solution above. Must be stored under hydrogen for long-term stability; otherwise it must be titrated against the $K_2Cr_2O_7$ solution each day.

Diphenylamine indicator. Dissolve 0.5 gm in 100 ml of concentrated H_2SO_4 and pour carefully into 20-ml cold distilled water.

85% H_3PO_4 solution.

Procedure

1. Transfer 0.500 gm of ground dried sediment to a 500-ml Erlenmeyer flask, add 10 ml $K_2Cr_2O_7$ solution, followed by 20 ml of H_2SO_4-Ag_2SO_4 solution.

2. Heat the flask gently to 150°C within 1 minute and allow to cool.

3. Excess $K_2Cr_2O_7$ is measured by back-titration with $FeSO_4$ solution after dilution to 200 ml and addition of 10 ml 85% H_3PO_4, 0.2 gm NaF, and 30 drops of diphenylamine indicator. The endpoint is indicated by blue to bright green.

4. A blank analysis should be run with each series of samples. The same procedure is followed except that no sample is added to the flask.

5. The results are calculated as follows:

percent organic carbon =

$$\frac{(mlFeSO_4)blank - (mlFeSO_4)\,sample \times 0.3}{weight\ of\ sample}$$

This is also reported as oxidizable carbon (Jackson, 1958, p. 206) or as reducing capacity (Trask, 1939, p. 432; Frost, 1962).

References

Allison, 1935; Walkley, 1947; Van Slyke, 1954; Jackson, 1958, pp. 214–222; Black et al., 1965, pp. 1367–1378.

CARBONATE BY WEIGHT LOSS AFTER ACIDIFICATION

Apparatus

Hot plate (or infrared heaters) and 250-ml tall-form Pyrex beakers.

Reagents

Dilute HCl (such as 0.1 N), pH indicator paper.

Procedure

1. For sediment containing more than 1% $CaCO_3$ of ground sediment dried at 110°C, transfer approximately 2 gm to a preweighed tall-form 250-ml beaker; cover with watch glass or inverted petri dish. For sediment with more than 1% $CaCO_3$, smaller samples may be used.

2. Add 10 ml of distilled water and 5 ml of 0.1 N HCl and stir to promote reaction. After initial reaction has ceased, add additional 2 ml 0.1 N HCl and continue to add acid until no further reaction is observed or until the solution is strongly acidic to pH indicator paper.

3. Decant excess acid and carefully wash sample in distilled water to remove $CaCl_2$.

4. Evaporate to dryness and heat to 110°C before transferring to desiccator to cool.

5. Reweigh and calculate weight loss.

6. Results may be calculated and reported as follows:

$$\text{percent carbonate-carbon} = \frac{100 \times \text{weight loss} \times 0.12}{\text{initial dry weight of sample}}$$

assuming that all the carbonate occurred as $CaCO_3$.

Reference

Black et al., 1965, pp. 1388–1389.

CARBONATE-CARBON BY SIMPLE VOLUMETRIC TECHNIQUES

Apparatus

Graduated cylinder, 500 ml, 500-ml Erlenmeyer flask, 125-ml Erlenmeyer flask, glass and plastic tubing, 8-ml shell vials (see Fig. 4).

Reagents

HCl, concentrated.
$CaCO_3$, reagent-grade, for calibration purposes.

Procedure

1. Transfer 2 gm of dried, ground sample (sample containing less than 50% $CaCO_3$) to a 125-ml conical flask and, using forceps, set a vial containing 3 ml concentrated HCl in the flask.

2. Connect the flask containing the sample and the acid to a 500-ml flask (Fig. 4) containing an indicator solution of 355 gm anhydrous $CaCl_2$ in 500 ml water containing a few drops HCl and methyl red indicator.

3. After checking all connections for tightness, tip the flask to cause the vial to tip over, releasing the acid to react with the sample.

4. CO_2 given off in the reaction will displace indicator fluid from the large flask, causing it to flow into the graduated cylinder. The volume of fluid displaced is a measure of the amount of CO_2 evolved.

5. Results may be calculated as follows:

(a) percent carbonate-carbon =

$$\frac{100 \times \text{volume of water displaced by sample} \times 0.12}{\text{volume of water displaced by equal weight of } CaCO_3}$$

(This calculation involves no assumptions about the form of the carbonate phase.)

Derivation of the factor: $\quad 0.12 = \dfrac{\text{weight of carbon}}{\text{weight of } CaCO_3} = \dfrac{12}{100}$

(b) percent $CaCO_3 =$

$$\frac{100 \times \text{volume of water displaced by sample}}{\text{volume of water displaced by equal amount of } CaCO_3}$$

(This calculation involves assumption that all the carbonate occurs as $CaCO_3$)

This technique should be calibrated at the beginning and end of each series of analyses by analyzing a standard amount of reagent grade $CaCO_3$. Changes in atmospheric pressure affect the volume of the CO_2 given off during the reaction. For samples containing carbonate equivalent to more than 20% $CaCO_3$, the results are reported to agree within 1% with those obtained by titration analyses (Jackson, 1958, pp. 79–80).

References

Jackson, 1958, p. 80; Black et al., 1965, pp. 1389–1396. Hülsemann, 1966, described a more elaborate but inexpensive laboratory instrument.

CARBONATE-CARBON BY TITRATION TECHNIQUE

Apparatus

Volumetric pipettes 50 ml, 50-ml buret (glass-bead type), 250-ml beakers, hot plate, magnetic stirrer and stirring bars, pH meter desirable.

Reagents

Hydrochloric acid solution, 0.5 N, standardized.
Sodium hydroxide solution, 0.25 N, standardized.
Potassium acid phthalate. Reagent grade.

Standardization Procedure

Standardizations for both the acid and the base should be made in triplicate. The NaOH solution is standardized in the following manner: Weigh out about 0.8 gm of potassium acid phthalate and record the weight to the nearest milligram. Place the acid in a beaker with 150 ml of distilled water. Add 3 or 4 drops of phenolphthalein (0.1% in ethyl alcohol) and titrate with NaOH to a pink color. If the potentiometric procedure is used, merely titrate to pH 7. The normality is found by use of the following equation:

$$N_{NaOH} = \frac{\text{wt. potassium acid phthalate}}{0.2042 \times \text{ml NaOH}}$$

In order to standardize the H_2SO_4 pipette 25 ml of the acid into a beaker containing about 100 ml of distilled water. Add 3 to 4 drops of phenolphthalein and titrate with standardized NaOH to a pink color or titrate to pH 7, using the potentiometric method. The following equation is used to calculate the normality of the acid:

$$ml_{H_2SO_4} \times N_{H_2SO_4} - ml_{NaOH} \times N_{NaOH}$$

(For more information about standardizing solutions, see Hillebrand et al., 1953.)

Analytical Procedure

1. Transfer an accurately weighed sample (approximately 1 gm) which has been ground and dried at 110°C to a 250-ml beaker and add by pipette 50 ml of a previously standardized HCl solution.

2. Heat to about 90°C for 20 minutes and then test with pH indicator paper. If the pH is greater than 2, add 50 ml of HCl solution and heat an additional 20 minutes.

3. When the pH remains less than 2 after heating, fill the beaker about half-full with distilled water and back-titrate with a previously standardized NaOH solution to pH 7, stirring constantly, preferably with a magnetic stirrer.

4. If no pH meter is available, add 3 or 4 drops of phenolphthalein and back-titrate to the phenolphthalein endpoint. Stir constantly.

5. The results are calculated as follows:

(a) percent carbonate-carbon =

$$100 \times 0.006 \times (ml_{HCl} \times N_{HCl}) - (ml_{NaOH} \times N_{NaOH})$$

(This involves no assumptions about the forms of the carbonate phases.)

(b) percent $CaCO_3$ =

$$100 \times 0.05 \times (ml_{HCl} \times N_{HCl}) - (ml_{NaOH} \times N_{NaOH})$$

(This assumes that all the carbonate occurs as calcium carbonate.)

References

Herrin et al., 1958; Black et al., 1965, p. 1387.

REFERENCES

Allison, L. E., 1935, Organic soil carbon by reduction of chromic acid; *Soil Sci.*, **40**, 311–320.

American Public Health Association, 1965, Standard methods for the examination of water and wastewater including bottom sediments and sludges, 12th ed., 769 pp.

Black, C. A., Editor-in-Chief, et al., 1965, Methods of soil analysis, American Society of Agronomy, Pts. 1 and 2, 1572 pp.

Bowen, H. J. M., 1966, Trace elements in biochemistry, Academic Press, 241 pp.

Degens, E. T., 1965, Geochemistry of sediments: a brief survey, Prentice-Hall, 342 pp.

Frost, I.C., 1962, Evaluation of the use of dichromate to estimate the organic carbon content of rocks, U.S. Geol. Surv. P. P. 424–C, C376–377.

Griffiths, J. C., 1967, Scientific method in analysis of sediments, McGraw-Hill Book Co., 508 pp.

Herrin, E., H. S. Hicks, and H. Robertson, 1958, A rapid volumetric analysis for carbonate in rocks, *Field and Laboratory,* **26,** 139–144.

Hillebrand, W. F., and G. E. Lundell, 1953, Applied inorganic analysis, 2nd ed., John Wiley and Sons, 1034 pp.

Hülsemann, Jobst, 1966, On the routine analysis of carbonates in unconsolidated sediments, *J. Sed. Pet.,* **36,** 622–625.

Jackson, M. L., Soil chemical analysis, Prentice-Hall, 498 pp.

Krumbein, W. C., and F. J. Pettijohn, 1938, Manual of sedimentary petrography, Appleton-Century Crofts, 549 pp.

Menzel, D. W., and R. F. Vaccaro, 1964, The measurement of dissolved organic particulate carbon in seawater, *Limnology and Oceanography,* **9,** 138–142.

Mortenson, J. L., and F. L. Himes, 1964, Soil organic matter, pp. 206–241, *in* F. E. Bear, ed., Chemistry of the soil, 2nd ed., Reinhold, 515 pp.

Schopf, T. J. M., and F. T. Manheim, 1967, Chemical composition of Ectoprocta (Bryozoa), *J. Paleon.,* **41,** 1197–1225.

Simons, E. L., J. E. Fagel, E. W. Balis, and L. P. Pepkowitz, 1955, High-frequency combustion — volumetric determination of carbon in metals, *Analytical Chemistry,* **27,** 1119.

Strickland, J. D. H., 1960, Measuring the production of marine phytoplankton, Fisheries Research Board of Canada, Bull. 122, 172 pp.

Trask, P. D., 1932, Origin and environment of source sediments of petroleum, Gulf Publishing Co., 323 pp.

_____, 1939, Organic content of recent marine sediments, pp. 428–453, *in* P. D. Trask, ed., Recent marine sediments: a symposium, Am. Assoc. Petrol. Geologists, Tulsa, Oklahoma, 736 pp.

_____, and H. W. Patnode, 1942, Source beds of petroleum, Am. Assoc. Petrol. Geologists, Tulsa, Oklahoma, 566 pp.

Van Slyke, D. D., 1954, Wet Carbon combustion and some of its applications, *Analytical Chemistry,* **26**, 1706–1712.

Walkley, Allan, 1947, A critical examination of a rapid method for determining organic carbon in soils — effect of variations in digestion conditions and of inorganic soil constituents, *Soil Sci.,* **63**, 251–264.

CHAPTER 26

Eh-pH DETERMINATION

DONALD LANGMUIR

The Pennsylvania State University, University Park, Pennsylvania

Eh is a measure of the aqueous electron concentration,[1] and pH a measure of the aqueous proton or hydrogen ion concentration. Because electrons neutralize protons, a great many natural reactions are both Eh and pH dependent. For the same reason, high Eh values (low electron content) are generally found in conjunction with low pH values (high proton content) and conversely (Fig. 1).

The reactions controlling Eh involve elements present in more than one oxidation state. In sedimentary environments these elements are iron, sulfur, carbon, nitrogen, oxygen, and, less importantly, manganese. In most cases Eh is determined by photosynthesis, by respiration, and by reactions associated with the oxidation or reduction of iron and sulfur (Baas Becking et al., 1960).

---

[1]The electron concentration may also be described using the simpler concept of pE, where by analogy to pH, $pE = \log(e)$ (see Stumm, 1966, p. 2).

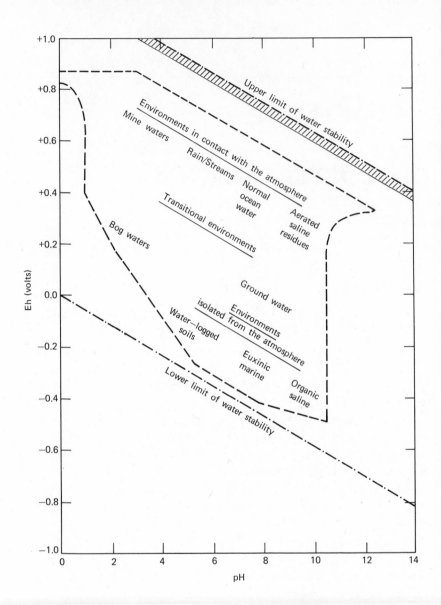

Fig. 1 Approximate position of some natural environments in terms of Eh and pH, modified after Garrels (1960, p. 201). The dashed line represents the limits of measurements in natural environments as indicated by Baas Becking et al. (1960). Crosshatched area represents the Eh and pH of waters containing more than 0.01 ppm dissolved oxygen based on reaction (22).

In organic-rich environments pH may also be controlled by the processes of photosynthesis and respiration. However, in the absence of active organic materials pH is most often determined by the weathering and alteration of aluminum silicate minerals, the solution and precipitation of carbonates, and cation exchange reactions with fine-grained sediments such as clays.

Many geologists, biologists, and soil scientists have used Eh and pH measurements as qualitative tools to characterize environments within the hydrosphere (e.g., Zobell, 1946; Baas Becking et al., 1960). A few others have attempted to relate measured Eh values to the activities of species in the aqueous environment through the Nernst equation (e.g., Back and Barnes, 1965). With care it is frequently possible to measure a field pH that is stable and reproducible and quantitatively meaningful in terms of the activities of natural species present. However, only under exceptional conditions are stable and reproducible Eh measurements possible, and only when such measurements are possible is there a chance that they can be closely related to the activities of species present. The quantitative approach is restricted further by the fact that Eh-related reactions in natural environments are usually irreversible, and that measured Eh is often a mixed potential as a result of several redox reactions. Thus, in general, Eh measurements should be considered primarily as a descriptive tool.

In this chapter we are concerned with how to measure accurately Eh and pH in sedimentary environments. Space does not permit a detailed discussion of the theory and interpretation of Eh and pH measurements. Some theory has been included, however, when it is felt it contributes to a better understanding of measurement practices and problems.

DEFINITION OF Eh AND pH

The general expression for a reduction reaction may be written[2]

$$aA + bB + ne^- = cC + dD \qquad (1)$$

where A and B are species in the oxidized state, the symbol e^- represents the electrons, C and D are in the reduced state, and a, b, c, d, and n are mole numbers. If expression (1) is a reversible reaction, then at equilibrium the rate of oxidation equals the rate of reduction, and Eh is given by the Nernst equation

$$\text{Eh} = E° + \frac{RT}{nF} \ln \frac{[A]^a [B]^b}{[C]^c [D]^d} \qquad (2)$$

In this expression R is the gas constant, 0.001987 kcal per degree, T the absolute temperature in degrees Kelvin, n the number of electrons, and F, the Faraday constant, is 23.06 kcal/(volt)(gram) equivalent. The last term on the right is the natural logarithm of the activity product for the reaction, neglecting electrons, with oxidized species in the numerator, reduced species in the denominator, and the activity of each species raised to the power of its respective stoichiometric coefficient. $E°$ is the standard potential of the reaction in volts and may be calculated from the relationship

$$E° = -\frac{\Delta G°}{nF} \qquad (3)$$

where $\Delta G°$ is the standard Gibbs free energy of the reaction at the temperature of interest.

It is not possible to measure absolute potentials, only differences in potential. Thus electrometric measurements of both Eh and pH are based on the potential difference between a reference electrode of constant voltage and an indicator electrode whose voltage depends on the concentration of electrons or protons.

---

[2]The Stockholm convention for writing Eh and redox expressions will be used throughout this chapter (see Ives and Janz, 1961, p. 26).

The ultimate reference for Eh and pH measurements is the hydrogen electrode which operationally corresponds to the hydrogen gas-hydrogen ion couple

$$\frac{1}{2} H_2 (g) = H^+ + e^- \tag{4}$$

This electrode may be formed by bubbling absolutely pure water-saturated hydrogen gas over a platinum surface which has been coated with a catalytic layer of platinum black. The reference state is fixed by the assumption that E_H° for the hydrogen gas-hydrogen ion couple is zero at all temperatures. The potential of the electrode is then given by

$$E_H = -\frac{RT}{F} \ln \frac{[H^+]}{[P_{H_2}]^{1/2}} \tag{5}$$

The standard hydrogen electrode is formed by placing the hydrogen electrode in an acid solution in which $[H^+] = 1$. Thus with $P_{H_2} = 1$, $E_H = 0$. For Eh measurements a standard hydrogen electrode is connected to the sample solution by a salt bridge, and a second indicator electrode, which must be chemically inert and which acts as an electron donor or acceptor, is placed in the sample solution to complete the circuit. The indicator electrode is usually made of platinum. If the sample solution contains ferric and ferrous ions, then

$$Fe^{3+} + e^- = Fe^{2+} \tag{6}$$

The total redox reaction is given by the sum of the individual redox couples or

$$\frac{1}{2} H_2 (g) + Fe^{3+} = H^+ + Fe^{2+} \tag{7}$$

The Eh is

$$\text{Eh} = E^\circ_{Fe^{3+},Fe^{2+}} + \frac{RT}{F} \ln \frac{[Fe^{3+}]}{[Fe^{2+}]} - E^\circ_H - \frac{RT}{F} \ln \frac{[H^+]}{[P_{H_2}]} \qquad (8)$$

But $E^\circ_H = 0$, $[H^+] = 1$, and $P_{H_2} = 1$. Thus (8) simplifies to

$$\text{Eh} = E^\circ_{Fe^{3+},Fe^{2+}} + \frac{RT}{F} \ln \frac{[Fe^{3+}]}{[Fe^{2+}]} \qquad (9)$$

which is the Eh dependence of the ferric-ferrous iron couple. The Eh of other redox couples may be similarly defined.

The hydrogen electrode is also the ultimate reference for pH measurements. If the electrode is placed in a solution of variable pH, with $P_{H_2} = 1$ constant, then the response of the electrode as given by expression (5) reduces to

$$E_H = -\frac{RT}{F} \ln [H^+] \qquad (10)$$

By definition pH = -log $[H^+]$. Thus (10) simplifies to

$$E_H = 1.984 \times 10^{-4}\, T \times \text{pH} \qquad (11)$$

At 25°C this is equivalent to

$$\text{pH} = \frac{E_H}{0.05916} \qquad (12)$$

In other words a one unit increase in pH corresponds to an increase of 59.16 mv in the measured potential at 25°C. If we connect a standard hydrogen electrode to the test solution through a salt bridge, but instead of an inert platinum electrode place a second hydrogen electrode in the solution, then, because $E_H = 0$ for the standard hydrogen electrode, the pH of the solution will be given by expression (12) at 25°C.

Although Eh and pH measurements are based ultimately on the potential of the hydrogen electrode, it is not practical for use in routine Eh-pH determinations. Proper operation of the hydrogen electrode requires large volumes of pure H_2 gas of accurately known partial pressure and the total exclusion of oxygen. The electrode cannot be used as a pH or redox indicator in the presence of redox-sensitive species because of its strong reducing action. Electrode response is slow, particularly in poorly buffered solutions, and a single measurement may take an hour or more. Response is further inhibited by H_2S and other species that poison the platinum surface (see Bates, 1964, p. 231).

Eh AND pH DETERMINATION

Colorimetric Methods

The routine measurement of Eh and pH under a wide variety of conditions has been made possible by the development of secondary reference and indicator electrodes, which are more versatile than the hydrogen electrode, and through the use of colorimetric Eh and pH indicators. Colorimetric methods depend on the fact that certain redox and acid-base indicators exhibit different colors in solutions of different Eh or pH, respectively. Eh or pH measurement involves adding a colored redox or pH-sensitive indicator to the sample, and visually or photometrically comparing the color that develops to that of standards containing the same indicator at known Eh or pH.

There are many limitations to using colorimetric methods in natural environments. First, colored or turbid samples make it impossible to determine precisely the color of the indicator. Second, each indicator gives its optimum accuracy (±3 to 6 mv, ± 0.05 to 0.1 pH units) over a range of only a few 100 mv or 1 or 2 pH units, so that several indicators may be required to cover the range of interest. In addition, the indicator itself has a capacity for electrons or protons, and may therefore change the sample being measured if the sample is poorly poised or buffered. Insofar as most Eh and pH measurements today are done electrometrically we will not comment further on colorimetric methods. The interested reader is referred to Zobell (1946, p. 489) for a review of colorimetric Eh determinations. Bates (1964, p. 131) and Willard et al. (1951, p. 184) discuss colorimetric pH techniques in some detail.

Electrometric Methods

The most widely used secondary reference electrodes for Eh and pH measurement are the calomel, or mercury-mercurous chloride electrode, and the silver-silver chloride electrode. The Eh indicator electrode is most often made of bright platinum. Gold electrodes have also been suggested for this purpose, but should not be used in poorly poised natural waters (see Barnes and Back, 1964, and Manheim, 1961). The pH indicator electrode that is now employed almost exclusively is the glass electrode. The quinhydrone electrode is capable of greater precision and accuracy than the glass electrode, but its useful operation is limited to the pH range from 0 to 8. Measurement of pH includes dissolving quinhydrone in the sample, a procedure that can alter the pH of poorly buffered samples. The quinhydrone electrode is considerably less versatile than the glass electrode for measurements in natural environments, and will not be considered further here (see Bates, 1964, p. 256; Mattock and Taylor, 1961, p. 86).

Electrometers

An extensive review of electrometers for Eh and pH measurement is beyond the scope of this chapter; however, a few remarks concerning the capabilities of such instruments seem appropriate. Basically the two kinds of meters suitable for Eh and pH measurement are the null-balance potentiometer and the direct reading meter. The null meter can provide somewhat greater accuracy and sensitivity because of its extremely low current demand at the balance point. However, recent advances in circuit design have greatly improved the capabilities of direct reading meters. The best of both modern types now provide repeatabilities and relative accuracies of better than ±1 mv or ±0.01 pH units.

For maximum convenience and utility, a meter for field use should be battery operated, have scales permitting the direct reading of pH and millivolts, and be provided with a manual thermal compensator, preferably with one degree settings from 0 to 50°C (see p. 626). The direct reading type is the most convenient if the meter is to be used for alkalinity or redox titrations. A transistorized instrument is superior to one employing vacuum tubes with regard to speed of warmup and freedom from drift and electrical interferences (see p.   ). The reader will find extensive discussions of the principles and problems of meter operation in Mattock and Taylor, (1961, p. 198) and Bates (1964, p. 339).

Reference Electrodes

Liquid Junction Effects. The operation of secondary reference electrodes involves the use of a salt bridge to connect the internal electrochemical reference cell to the solution in which the electrode is immersed (Fig. 2). At the point where mixing of electrolyte and sample takes place, a liquid junction is formed. Cations and anions migrate across the junction in response to concentration gradients. In some cases this effect and physical flow can lead to significant contamination of dilute samples by concentrated electrolytes and conversely. In addition, the liquid junction potential, E_j, which is formed may lead to an appreciable error in the measured potential.

The size of E_j depends chiefly on the relative mobilities and concentrations of the major ions in the electrolyte and the sample, the rate of flow of the electrolyte, and the physical design of the junction itself. H^+ and OH^- move several times faster than other common ions so that strong acids or bases produce junction potentials up to several tens of millivolts. The salts used in bridge solutions (KCl, K_2SO_4, KNO_3, NH_4NO_3) are chosen so that corresponding cations and anions have nearly the same mobilities. Thus the electrolyte itself contributes minimally to the junction potential.

The danger of junction potentials caused by ions from the sample migrating into the salt bridge can be avoided largely by using a relatively concentrated electrolyte to swamp out the effect of sample ions, and by establishing a constant flow of electrolyte into the sample (see Dole, 1941, p. 111). Generally, proper flow is assured with an open electrolyte filling hole (see Fig. 2) and a constant height of electrolyte above the level of the solution in which the electrode is immersed during calibration and measurement.

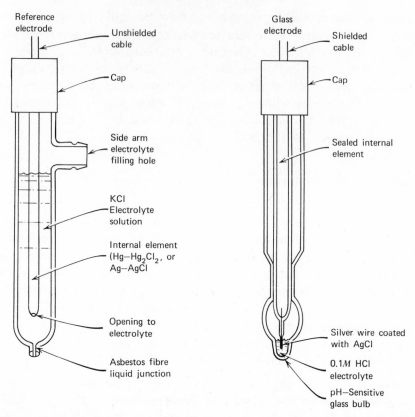

Fig. 2   Basic design of a typical commercial reference electrode and glass electrode.

Liquid junction effects are most serious in muds or other sediment-rich samples. In order to maintain adequate flow and to avoid junction clogging in such systems it may be necessary to apply external pressure to the electrolyte through an elevated electrolyte reservoir connected to the electrode-filling tube. When measurements are made in fine-grained sediments, the so-called suspension effect may lead to negative errors of several hundred millivolts or of several pH units. The effect is caused apparently by highly charged colloidal materials in the sediment which interfere with the relative mobilities of K^+ and Cl^- ions from the reference-electrode salt bridge. The error is most serious in very fine-grained and poorly aggregated materials at pHs below 7. When Eh or pH measurements are made near the interface between a sediment and overlying body of water, the magnitude of the effect can be estimated by measuring the potential difference between one reference electrode inserted into the sediment and a second one placed above the first in the overlying solution. In open or interstitial waters of constant sodium or potassium ion content, a sodium or potassium glass electrode may be used as a reference, thus avoiding problems owing to liquid junction effects (Bates, 1964, p. 274; Wilde and Rodgers, 1968).

Table 1 presents a summary of junction characteristics and the importance of liquid junction design.

The Calomel and Silver-Silver Chloride Electrodes. Operation of the calomel electrode is based on the half-cell reaction

$$Hg_2Cl_2 \ (s) + 2e^- = 2Hg \ (1) + 2Cl^-$$

(13)

calomel

The potential due to this reaction is

$$E_c = E_c^\circ - \frac{RT}{F} \ \ln \ [Cl^-]$$

(14)

where E_c° is the standard potential of the reaction. The internal element of the electrode contains a pool of liquid mercury which is connected by a platinum wire to the external circuit. The mercury is in contact with a paste of finely divided mercury and calomel. A KCl solution saturated with Hg_2Cl_2 and in contact with both the mercury and calomel acts as a salt bridge providing electrical connection to the test solution (Fig. 2). Because the salt bridge involves a liquid junction, the measured voltage of the calomel electrode is given by

$$E_c' = E_c + E_j$$

(15)

where E_j is the liquid junction potential.

TABLE 1 Common reference electrode liquid junction designs and their characteristics[a]

	Palladium Wire Annulus	Asbestos Fiber	Porous Ceramic Plug	Ground-Glass Sleeve
Typical flow rates (ml/day)	0.01	0.1–1	1–3	3–5
Stability of E_j per day	±0.2 mv (±0.003 pH)	±2 mv (±0.03 pH)	Relatively stable, but figures not available.	±0.06 mv (±0.001 pH)
Effect of pressure	Flow to 5 ml/day at 10 psi	Flow to 10–50 ml/day at 10 psi	May be pressurized, but flow data not available.	Flow too high for pressure operation.
Clogging characteristics, limitations	Little tendency to clog in viscous material or suspensions. Cannot be used in strongly oxidizing or reducing media.	Tendency to clog in viscous material or suspensions. Pressurization is not advised.	Less subject to clogging than fiber junction. Electrolyte leakage may be excessive in small, dilute samples.	Difficult to clog. Electrolyte leakage may be excessive in small, dilute samples.
Recommended use	Unsuitable for use in natural samples unless they are relatively free of redox species.	Recommended for use in clear and dilute samples.	Best choice for measurements in sediment-rich samples where external pressure is required.	May be used in samples containing suspended material.

[a]Data in large part from *The Industrial pH Handbook*, 1957, Beckman Instruments, Inc. See also Mattock and Taylor (1961, p. 180).

The Ag-AgCl electrode is often used as the internal reference in glass electrodes (see p. 625) and less frequently as a separate reference electrode. As a separate reference, the electrode consists commonly of metallic silver which has been electrodeposited on a platinum wire and coated with solid AgCl in contact with a KCl electrolyte which also acts as a salt bridge. The potential of the electrode is based on the half-cell reaction

$$AgCl(s) + e^- = Ag(s) + Cl^- \qquad (16)$$

The measured voltage of the Ag-AgCl electrode and salt bridge is

$$E'_s = E_s + E_j \qquad (17)$$

where E_s, the potential due to the half-cell reaction, is

$$E_s = E°_s - \frac{RT}{F} \ln [Cl^-] \qquad (18)$$

Standard potentials, $E°_c$ and $E°_s$, relative to the standard hydrogen electrode from 0 to 50°C are given in Table 2.

Tabulated values of E_c and E'_c are available in the literature (e.g., Ives and Janz, 1961, p. 161; Bates, 1964, p. 278). Insofar as measured potentials correspond to E'_c, these data are most useful and have accordingly been listed in Table 3 for several KCl concentrations from 0 to 50°C. At 25°C, E'_c includes a liquid junction potential from +2 to +4 mv. Because E_j is responsive to H$^+$ and OH$^-$ ions, below pH 2, E'_c is about 1 mv more positive than values in Table 2, and above pH 12, 1 mv less than these values.

Values of E'_s at temperatures other than 25°C are not readily available in the literature. However, E'_s may be calculated from the data in Tables 2 and 3. If a calomel and an Ag-AgCl electrode, both containing the same KCl electrolyte, are placed in the same solution, then because [Cl$^-$] and E_j are constant, the potential difference between the two electrodes is

$$E'_s - E'_c = E_s - E_c = E°_s - E°_c \qquad (19)$$

The potential of the Ag-AgCl electrode including E_j is then

$$E'_s = E'_c + (E^\circ_s - E^\circ_c) \tag{20}$$

Thus at 25°C, E'_s for a 3N KCl electrolyte is

$$E'_s = 0.2549 - 0.0459 = 0.2090 \text{ volt} \tag{21}$$

The other data in Table 4 have been calculated in a similar manner.

Several calomel and AgCl reference electrodes are briefly characterized in Table 5. Calomel electrodes are best limited to use between −5 and 70°C because of calomel decomposition at higher temperatures. Ag-AgCl electrodes may be used from −5 to well above 100°C. Below 70°C both electrodes are about equally suitable, and have roughly the same limitations in natural environments. E_c and E_s are affected by dissolved oxygen in acid waters. Bromide poisoning also affects both electrodes, and may be a problem in brines and sea water (65 mg/1 Br). Bates (1964, p. 280) notes an error of −0.1 to −0.2 mv in E_s due to about 30 mg/1 Br. Sulfide can also alter E_s.

TABLE 2   Standard potentials of the mercury-mercurous chloride (calomel) and silver-silver chloride electrodes from 0 to 50°C, in volts

$T(°C)$	E°_c Hg-Hg$_2$Cl$_2^a$	E°_s Ag-AgCl^b	$-(E^\circ_s - E^\circ_c)$
0	0.274	0.2366	0.0374
5	0.2729	0.2341	0.0388
10	0.2719	0.2314	0.0405
15	0.2708	0.2286	0.0422
20	0.2696	0.2256	0.0440
25	0.2682	0.2223	0.0459
30	0.2666	0.2190	0.0476
35	0.2649	0.2157	0.0492
40	0.2630	0.2121	0.0509
45	0.2610	0.2084	0.0526
50	0.2584	0.2045	0.0539

[a] Ives and Janz, 1961, p. 138.

[b] Bates, 1964, p. 285.

TABLE 3 Standard potentials (E_c') of calomel electrodes
from 0 to 50°C, in volts, as a function of KCl normality[a]

$T(°C)$	0.1 N[b]	1 N[c]	3 N	3.5 N[d]	Saturated (4.16 N at 25°C)
0	(0.337)	0.2883	(0.262)	–	0.2602[c]
5	(0.337)	(0.2877)	(0.261)	–	(0.2574)
10	(0.3363)	0.2868	0.2602	0.2556	0.2543
15	0.3361	(0.2857)	(0.2586)	(0.2539)	0.2511
20	0.3358	0.2844	0.2569	0.2520	0.2479
25	0.3356	0.2830	0.2549	0.2501	0.2444
30	0.3354	0.2815	0.2530	0.2481	0.2411
35	0.3351	(0.2799)	(0.2509)	(0.2460)	0.2376
40	(0.3349)	0.2782	0.2487	0.2439	0.2340
45	(0.3346)	(0.2763)	(0.246)	(0.242)	(0.2306)
50	(0.334)	0.2745	(0.244)	(0.240)	0.2272[c]

[a] All data from Bates (1964, p. 278) unless otherwise noted. Values in parentheses are based on graphic interpolation or extrapolation.

[b] These potentials are for an electrode with double salt bridge, a 0.1 N internal electrolyte, and KCl saturated external bridge.

[c] Ives and Janz (1961, Table 7, p. 161). 1 N values calculated from data same table assuming $E_j = +2.9$ mv.

[d] The 3.5 N electrode is saturated below 7°C.

Reference electrode poisoning is rarely significant during an individual measurement. However, poisoning effects are usually cumulative and can lead to a shift of 10 mv or more in potential after a few months of electrode use. If calibration of an electrode indicates an error of more than ±5 mv (see Table 7), the cause may be electrolyte contamination or poisoning of the internal element (Fig. 2). The electrolyte reservoir should be drained, rinsed several times, and refilled with fresh electrolyte. If this does not improve response, the electrode should be discarded.

As noted in Table 5 the only commercially available filling solutions are the saturated KCl solution for calomel electrodes and the 4 N KCl solution saturated with AgCl for Ag-AgCl electrodes. All electrolytes more concentrated than about 3.7 N become saturated with KCl at temperatures above 0°C. The 3 N electrolyte is

TABLE 4   Standard potentials (E_s') of silver-silver chloride electrodes from 0 to 50°C, in volts, as a function of KCl normality[a]

T(°C)	3 N	Saturated (4.16 N at 25°C)	At 25°C N	E_s'
0	0.225	0.2228	0.1	0.2897
5	0.222	0.2186	1.0	0.2371
10	0.2197	0.2138	3.5	0.2052
15	0.2164	0.2089	4.0	0.200[b]
20	0.2129	0.2039		
25	0.2090	0.1985		
30	0.2054	0.1935		
35	0.2017	0.1884		
40	0.1978	0.1831		
45	0.193	0.1780		
50	0.190	0.1733		

[a] See text for method of calculating E_s'.

[b] Determined by graphic interpolation.

unsaturated and probably the best choice for general use in either the calomel or AgCl reference electrodes.

When samples would be contaminated seriously by chloride ions, it may be desirable to use a reference electrode with a double salt bridge. These electrodes are available commercially from major suppliers, as are cheaper external salt bridge tubes designed to accomodate a standard reference electrode. The outer salt bridge may be filled with an electrolyte (usually saturated) of K_2SO_4, KNO_3, or NH_4NO_3.

Eh Measurement

Significance and Limitations. In waters in equilibrium with atmospheric oxygen, there is no correlation between the measured Eh and dissolved oxygen (DO) concentration. The reaction

$$\frac{1}{2} O_2 + 2H^+ + 2e^- = H_2O \tag{22}$$

which represents the upper stability limit for water (Fig. 1) is too slow to control measured Eh. In fact, based on reaction 22, the

TABLE 5 Characteristics of some calomel and silver-silver chloride reference electrodes in terms of KCl-normality

Electrode	Electrolyte	Advantages	Disadvantages	Remarks
Calomel	Saturated KCl (4.16 N at 25°C)	Electrolyte easy to maintain and least vulnerable to poisoning by sample. Has smallest E_j.	Potential not as reproducible as that of unsaturated electrodes. Thermal hysteresis due to solution and recrystallization of KCl. Tendency for KCl crystals to clog junction.	This is the most popular reference electrode. KCl clogging can be minimized by filling with electrolyte at temperature of operation.
Calomel	3 N KCl	No KCl crystal clogging problems. More rapid adjustment to temperature changes than saturated calomel. E_c' less temperature dependent than for saturated electrolyte.	Electrolyte must be made up, and is slightly more difficult to maintain than the saturated calomel.	This electrode is superior in overall performance to the saturated calomel.
Ag-AgCl	Saturated KCl (4.16 N at 25°C)	Same as saturated calomel.	Same as saturated calomel. Also electrolyte must be presaturated with AgCl.	The 4 N electrolyte is the only one sold, but it may be saturated by adding a few crystals of a KCl-AgCl mixture.[a]

[a]Beckman Instruments sells this mixture which may also be used to make up less concentrated electrolytes.

upper limit of Eh measurements in environments in contact with the atmosphere, as shown in Fig. 1, corresponds to an oxygen partial pressure of less than 10^{-10} atmosphere. Oxygen pressures in stagnant surface waters may sometimes be 10^{-10} atmosphere or lower. However, in most surface waters at 25°C, P_{O_2} exceeds $10^{-3.6}$ atmosphere, which is equivalent to a DO content of 0.01 ppm. If reaction 22 controlled Eh, Eh and pH measurements in such waters should lie within the narrow crosshatched area in Fig. 1. To explain this seeming disparity, Sato (1960) has proposed that Eh measurements in oxygenated surface waters are controlled not by reaction 22 but by the faster reaction

$$O_2 + 2H^+ + 2e^- = H_2O_2 \qquad (23)$$

In any case, most surface waters contain more than 0.01 ppm DO. Their state of oxidation is then most usefully determined by a DO analysis using the Winkler method (Rainwater and Thatcher, 1960) or any one of the commercially available oxygen electrodes. The measurement of Eh should be considered only if DO is below the limits of detection (about 0.01 ppm).

Eh has been defined as the equilibrium potential of an oxidation-reduction reaction relative to the potential of a standard hydrogen electrode. Garrels and Christ (1965) and Cloke (1966) among others have shown how to calculate the Eh dependence of any redox reaction from free energy data. Although such calculations may be qualitatively instructive, they are often without operational significance, for only if the reaction is readily reversible can the measured Eh equal the Eh calculated through the Nernst equation. However, most redox reactions in natural environments are not readily reversible. This applies to reaction 22, and it is also true of biologically mediated processes such as photosynthesis and respiration. When there are substantial concentrations of active organic material, the Eh is ultimately controlled by the rate and status of such reactions, and tends to drift continuously as the concentration of reactants decreases.

Eh readings free of drift are only possible in well-poised systems. Such systems, which are also described as having high redox capacity, tend to resist changes in Eh, just as systems which have high pH buffer capacity resist changes in pH (see p. 621). What we are concerned with here is not the total oxidation or reduction capacity (see Zobell, 1946, p. 492) but the redox capacity at the Eh of the natural sample. The redox capacity, ρ, at any Eh is defined as the quantity of strong reductant in moles, which must be added to a liter of the sample solution to lower the Eh by 1 volt (Nightingale, 1958, p. 268). That is

$$\rho \equiv \frac{dC_r}{d(\text{Eh})} \tag{24}$$

(Strictly speaking ρ is the oxidation capacity. The reduction capacity similarly defined is equal but opposite in sign to ρ.) In well-poised systems ρ is large, and Eh measurements are relatively easy and reproducible.

The concept of redox capacity may be illustrated in terms of the ferrous-ferric iron couple as expressed by Equation (6). At 25°C the Eh of this couple simplifies to

$$\text{Eh} = 0.771 + 0.0592 \log \frac{[Fe^{3+}]}{[Fe^{2+}]} \tag{25}$$

The redox capacity is greatest when $[Fe^{3+}] = [Fe^{2+}]$ and the total iron concentration is large. The system is poorly poised (ρ small) when $[Fe^{3+}]/[Fe^{2+}]$ is less than 10^{-5} or greater than 10^5 or when both Fe^{3+} and Fe^{2+} are less than about 10^{-6} molar.

W. Stumm (1966, p. 10) describes redox capacity in terms of the net current, which equals zero for the sum of the oxidation and reduction reactions at equilibrium. In poorly poised systems the net current is close to zero for a wide range of Eh values in the vicinity of true equilibrium, with the result that it may not be possible to measure a stable or meaningful Eh.

As a further complication the Eh of natural systems can represent a mixed potential. The net current at the electrodes is the algebraic sum of anodic and cathodic currents produced by all active oxidation and reduction reactions taken together (Charlot et al., 1962, p. 11). Eh is the potential at which this net current equals zero, but as a mixed potential it will not correspond to the equilibrium potential for any one redox couple.

TABLE 6 The effect of environment on Eh response during measurement

Types of Environment	Redox Characteristics	Eh Response
Oxygenated waters; rain, streams, shallow lakes and ocean, some ground waters.	Poorly poised; electroactive, reduced species may be absent	Reproducible Eh may be impossible to measure due to drift. Dissolved oxygen measurements are simpler and more meaningful than Eh under these conditions.
Some acid-mine waters, iron-rich ground waters at pHs above 4.	May be well poised; Eh controlled by reversible redox reactions	Eh relatively easy to measure with rapid response and little or no drift. Such systems are uncommon in natural environments.
Organic-rich systems, bog waters, waterlogged soils, and fine-grained sediments; heavily polluted or stagnant waters.	May be poorly poised due to absence of oxidized species. Eh possibly a mixed potential controlled by irreversible redox reactions (especially biologically mediated)	Rapid to slow initial response followed by slow drift. Absolute equilibrium never attained in the presence of living organisms. The lowest Eh values are found in these environments.
Bodies of water or sediment subject to frequent mixing or turnover; heterogeneous environments.	Eh variable and unstable	A reproducible Eh may be impossible to measure.

TABLE 7   Calibration procedures for reference electrodes and the platinum electrode-reference electrode combination

Reference Electrodes. Compare the questionable electrode with one known to be reliable. For the same KCl electrolytes; 2 calomels immersed in the same solution should agree within ±5 mv, a calomel and an Ag-AgCl electrode within 46±5 mv at 25°C (see Table 2).

Platinum Electrode-Reference Electrode. Check the platinum electrode against a reference electrode in Zobell solution (Garrels, 1960, p. 68). Zobell solution is made by dissolving 1.2672 gm (0.003M) potassium ferrocyanide, 0.9878 gm (0.003M) potassium ferricyanide, and 7.4557 gm (0.1M) potassium chloride in 1 liter of solution. Near 25°C.

$$Eh_{Zobell} = 0.429 + 0.0024(25 - t)$$

where t is in degrees Celsius. For the saturated calomel electrode

$$E(observed) = 0.185 + 0.00164(25 - t)$$

For the saturated Ag-AgCl electrode

$$E(observed) = 0.231 + 0.0013(25 - t)$$

In spite of the foregoing difficulties, in acid, oxygenated waters, and in transitional or isolated environments at pHs above 4 (see Fig. 1), Eh measurements are sometimes relatively fast and reproducible, and appear to have thermodynamic significance. This probably reflects the common occurrence of iron species under such conditions. Redox reactions among carbon, nitrogen, and sulfur species are generally too slow to affect an Eh reading when significant concentrations of oxidized and reduced iron species are present. In acid and oxygenated waters, the ferric-ferrous iron couple can give the water a high redox capacity. At pHs above 4, many anoxic waters are in equilibrium with both dissolved ferrous iron and solid ferric oxhydroxides. Stumm (1967) has found that the reaction

$$Fe(OH)_3 + 3H^+ + e^- = Fe^{2+} + 3H_2O \qquad (26)$$

is electroactive and can produce a stable and reproducible Eh at pH = 7 for Fe^{2+} concentrations as low as 0.6 ppm. Doyle (1968) concluded that reaction (26) is catalyzed by a layer of ferric oxyhydroxide, which tends to form on the surface of the platinum electrode during an Eh measurement (see Langmuir, 1969). (An analogous reaction may also occur in waters which contain Mn^{2+} ion and the higher oxides of manganese.)

Thus, because of reactions (6) and (26), Eh measurements are sometimes possible which are consistent with the activities of iron species present. Frequently, however, an unambiguous quantitative interpretation of Eh measurements is not possible. The quantitative approach has been tried with mixed success by Sato (1960), Berner (1963), and Back and Barnes (1965), among others, in studies of the occurrence and behavior of inorganic species of iron and sulfur. In general, however, Eh should be thought of as a qualitative expression of the state of oxidation or reduction of a natural system (see Stumm, 1966; Morris and Stumm, 1967).

The effect of the redox characteristics of the environment on Eh response during measurement is summarized in Table 6. The first three categories describe conditions in homogeneous environments. The fourth represents the unstable response which may be encountered in physically and chemically heterogeneous systems.

The Platinum Electrode. The potential measured with a platinum electrode and a secondary reference electrode must be corrected to give true Eh. For the calomel and AgCl electrodes the correction consists of adding the standard half-cell potential of the reference electrode to the measured potential (see Tables 3 and 4). Thus, for example,

$$Eh = E \text{ (measured)} + E'_c \tag{27}$$

The rate of a redox reaction at the platinum electrode is proportional to the electrode surface area (Charlot et al., 1962, p. 118). Thus laboratory Eh determinations may be expedited with a platinized platinum electrode which has a surface area of up to several hundred times that of polished platinum. Unfortunately, in field studies platinized platinum is less convenient and more vulnerable to poisoning than bright platinum. As a compromise, the thimble-type bright platinum electrode, which has a larger surface area than the inlay or loop types, is recommended for solution Eh studies. For measurements in muds and waterlogged soils, when sample disturbance must be avoided, a bright platinum wire electrode is probably the best choice (Zobell, 1946, p. 495).

Poisoning of the platinum electrode is a common occurrence during field Eh work. Poisoning can reduce the effective electrode surface area to 1% of its original value, thus making accurate Eh readings difficult or impossible (Ives and Janz, 1961, p. 277). In aerated waters there is a tendency for dissolved oxygen to react with the electrode, coating it with a film of platinum oxide or hydroxide (Charlot et al., 1962, p. 147; Watanabe and Devanathan, 1964). At low Ehs, H_2S may complex with the platinum surface. H_2, CH_4, and CO have also been found to interfere with electrode function, as have films of organic material or ferric oxyhydroxides.

To minimize the danger of poisoning, platinum electrodes should be carefully cleaned between measurements. This may be accomplished by washing in detergent solution and burnishing the tip with scouring powder, jeweler's rouge, or crocus cloth. Doyle (1968) recommends cleaning the platinum electrode with a paste of powdered pyrex glass and glycerine, and then rinsing with distilled water.

Before and after an Eh measurement, operation of the platinum and reference electrode combination and meter should be checked in Zobell solution (see Table 7). If the check is in error by more than ±5 mv with a freshly cleaned platinum electrode, the reference electrode may be poisoned and should be compared to a second reference electrode that is known to be reliable (see Table 7).

As a general practice, Zobell (1946, p. 500) and Garrels and Christ (1965, p. 136) recommend alternating two platinum electrodes in the measuring circuit during an Eh determination. In this way electrode poisoning is less likely to go undetected, and the reproducibility of a single measurement is evident from the difference in potential between readings with the two electrodes.

Stable and reproducible Eh readings are possible only if the potential registered by the electrodes and meter reflects solely the redox state of the unaltered sample. Unfortunately, this is rarely true. Drift and instability of Eh readings in natural samples are more often the rule than the exception. We have noted that drift may occur as a result of low redox capacity of the sample or poisoning of the electrodes. Other causes of instability and drift, like poisoning, can be largely eliminated if suitable precautions are taken. Electrode cables, measuring vessel, and millivoltmeter should be carefully shielded and grounded. This will minimize electrical interferences caused by moving objects such as pumps and motors, power lines, rain, and wind (see Back and Barnes, 1961). The electrodes and Zobell solution should be brought to the temperature of the sample ($\pm 1°C$, if possible), and the meter should be properly warmed up and stabilized at ambient temperature. All measuring equipment should be kept out of direct sunlight. Otherwise changes in the potential of reference electrodes and in meter characteristics due to thermal effects may cause appreciable Eh drift. Whatever detailed measuring procedure is used, it should not permit introducing oxygen or other contaminants into the sample. Because some oxidation is almost unavoidable in studies of reduced environments, the lowest measured Eh is usually closest to the true value (Sato, 1961).

Several authors have suggested that Eh be recorded continuously to show the existence or absence of drift (e.g., Berner, 1963, p. 569). Because drift is the rule, the observer must generally decide what drift rate is negligible for his purposes. In poorly poised waters appreciable drift may continue for hours (Back and Barnes, 1961). In muds, Eh may become relatively stable in less than 10 minutes, although slow drift will usually continue indefinitely in organic-rich sediments (e.g., Zobell, 1946, p. 499). Sometimes Eh can be measured in well-poised systems with a reproducibility of ± 5 mv, but in poorly poised systems the reproducibility is often no better than ± 50 mv.

When there is relative movement of electrodes and sample, a flowing potential results and may cause a negative error of several tens of millivolts (e.g., Back and Barnes, 1961; MacInnis, 1961, p. 437). Relative movement of electrodes and solution is helpful in speeding up the stabilization of Eh readings in poorly poised samples, but should be stopped during readings to eliminate flowing potential.

Eh readings in fine-grained sediments are susceptible to errors up to several hundred millivolts because of the suspension effect on the liquid junction potential of the reference electrode. This problem has already been discussed in detail on p. 606, and need not be considered further.

If at all possible, Eh should be measured *in situ*. Eh changes during collection, transport, and storage of samples as a result of changes in P_{O_2}, temperature, biological activity, and so on, are often considerable. The danger of oxidation is particularly great in samples of very low Eh, which are usually poorly poised. Water especially should be measured *in situ*, or errors (usually positive) up to several hundred millivolts may result (Lisitsin, 1963). The Eh of muds or waterlogged soils is also best measured in place. If sample removal is necessary, Zobell (1946, p. 507) suggests cooling the sample to just above the freezing point and storing in a nitrogen gas atmosphere. These procedures inhibit biological and chemical reactions which can substantially raise or lower Eh. For best results samples should be measured within a few hours after collection.

Specific field Eh measuring techniques, which have been used in different environments, are described briefly in Table 9. The reader is referred to cited references for more detail on exact methods and procedures.

pH Measurement

Significance and limitations. Stable and reproducible pH measurements are possible only in well-buffered systems. The reasons are analogous to those mentioned in the discussion of Eh measurements in well-poised systems. The hydrogen ion concentration in most natural waters is ultimately buffered by reactions with solid mineral substances and organic substances. However, only reactions between dissolved aqueous species, and other dissolved or colloidal-sized aqueous species, are fast enough to buffer effectively pH during a measurement.

A pH buffer resists pH change caused by the addition of an acid or base. The effective buffer capacity or buffer index (β) of a solution at any pH is defined as the number of moles of strong base which must be added to 1 liter of solution to raise the pH by one unit (Weber and Stumm, 1963; Butler, 1964, p. 150). That is,

$$\beta \equiv \frac{dC_b}{d(\text{pH})} \tag{28}$$

TABLE 8 The pH from 0 to 50°C of several buffers which are commercially available
Data are from Bates (1964, p. 76) unless otherwise noted [a, b, d]

T(°C)	0.05m KHC$_8$H$_4$O$_4$ (phthalate)	0.025mKH$_2$PO$_4$ 0.025Na$_2$HPO$_4$ (phosphate)	0.0253mKH$_2$PO$_4$ 0.0413mNa$_2$HPO$_4$ (phosphate)[c]	0.008695mKH$_2$PO$_4$ 0.03043mNa$_2$HPO$_4$ (phosphate)	0.01mNa$_2$B$_4$O$_7$·10H$_2$O (borax)
0	4.003	6.984	7.12	7.534	9.464
5	3.999	6.951	7.09	7.500	9.395
10	3.998	6.923	7.06	7.472	9.332
15	3.999	6.900	7.04	7.448	9.276
20	4.002	6.881	7.02	7.429	9.225
25	4.008	6.865	7.00	7.413	9.180
30	4.015	6.853	6.99	7.400	9.139
35	4.024	6.844	6.98	7.389	9.102
40	4.035	6.838	6.98	7.380	9.068
45	4.047	6.834	6.97	7.373	9.038
50	4.060	6.833	6.97	7.367	9.011

[a] All except the pH '7' buffer are NBS primary standards.
[b] Coleman Instruments offers pH 2.00 and 3.00 buffers, and Beckman Instruments the pH 12.45 buffer (saturated Ca(OH)$_2$ solution).
[c] The pH '7' buffer generally has the compositon shown and is based on Sorenson (Welcher, 1942, p. 330).
[d] At 25°C, approximate ionic strengths of these buffers are, from left to right, 0.0533, 0.10, 0.125, 0.090, and 0.02, respectively.

TABLE 9 Field methods of Eh and pH measurement

Type of Environment	Nature of Measurement	Method	Reference
Surface waters	Eh and pH of lake water with depth	Lowered probe with thermistor and platinum, high-pressure glass, and side-arm reference electrodes. Reference (unsaturated) connected to an elevated reservoir at surface.	Kramer (1961)
	Eh and pH of ocean water with depth	Lowered probe with combination platinum-glass electrode. Separate calomel electrode (3N KCl) attached to float at surface.	Manheim (1961)
Ground waters	Eh and pH of pumped well-water[a]	Water circulated through plastic tubing into sealed glass jar. Platinum, glass, and saturated calomel electrodes sealed into jar lid.	Back and Barnes (1961)
Sediments	Eh and pH of mine-seepage	Probe with platinum, glass, and calomel electrodes inside perforated plastic tube, sealed into hole in wall rock.	Sato (1960). See also Germanov et al. (1958)
	Eh and pH of mud	Eh measured in wide-mouthed bottle using platinum wire electrode and fiber junction calomel electrode inserted through a rubber stopper.	Zobell (1946). See also Berner (1963), Semenovich (1963), Kuznetsov (1963)
	Eh and pH[b] of soil	Measured with platinum, glass, and calomel electrodes inside a pencil-shaped probe designed for forced insertion into soil.	Starkey and Wight (1945)

[a]Pumping water-table aquifers may cause mixing with oxygen-rich recharge (see Heidel, 1963).
[b]Pommer (1967) also discusses soil pH measurements.

623

Addition of a strong acid has the reverse effect so that

$$\beta = -\frac{dC_a}{d(\text{pH})} \qquad (29)$$

The pH of most natural waters lies between 4.5 and 8.2. In this range β is generally a function of the relative and absolute concentrations of H_2CO_3 and HCO_3^-. The first dissociation constant of carbonic acid is

$$K_1 = \frac{[H\pm]\ [HCO_3^-]}{[H_2CO_3]} \qquad (30)$$

Taking logarithms and introducing the definition of pH and the value for K_1 at 25°C (Langmuir, 1968), we can write

$$\text{pH} = 6.36 + \log\frac{[HCO_3^-]}{[H_2CO_3]} \qquad (31)$$

For a given total alkalinity, buffer capacity is controlled by the HCO_3^- / H_2CO_3 ratio and is greatest at a pH of about 6.4 when this ratio is unity, and least when it is about 10^2 at pH 4.5, or 10^{-2} at pH 8.2[3]. Below pH 4.5 and above 8.2, β increases rapidly due to increasing concentrations of hydrogen ions and carbonate and hydroxyl ions, respectively. Low buffer capacity makes accurate pH measurement a problem in the vicinity of pH 4.5 and 8.2, particularly in waters containing less than about 50 mg/1 HCO_3^- at temperatures below 25°C. Thus pH measurements of rainwater and surface and ground waters from noncarbonate rock terrains may be difficult to measure more accurately than to about ± 0.1 pH units. In well-buffered samples, with proper care, pH may be determined with a reproducibility of about ± 0.02 units and an accuracy of ± 0.05 units in the field.

---

[3]The endpoints of titrations for CO_3^{2-} and HCO_3^- alkalinity with a strong acid are pH 8.2 and 4.5, respectively (see Rainwater and Thatcher, 1960, p. 94).

The glass electrode. The design of a typical glass electrode is shown in Fig. 2. The electrode usually contains an Ag-AgCl internal reference cell in contact with a buffered chloride electrolyte such as 0.1 molar HCl. The chloride ion activity of the electrolyte fixes the reference cell potential. The electrode tip is made of H^+ ion sensitive glass. Changes in electrode potential are caused chiefly by changes in hydrogen ion activity at the external surface of the glass membrane relative to its activity in the buffered internal electrolyte. (See Eisenman, 1967, p. 133, for a discussion of the detailed mechanisms of glass electrode response.)

Ideally, the potential across the glass membrane should be zero when $[H^+]$ is the same at both sides. This, however, is generally not the case because of dissimilar behavior of the glass inside and outside the pH-sensitive bulb. This residual potential is called the *asymmetry potential.*

Within the pH range of most natural waters the glass electrode exhibits a Nernst or near-Nernst response to changes in pH; that is, a change in the measured potential in volts is related to pH change by the factor $(RT \ln 10)/F = 1.984 \times 10^{-4}\ T$. (See Equation (2)). Thus at 0, 25, and 50°C, 1 pH unit corresponds to 54.2, 59.2, and 64.1 mv, respectively. Additional values of $(RT \ln 10)/F$ from 0 to 100°C are given by Bates (1964, p. 403).

Because the glass electrode must be calibrated in buffers of known pH before use, it has been found convenient to define an operational pH in terms of the calibration procedure or

$$pH_x = pH_b + \frac{(E_x - E_b)}{1.984 \times 10^{-4}\ T} \tag{32}$$

where the subscripts x and b represent the sample and buffer solutions, respectively. At 25°C this expression becomes

$$pH_x = pH_b + \frac{(E_x - E_b)}{0.05916} \tag{33}$$

As indicated by expression (33), the potential of the external reference electrode need not be known so long as it remains constant during calibration and pH measurement.

The pH of several commercially available buffers from 0 to 50°C is presented in Table 8. The pH values of National Bureau of Standards primary standards are given to four significant figures. However, in view of the approximations involved in buffer standardization these values may be no more accurate than about ±0.01 pH units (Mattock and Band, 1967, p. 34). Buffers should be prepared following the manufacturer's instructions, and distilled water should be used with a specific conductance of less than about 5 micromhos. Buffer solutions kept tightly capped and in a cool, dark place will often maintain their pH to within ± 0.01 units for six months or more. With moderate precautions at constant temperature any two of the buffers listed in Table 8, differing by 2 or 3 pH units and freshly made, should agree with each other in the laboratory to within ± 0.02 pH units. In the field, agreement should be within ± 0.05 pH units. If greater accuracy or precision is required, additional precautions must be taken (see Bates, 1964, p. 378). For more detailed information on the preparation and use of pH buffers see Clark (1960, p. 249), Mattock and Taylor (1961, p. 39), and Bates (1964, p. 124).

Most pH meters are provided with a manual thermal compensator, which is essentially a variable resistor, to adjust pH-mv response with temperature in accordance with the Nernst relationship. Many pH meters are also equipped for automatic thermal compensation. This is accomplished with a resistance thermometer which is immersed in sample and buffer solutions during measurement together with the glass and reference electrodes. Changes in temperature alter the potential of the internal reference cell and the asymmetry potential of the glass electrode, as well as the liquid junction potential and cell potential of the reference electrode. The total effect of these factors on the scale position of a particular pH value after temperature change is usually difficult to evaluate (see Bates, 1964, p. 365). Thermal compensation corrects for changes in scale length, but not changes in scale position. Thus the compensator cannot be used to adjust for differences in temperature between the buffer and the sample, or large errors may result. The only acceptable procedure is to set the manual thermal compensator (or immerse the automatic thermal compensator) at the temperature of the sample, and to make both the buffer check and pH measurement at this same temperature.

Inherent in the operational definition of pH is the assumption that the asymmetry potential of the glass electrode and the liquid junction potential, E_j, of the reference electrode are the same in the buffer solution and sample solution, and thus cancel out. This reasoning is valid as long as calibration and measurement are made at nearly constant temperature, and the ionic strength and pH of the buffer and sample do not differ greatly. The dissolved solids content of most natural waters is close enough to that of the typical buffer so that the effect of changes in ionic strength on E_j, and thus on pH, generally amounts to less than ±0.01 pH units. The error does become significant in sea water or more concentrated solutions where it may be ±0.01 pH units or more, but can be reduced if pH buffers are prepared in a synthetic medium of the same ionic strength as the sample (see Pytkowicz et al., 1967, p. 417).

The pH effect on E_j and thus on measured pH in acid and alkaline solutions has already been discussed (see p. 605). Of concern also is the non-Nernst response of the glass electrode in acid and alkaline solutions. In most solutions between pH 1 and 9, this error is much less than ± 0.01 pH units (MacInnes, 1961, p. 268). Below pH 1, glass electrode response exceeds 59.16 mv per pH unit at 25°C, possibly due to changes in water activity (Willard et al., 1951, p. 199). Unfortunately, there is no routine procedure to correct for acid errors. Above pH 9, glass electrode response becomes less than Nernst, usually because of interference from the high sodium ion concentrations commonly associated with alkaline solutions. The alkaline error of most general-purpose electrodes is negligible in sea water (pH 8.15, Na 0.5N), but may lead to a pH reading several tenths of a unit too low in some alkali lakes and brines. All major electrode manufacturers supply high-pH, low-sodium error electrodes which should be used in these cases.[4]

---

[4] Corning Glass Works makes an electrode with an error of less than 0.01 pH units at 50°C, pH = 10, and Na = 10N.

Alkaline errors increase appreciably with temperature, although acid errors do not. Nomograms provided with most electrodes may be used to approximate the alkaline pH error at any temperature and pH as a function of sodium ion concentration. For further discussion of acid and alkaline errors, see MacInnes (1961, p. 268), Mattock and Taylor (1961, p. 105), and Bates (1964, p. 316).

With a dc resistance of 50 to 500 megohms at room temperature, the membrane of the glass electrode is the principal source of electrical resistance in the pH measuring circuit. The magnitude of this resistance has a pronounced effect on pH measuring response and stability. At $50°C$ electrode resistance is usually less than 50 megohms, and response time may be only a few seconds. Below $50°C$ resistance increases exponentially. Thus at $10°C$ response is quite slow, and readings may be somewhat erratic, whereas near $0°C$ accurate pH measurement may become practically impossible in moderately or poorly buffered solutions. Low-resistance electrodes are the best choice for measurements below ambient temperature. Unfortunately, these electrodes have a relatively thin glass bulb which is easily broken and thus requires careful handling. The lower resistance of glass above $25°C$ permits high-temperature electrodes to be made with a thicker and more durable glass bulb. The practical temperature range of commercial glass electrodes is usually specified in literature provided by the manufacturer.

Improper use and performance of the glass electrode are major causes of pH measurement problems. The glass membrane should be hydrated before use by soaking for an hour or more in distilled water or a buffer with a pH close to that of the sample. Conditioning may be accomplished in transit by slipping a protective rubber or plastic cup filled with distilled water or buffer over the glass bulb. Unless the electrode is to be used on successive days, it may be stored dry.

Aging of the glass membrane leads to longer response times and instability of readings. Sometimes proper operation can be restored by dry storage for several weeks. Various rejuvenation procedures are suggested by manufacturers in their electrode brochures, but these are effective only occasionally.

To minimize the risk of breaking or scratching the glass electrode during insertion into muds or other sediment rich samples, special high-strength penetration electrodes with a tapered conical tip should be used. In general, to avoid scratching the bulb should be blotted dry with soft, absorbent tissue rather than wiped after rinsing. If the pH of a number of similar water samples is being measured, the electrode should be rinsed with successive samples rather than with distilled water and wiped between samples. This procedure avoids wiping damage, saves time, and improves electrode response.

The best test for proper operation of the electrodes and meter prior to a field measurement is a double buffer check. The check requires two buffers that differ by 2 or 3 pH units and whose pH values bracket that of the sample. The meter is set at the pH of the first buffer, and the pH reading in the second buffer compared with its known pH. The difference should be within ± 0.05 pH units. The most common reasons for an unsatisfactory check are dirty or damaged electrodes, lack of thermal equilibrium between buffers and electrodes, and inaccurate buffers. It may be necessary to repeat the double buffer check several times before obtaining a satisfactory calibration. The necessity for repeating the check between measurements depends on the stability of the pH meter, changes in sample and ambient temperature, and the desired precision and accuracy of measurement. These and other measurement problems are considered in some detail by Barnes (1964) and Bates (1964, p. 374). Similar discussion is also found in manufacturers' electrode and meter brochures.

The glass electrode will respond with greater precision and speed if successive samples are close to each other in temperature, pH, and buffer capacity. The accuracy and speed of readings in poorly buffered samples can be improved if the sample is flowing, or can be sloshed or stirred around the electrodes for a few seconds before a reading is taken. The sample should be static during the actual reading, or a flowing potential error results (see p. 620). Thus, Barnes (1964) recorded an error of -0.1 pH units in a stream moving 1 ft/sec. For best results measurement should be repeated several times with fresh batches of the sample until the pH of successive batches remains constant.

For routine field pH measurements in water samples, the author strongly recommends combination pH-reference electrodes (Langmuir, 1969). Their many advantages relative to separate glass and reference electrodes include (a) the reference electrolyte shields the glass electrode internal and thus reduces electrical interferences;

(b) errors due to differences in temperature are less likely than with separate electrodes; (c) cleaning and manipulation are simpler; (d) buffers may be kept in small-mouthed plastic bottles and need not be poured out during calibration, thus reducing the danger of errors due to contamination, evaporation, or temperature change.

Some examples of pH measurement techniques used in different environments are reviewed briefly in Table 9. In addition, Barnes (1964) has outlined the precautions and procedures which should be followed when studying the field pH of surface waters. Pressure-compensated glass electrode cells for pH measurement in deep wells or the open ocean are described by Kunkler et al. (1967) and Disteche (1959; 1962).

Like Eh, the pH of samples which are removed from the natural environment will change due to changes in temperature, pressure, loss or gain of dissolved gases, reactions with sediments and sample bottle, precipitation of solids, and organic processes. These effects will usually alter pH by ± 0.2 to ± 1 units after a few weeks of storage at ambient temperature (see Roberson et al., 1963; Back, 1963). Surface waters which are in equilibrium with the atmosphere at the time of collection are much less subject to change than ground waters. Changes in pH are most rapid and pronounced in poorly buffered samples, particularly if they are stored in plastic bottles which are usually quite permeable to O_2 and CO_2.

If samples are clear and well-buffered, it is sometimes possible to estimate the original pH to ± 0.1 units from a laboratory measurement made a few hours or days after collection. The sample should be stored at field temperature in a completely filled and tightly sealed hard-glass bottle, and brought to ambient temperature a few hours before measurement. The original pH may be estimated if the field to laboratory pH change is due to changes only in the dissociation constants of carbonic acid with temperature, and no carbonate or other mineral precipitation occurs. Tabulated corrections based on these assumptions for pure carbonate waters are given by Langelier (1946). Harvey (1957) lists similar corrections for the change in sea water pH with temperature and pressure. The reliability of such procedures should be spot-checked frequently against actual field pH measurements.

REFERENCES

Baas Becking, L. G. M., I. R. Kaplan, and D. Moore, 1960, Limits of the natural environment in terms of pH and oxidation-reduction potentials, *J. Geol.,* **68,** 243–284.

Back, W., 1963, Preliminary results of a study of calcium carbonate saturation of ground water in Central Florida, *Inter. Assoc. Scientific Hydrol.,* **8,** No. 3, 43–51.

———, and I. Barnes, 1961, Equipment for field measurement of electrochemical potentials, U.S. Geol. Surv. P.P. 424-Q, C366–C368.

———, 1965, Relation of electrochemical potentials and iron content to ground-water flow patterns, U.S. Geol. Surv. P.P. 498-C, 16 pp.

Barnes, I., 1964, Field measurement of alkalinity and pH, U.S. Geol. Surv. Water-Supply Paper 1535-H, 17 pp.

———, and W. Back, 1964, Geochemistry of iron-rich ground water of Southern Maryland, *J. Geol.,* **72,** 435–447.

Bates, R. G., 1964, Determination of pH, theory and practice, John Wiley and Sons, 435 pp.

Berner, R. A., 1963, Electrode studies of hydrogen sulfide in marine sediments, *Geochim. et Cosmochim. Acta,* **27,** 563–575.

Butler, J. N., 1964, Ionic equilibria, Addison-Wesley Publishing Co., 547 pp.

Charlot, G., J. Badoz-Lambling, and B. Tremillon, 1962, Electrochemical reactions, Elsevier Publishing Co., 376 pp.

Clark, W. M., 1960, Oxidation-reduction potentials of organic systems, The Williams and Wilkins Co., 584 pp.

Cloke, P. L., 1966, The geochemical application of Eh-pH diagrams, *J. Geol. Education,* **14,** 140–148.

Disteche, A., 1959, pH measurements with a glass electrode withstanding 1500 kg/cm^2 hydrostatic pressure, *Rev. Scientific Instruments,* **30,** 474–478.

———, 1962, Electrochemical measurements at high pressures: *J. Electrochemical Soc.,* **109,** 1084–1092.

Dole, M., 1941, The glass electrode, John Wiley and Sons, 332 pp.

Doyle, R. W., 1968, The origin of the ferrous ion-ferric oxide Nernst potential in environments containing dissolved ferrous iron, *Am. J. Sci.*, **266**, 840–859.

Eisenman, G., 1967, The origin of the glass-electrode potential, pp. 133–173, *in* Glass electrodes for hydrogen and other cations, George Eisenman, ed., Marcel Dekker, 582 pp.

Garrels, R. M., 1960, Mineral equilibria at low temperature and pressure, Harper and Brothers, 254 pp.

_____, and C. L. Christ, 1965, Solutions, minerals and equilibria, Harper and Row, 450 pp.

Germanov, A. I., G. A. Volkov, A. K. Lisitsin, and V. S. Serebrennikov, 1959, Investigation of the oxidation-reduction potential of ground waters, *Geochemistry,* No. 3, 322–329.

Harvey, H. W., 1957, The chemistry and fertility of sea waters, The Cambridge Univ. Press, 234 pp.

Heidel, S. G., 1965, Dissolved oxygen and iron in shallow wells at Salisbury, Md., *J. Am. Water Works Assoc.*, **57**, 239–244.

Ives, D. J. G., and G. J. Janz, 1961, Reference electrodes, Academic Press, 651 pp.

Kramer, J. R., 1961, Chemistry of Lake Erie, Great Lakes Res. Division, Inst. Sci. and Technology, The Univ. of Michigan, Publ. No. 7, 27–56.

Kunkler, J. L., F. C. Koopman, and F. A. Swenson, 1967, An instrument for measuring pH values in high-pressure environments, U.S. Geol. Surv. P.P. 575-B, B250-B253.

Kuznetsov, S. I., 1963, The oxidation-reduction potential of surface layers of bottom muds in different types of lakes, *Doklady Akad. Nauk SSSR.,* **151**, 180–182.

Langelier, W. F., 1946, Effect of temperature on the pH of natural waters, *J. Am. Water Works Assoc.*, **38**, p. 179–185.

Langmuir, D., 1968, Stability of calcite based on aqueous solubility measurements, *Geochimica et Cosmochimica Acta*, **32**, No. 8, 835–851.

_____, 1969, Geochemistry of iron in a coastal-plain ground water of the Camden, New Jersey, area, U.S. Geol. Surv. P.P. 650-C, C224-C235.

Lisitsin, A. K., 1963, Description of media in hydrogeochemical investigations, *Geochemistry*, 165—173.

MacInnes, D. A., 1961, The principles of electrochemistry, Dover Publications, 478 pp.

Manheim, F., 1961, In situ measurement of pH and Eh in natural waters and sediments, *Stockholm Contributions Geol.*, **8**, 27—36.

Mattock, G., and D. M. Band, 1967, Interpretation of pH and cation measurements, *in* Glass electrodes for hydrogen and other cations, George Eisenman, ed., Marcel Dekker, pp. 9—49.

———, and G. R. Taylor, 1961, pH measurement and titration, Macmillan, 406 pp.

Morris, J. C., W. Stumm, 1967, Redox equilibria and measurements of potentials in aquatic environment, pp. 270—285, *in* Equilibrium concepts in natural water systems, Robert F. Gould, ed., Am. Chemical Soc., Washington, D. C.

Nightingale, E. R., Jr., 1958, Poised oxidation-reduction systems, *Analytical Chemistry*, **30**, 267—272.

Pommer, A. M., 1967, Glass electrodes for soil waters and soil suspensions, 362—411, *in* Glass electrodes for hydrogen and other cations, George Eisenman, ed., Marcel Dekker, 582 pp.

Pytkowicz, R. M., D. R. Kester, and B. C. Burgener, 1967, Reproducibility of pH measurements in sea water, *J. Oceanography and Limnology*, **11**, 417—419.

Rainwater, F. H., and L. L. Thatcher, 1960, Methods for collection and analysis of water samples, U.S. Geol. Surv. Water-Supply Paper 1454, 301 pp.

Roberson, C. E., J. H. Feth, P. R. Seaber, and P. Anderson, 1963, Differences between field and laboratory determinations of pH, alkalinity and specific conductance of natural water, U.S. Geol. Surv. P.P. 475-C, C212-C215.

Sato, Motoaki, 1960, Oxidation of sulfide ore bodies, 1. Geochemical environments in terms of Eh and pH, *Econ. Geol.*, **55**, No. 5, 928—961.

Semenovich, N. I., 1963, The oxidation-reduction potential and pH of the sediments of Lake Ladoga, *Geochemistry*, No. 2, 183—184.

Starkey, R. L., and K. M. Wight, 1945, Anaerobic corrosion of iron in soil, Am. Gas Assoc., 108 pp.

Stumm, W., 1966, Redox potential as an environmental parameter; conceptual significance and operational limitation, 3rd Inter. Conference on Water Pollution Res. Section 1, No. 13, Water Pollution Control Federation, Washington, D. C., 15 pp.

———, 1967, Discussion of Chemical oxidation and reduction of metals and ions in solution, by T. E. Larson *in* S. D. Faust and J. V. Hunter, eds., Principles and applications of water chemistry, John Wiley and Sons, 448 pp.

The Industrial pH Handbook, 1957, T. J. Kehoe, ed., Beckman Instruments, 74 pp.

Watanabe, N., and M. A. V. Devanathan, 1964, Reversible oxygen electrodes, *J. Electrochemical Soc.,* **111**, 615—619.

Weber, W. J., and W. Stumm, 1963, Mechanism of hydrogen ion buffering in natural waters, *J. Am. Water Works Assoc.,* **55**, 1553—1578.

Welcher, F., 1942, Chemical solutions, D. Van Nostrand Co., 404 pp.

Wilde, P., and P. Rodgers, 1968, Electrochemical measurements in the marine environment utilizing sea water as a 'spiked' solution (abstract), *Trans. Am Geophysical Union,* **49**, No. 1, 365—366.

Willard, H. H., L. L. Merrit, Jr., and J. A. Dean, 1951, Instrumental methods of analysis, D. Van Nostrand Co., 179—205.

Zobell, C. E., 1946, Studies on redox potential of marine sediments, *Bull. Am. Assoc. Petrol. Geologists,* **30**, 477—513.

SELECTED GENERAL REFERENCES

Bouma, A. H., 1969, Methods for the study of sedimentary structures, John Wiley and Sons, 458 pp.

Folk, R. L., 1968, Petrology of sedimentary rocks, 2nd ed., Hemphill's, 170 pp.

Griffiths, J. C., 1967, Scientific method in analysis of sediments, McGraw-Hill Book Co., 508 pp.

Holmes, Arthur, 1930, Petrographic methods and calculations, revised ed., Thomas Murby & Co., London, 515 pp.

Krumbein, W. C., and F. J. Pettijohn, 1938, Manual of sedimentary petrography, Appleton-Century-Crofts, 549 pp.

, and F. A. Graybill, 1965, An introduction to statistical models in geology, McGraw-Hill Book Co., 475 pp.

Milner, H. B., ed., 1962, Sedimentary petrography, 4th revised ed., Macmillan, Vol. I, 643 pp., Vol. II, 715 pp.

Müller, German, 1967, Sedimentary petrology, Part I, Methods in sedimentary petrology, translated by Hans-Ulrich Schmincke, Hafner Publishing Co., 283 pp.

Tickell, F. G., 1939, The examination of fragmental rocks, 1st revised ed., Stanford Univ. Press, 154 pp.

_____, 1965, The techniques of sedimentary mineralogy, Elsevier Publishing Co., 220 pp.

Twenhofel, W. H., and S. A. Tyler, 1941, Methods of study of sediments, McGraw-Hill Book Co., 183 pp.

Van der Plas, Leendert, 1966, The identification of detrital feldspars, Elsevier Publishing Co., 305 pp.

REFERENCE INDEX

Agterberg, F. P., 43
Albertson, M. L., 93
Alexander, L. E., 569
Ali, S. A., 527
Alimen, H., 146
Allen, J. R. L., 19, 248
Alling, H. L., 146
Amyx, J., 362
Anderson, G. E., 146
Anderson, P., 633
Angelucci, A., 177
Archie, G. E., 362
Armstrong, G. C., 19
Arnold, H. D., 92
Arthur, M. A., 214
Artini, E., 474
Aschenbrenner, B. C., 146, 362
Atherton, E., 146
Axelrod, J. M., 569
Ayer, N., 282

Baas Becking, L. G. M., 631
Back, W., 631
Badoz-Lambling, J., 631
Bailey, E. H., 527
Baker, R. A., 125
Balis, E. W., 595
Ballard, L. M., 451
Balzy, P., 527
Band, D. M., 633
Barnes, I., 631
Barr, J. L., 247
Barsdate, R. J., 449
Bartlett, W., 424
Bass, D., Jr., 362
Bates, R. G., 631
Baturin, V. P., 475
Be, A., 179
Beales, F. W., 247
Belding, H., 249

Bennett, C. A., 424
Berman, H., 509
Bernard, H. A., 214
Berner, R. A., 631
Bertholf, W. E., Jr., 451
Biederman, M., 177
Biggar, J. W., 282
Bissell, H. J., 247
Black, C. A., 595
Blanchard, L. R., 216, 284
Blatt, H., 449
Boardman, R. S., 215
Boggs, S., Jr., 146
Bonham, L. C., 310, 312, 333
Boswell, P. H. G., 475
Botty, M. C., 179
Bouma, A. H., 247, 282, 635
Bowen, H. J. M., 595
Bradley, W. A., 282
Bramer, H., 177
Bramlette, M. M., 475
Brammall, A., 475
Bredig, M. A., 568
Brewer, R., 146
Briggs, L. I., 147, 475
Brindley, G. W., 539, 568
Brison, R. J., 382
Brotherhood, G. R., 125
Brown, G., 568
Brown, W. E., 214
Browning, J. S., 449
Brunton, G., 568
Buchanan, H., 248
Buehler, E. J., 247
Buerger, M. J., 539
Bull, W. B., 126
Bunker, C. M., 282
Bunting, E. N., 363
Burgener, B. C., 633
Burger, J. A., 247

SUBJECT INDEX